Pharmaceutical Discovery, Development and Manufacturing Forum 2016

Core Programming Area at the 2016 AIChE Annual Meeting

San Francisco, California, USA
13 - 18 November 2016

ISBN: 978-1-5108-3445-3

Printed from e-media with permission by:

Curran Associates, Inc.
57 Morehouse Lane
Red Hook, NY 12571

Some format issues inherent in the e-media version may also appear in this print version.

Printed by Curran Associates, Inc. (2017)

For permission requests, please contact AIChE
at the address below.

AIChE
120 Wall Street, FL 23
New York, NY 10005-4020

Phone: (800) 242-4363
Fax: (203) 775-5177

www.aiche.org

Additional copies of this publication are available from:

Curran Associates, Inc.
57 Morehouse Lane
Red Hook, NY 12571 USA
Phone: 845-758-0400
Fax: 845-758-2633
Email: curran@proceedings.com
Web: www.proceedings.com

TABLE OF CONTENTS

(44a) Maximizing Poorly Water-Soluble API Nanoparticle Loading in Polymer Strip Films 1
Scott M. Krull, Meng Li, Ecevit Bilgili, Rajesh N. Davé

(44b) Computational Approaches for Modeling Fluid Bed Granulation Process 2
Gulsad Kucuk, Robert Schwabe, Leon Schultz, Ilgaz Akseli

(44c) Comparison of the Electrostatic Properties of Amorphous and Crystalline Lactose 3
Karolina Biegaj, Tim Lukas, Martin Rowland, Jerry Heng

(44d) Particle Engineering Research from the NSF Erc Research Experience for Teachers (RET) Program at Njit 4
Ecevit Bilgili, Meng Li, Lu Zhang, Mina Armani, Paul Orbe, Marie Aloia, Barbara Pielecha-Safira, Rajesh N. Dave

(44e) Study of DPI Carriers Design Parameters for Dry Powder Inhalation through Artificial Neural Network Modeling 5
Amir A. K. Farizhandi, Fengying Zhang, Raymond Lau

(44f) The Surface Properties of Organic Crystalline Solids and Their Interactions with Polymeric Excipients and Binders 6
Mark A. Isbell, Eftychios Hadjittofis, Steven J. Hinder, Geoff G. Z. Zhang, Jerry Heng

(44g) Narrow Tube Spray Drying with Precise Drying History Control for Particle Engineering 8
May Ginn Lee, Shahnaz Mansouri, Karen Hapgood, Meng Wai Woo

(44h) Soluble Tag-Assisted Liquid-Phase Peptide Synthesis Using Hydrophobic Benzyl Alcohols 16
Yingzi Liu, Yohei Okada, Kazuhiro Chiba, Hidehiro Kamiya

(67a) Biomanufacturing Science and Technology Consortium to Advance U.S. Manufacturing of Biopharmaceuticals 17
Seongkyu Yoon

(67b) Defining Challenges and Creating Solutions in Biopharmaceutical Manufacturing 18
Kelvin H. Lee

(67c) Technology Advances That Will Enhance Continuous Biomanufacturing 19
Melvin Koch

(67d) Energy-Based Medium Design for Mammalian Cell Fed-Batch Cultures 20
Ana L. Quiroga-Campano, Maria M. Papathanasiou, Efstratios N. Pistikopoulos, Athanasios Mantalaris

(67e) Evaluating the Potential of Miniature Scale Chromatography Process Characterization to Support Biomanufacturing Strategies and Regulatory Filings 21
John Welsh, James Pollock, Nihal Tugcu, David Roush, Jennifer Pollard

(67f) Ecoprime Twin – Continuous Two-Column Chromatography at the Process Scale 22
Kathleen Mihlbachler

(76a) Enzymatic Synthesis of Complex Natural Products 23
Chaitan Khosla

(76b) Quality-by-Control (QbC) Technologies for Pharmaceutical Process Development 24
Zoltan K. Nagy

(76c) Drug Product Continuous Manufacturing Control Strategy Considerations 25
Justin Pritchard, Kelly Swinney

(76d) Regulatory Approach for Supporting Emerging Pharmaceutical Technology 26
Rapti Madurawe, Sau Lee

(76e) The Integrated Quality Assessment: Office of Process and Facilities' Perspective on the Holistic Evaluation of Pharmaceutical Quality 27
David Doleski

(164a) A Phenomenological Model for the Prediction of Granule Drying Times in a Fluid Bed Dryer 28
Ariel R. Muliadi, Joseph Bullard

(164b) Practical Application of a Mechanistic Model to Scale up Fluid Bed Drying in Drug Product Development 29
Emmanuela Gavi, Pranay K. Ghosh, Susanne Page, Oskar Kalb

(164c) Modelling of Co-Rotating Twin-Screw Extruders in the Pharmaceutical Industry I: Single Component Model 30
Maria Fuentes-Gari, Pavol Rajniak, Niall Mitchell, John Robertson, Sean K. Bermingham, Tariq Islam

(164d) Modeling and Optimization of Agglomeration in Suspension through a Coupled Population Balance Framework 31
Ramon Pena, Zoltan K. Nagy, Doraiswami Ramkrishna, Christopher L. Burcham, Daniel Jarmer

(164e) Modelling of Water Absorption By a Pharmaceutical Tablet 33
Charalampos Christodoulou, Luca Mazzei, Eva Sorensen, Salvador García-Muñoz

(164f) Design Problem of Pharmaceutical Manufacturing Processes 34
Hirokazu Sugiyama

(165a) Multi-Scale Modeling in Pharmaceutical Drug Substance Manufacturing 35
José L.C. Santos, Rui C. Silva, Filipe A.P. Ataíde

(165b) A Simultaneous Approach for Kinetic Parameter Estimation and Curve Resolution Based on Spectroscopic Measurements 36
Lorenz Biegler, Weifeng Chen, Salvador García-Muñoz

(165c) Application of Computational Chemistry in Pharmaceutical Process Development 37
Dimitri Skliar, Antonio Ramirez

(165d) Systematic Optimisation of Continuous Separation for Artemisinin Recovery in Continuous Pharmaceutical Manufacturing ... 38
Hikaru G. Jolliffe, Dimitrios I. Gerogiorgis
(165e) Prediction of Liquid-Liquid Partitioning and Solubility of APIs with the SAFT-γ Mie Approach ... 40
Panatpong Hutacharoen, Simon Dufal, Vasileios Papaioannou, Claire S. Adjiman, George Jackson, Amparo Galindo
(165f) Use of a First Principles Model during QbD Studies to Assess Impurity Formation for a Coupling Reaction ... 41
Billie Kline, Conrad Fung, Elaine Lee, David Willcox
(165g) Verification and Use of CFD Modeling in Stirred Tank Reactors for Pharmaceutical Applications ... 42
Andrew Glace, Divya Seshadri, Fulya Akpinar, Benjamin Cohen
(165h) Product-Agnostic Computational Fluid Dynamic (CFD) Model for Single-Use Mixing System ... 43
Ahsan Munir, Sean Fitzgibbon, Kenroy Noicely, Alan Dominguez, Adam Sokolnicki, Cenk Undey
(169b) Continuous Biocompatible Reactive Extraction Concept for the Purification of Di-Carboxylic Acids from Dilute Aqueous Solutions ... 44
Jannick Gorden, Tim Zeiner, Christoph Brandenbusch
(169c) Combining PAT with Data Visualization and Machine Learning to Understand and Control Batch Cooling Crystallization ... 45
Daniel J. Griffin, Martha A. Grover, Yoshiaki Kawajiri, Ronald W. Rousseau
(169d) Stability Analysis of Solar Silicon Wafer Manufacturing By the Horizontal Ribbon Growth Process ... 46
Jiaying Ke, B. Erik Ydstie, Aditya S. Khair
(169e) Protein Enzymatic Hydrolysis Integrated with Ultrafiltration ... 47
Anna Trusek-Holownia
(169f) The Dynamics of Biodiesel Profile and Fungal Diversity of Activated Sludge During Lipid Accumulation on Acetic Acid ... 48
Dhan Lord Fortela, Rafael Hernandez, Mark Zappi, Andrei Chistoserdov, Rakesh Bajpai, Emmanuel Revellame, William Holmes
(226a) Computational Solid State Chemistry Applications in Pharmaceutical Industry ... 49
Yuriy Abramov
(226b) Process Optimization Using a Distillation Model Coupled with a Solubility Model for Increased Process Yield Following Multiple Distillations and Solvent Swap ... 50
Christopher J. Morrison, John Guo
(226c) Large-Scale Crystallization of a Pure Metastable Polymorph By Reaction Coupling ... 51
Tu Lee, Hong Yu Lin, Hung Lin Lee
(226d) PAT Enabled Drying Platform for API Process Development – Investigation of Solvent Trapping Associated with Form Transition and Crystal Collapse During Drying ... 52
David J. Lamberto, George Zhou, Rositza Petrova, Richard Varsolona
(233am) Modeling and Simulation Efforts in Scale up of Sustained Release Microspheres ... 53
Rishi Mehan, SVB Janardhan Garikipati, Birendra K. David, Yash Partani, RaviChandra Palaparthi
(233b) Combined Task Bayesian Optimization for Efficient Scale-up and Technology Transfer ... 54
Ryosuke Yoshizaki, Manabu Kano
(233v) Development and Optimization of a Mixed Suspension Mixed Product Removal (MSMPR) Crystallization Process Incorporating Wet-Milling ... 55
Yihui Yang, Charles D. Papageorgiou
(233f) Developing Process Models of Hydrogen Peroxide Decontamination in Pharmaceutical Manufacturing ... 56
Keisho Yabuta, Haruka Futamura, Koji Kawasaki, Masahiko Hirao, Hirokazu Sugiyama
(233w) Comparison of Raman and UV Spectroscopy for the in-Line Quantification of Active Ingredient in Pharmaceutical Semi-Solid and Liquid Formulations ... 57
Nils Bostijn
(233s) Process Development for Continuous Extrusion and Pelletization of a Heat and Shear-Sensitive Formulation ... 60
Theresa R. Hörmann, Gerold Koscher, Stephan Laske, Johannes G. Khinast
(233z) Wet Granulation-Induced Polymorphic Transitions in Piracetam: Raman Spectroscopy As a Potential in-Situ Process Analytical Tool ... 61
Gayathri Kollamaram, Catherine Kelly, Denise Croker, Gavin Walker
(233az) Dissolvable Microcarriers for Efficient Cell Production and Recovery ... 64
Yue Zhou, Jennifer L. Weber, Kathleen Krebs, Martial Hervy, Marylene Pecheul, Corinne Walerack, David Henry, Marie-Maud Bear, Stephen Caracci
(233af) The Effect of Particle Engineering on the Processing and the in Vitro Performance of Inhalation Blends in Dry Powder Inhalers ... 65
Eva Faulhammer, Sarah Zellnitz, Thomas Wutscher, Johannes G. Khinast, Amrit Paudel
(233aw) Feasibility Analysis of Flowsheet Models in Continuous Pharmaceutical Manufacturing Processes Considering the Effects of Noise ... 67
Zilong Wang, M. Sebastian Escotet-Espinoza, Ravendra Singh, Fernando J. Muzzio, Marianthi G. Ierapetritou
(233ag) Characterization of Compaction Behaviour of Powders in Solid-Dosage Manufacturing Using Finite Element Analysis ... 68
Ashish Kumar, Jens Dhondt, Johny Bertels, Didier Klingeleers, Krist V. Gernaey, Ingmar Nopens, Thomas De Beer
(233aq) Gender-Dependent PBPK Modeling ... 69
Ioannis P. Androulakis, Megerle Scherholz, Patrick J. Sinko, Marianthi Ierapertritou, Clara Hartmanshenn
(233l) Model-Based Process Analysis in Continuous Manufacturing of Pharmaceuticals: Calibration of a Fluid Bed Dryer Model ... 71
Michael Ghijs, Fien De Leersnyder, Daan Van Hauwermeiren, Valérie Vanhoorne, Jurgen Vercruysse, Philippe Cappuyns, Séverine T.F.C. Mortier, Thomas De Beer, Krist V. Gernaey, Ingmar Nopens

(233p) Development of a Model Predictive Tool for Continuous Manufacturing of Solid Oral Dosage Forms Via Roller Compaction ... 72
Barbara Schaller, Pishnamazi Mahboubah, Patrick Cronin, Ahmad Ziaee, Ahmad Albadarin, Denise Croker, Maurice Collins, Gavin Walker

(233q) Application of Model-Predictive Control to a Feeding Blending Unit ... 73
Jakob Rehrl, Julia Kruisz, Stephan Sacher, Martin Horn, Johannes G. Khinast, Isabella Aigner

(233ai) Systematic Technoeconomic Comparison of Separation Solvents for Green Continuous Pharmaceutical Manufacturing of Diphenhydramine ... 74
Samir Diab, Dimitrios I. Gerogiorgis

(233u) Continuous Wet Granulation in a Twin Screw Extruder: Effect of Process Variables on Residence Time Distribution and Granule Characteristics ... 76
Manuel Kuhs, Ahmad Albadarin, Gayathri Kollamaram, Shaza Darwish, Jacek Zeglinski, Rahamatullah Shaikh, Saeed Shirazian, Denise Croker, Gavin Walker

(233ar) Virtual High-Throughput Screening Pipeline: Limitation Characterization and Application to Different Targets ... 77
Jonathan J. Chen, Donald P. Visco

(233aa) Real-Time Monitoring and Ejection of out-of-Spec Tablets Via NIR Chemical Imaging ... 78
Patrick R. Wahl, Gregor Hauseder, Stephan Sacher, Matthias Kerschhaggl, Johannes G. Khinast

(233d) Multivariate Statistical Process Control for the Continuous Manufacturing of Solid Oral Dosage Forms ... 79
Salvador García-Muñoz, Zhenqi Shi, Joshua Hanson

(233g) Modelling Approaches to Spray Drying Process Development, Scale-up and Troubleshooting ... 80
Pedro Valente, Íris Duarte, Filipe Neves, Márcio Temtem

(233ab) Solid-State Reactivity of Drugs with Potential Reactive Excipient Impurities: In Silico and Experimental Studies ... 81
Corinna Gressl, Michael Brunsteiner, Adrian L. Davis, Margaret Landis, Klimentina Pencheva, Garry Scrivens, Gregory Sluggett, Geoffrey Wood, Johannes G. Khinast, Amrit Paudel

(233ap) Prediction of Formulation Effects of Continuously Formulated Solid Oral Dosage Forms on Bioavailability Using PBPK Modeling ... 82
Clara Hartmanshenn, Patrick J. Sinko, Marianthi G. Ierapetritou, Ioannis P. Androulakis

(233ao) Predicting the In Vivo Performance of Amorphous Solid Dispersions Based on Molecular Descriptors and Statistical Analysis ... 83
Íris Duarte, João Henriques, João F. Pinto, Márcio Temtem

(233r) Behaviour of Magnesium Stearate in Continuous Feeding ... 86
Raghu V. G. Peddapatla, Caroline A. Blackshields, Michael F. Cronin, Abina Crean

(233ac) Comparing and Contrasting the Solid-State Esterification of APIs with Excipient: Computational and Experimental Assessment ... 90
Corinna Gressl, Michael Brunsteiner, Adrian L. Davis, Margaret Landis, Klimentina Pencheva, Garry Scrivens, Gregory Sluggett, Geoffrey Wood, Johannes G. Khinast, Amrit Paudel

(233al) The Effects of Formulation Factors on the Aerosolization Performance of Metered Dose Inhalers ... 91
Denise S. Conti, Jay Holt, Poonam Sheth, Dennis Sandell, Anthony Hickey, Bhawana Saluja

(233ae) Exploiting Solvent and Impurity Effects to Modify Crystal Shape and Bulk Powder Properties of API and Intermediate Molecules ... 94
Anil Rane

(233t) Continuous Manufacturing of Co-Crystals in Twin Screw Extruder: A Case Study ... 95
Shaza Darwish, Rahamatullah Shaikh, Manuel Kuhs, Denise Croker, Gavin Walker

(233n) Fault-Tolerant Control Design and Risk MAP Based Resiliency Analysis of Continuous Solid Dose Manufacturing ... 96
Qinglin Su, Mariana Moreno, Sudarshan Ganesh, Arun Giridhar, Gintaras Reklaitis, Zoltan K. Nagy

(233h) A Comprehensive Statistics-Based Approach to Proactive Process Analysis Including Multivariate Analysis Methods ... 97
Sara Koynov, Paige Adack, Conor McCurdy, Orlaith Crowther, Michelle Cleary, Anthony Ruane, Dearbhla Carr, Eoin O'Brien, Jennifer Sun

(233ay) The Utilization of Amino Acid Consumption Rate Assessment As a Tool for CHO Clonal Stability Evaluation ... 98
Eric Brodean, Andrew Seletsky, Melissa Good, Wai Lam Ling

(233au) Powder Tester Comparison to Predict Discharge Rate from Hoppers and Filling Tablet Press Dies ... 99
Joerg Theuerkauf, Yi Fan, Andrew Horton, Madhusudhan Kodam, Karl Jacob

(233y) PAT Application in Expedited Development of Multi-Step Chemical Syntheses of Active Compounds ... 100
Dominique Hebrault, Nizar Haddad

(233ax) Process Innovation and Optimization of a Commercial Cell Culture Process ... 101
Sivaprakash Agastin, Simpson Gregoire, Sivashankar Sivakollundu, Jeffrey Savard

(233i) Design of Nonlinear Observer for State Estimation of a Bioreactor ... 102
Xinghua Pan, Tejasvi Jaladi, Jonathan Raftery, Melanie DeSessa, M. Nazmul Karim

(233e) Computational Fluid Dynamics (CFD) Guided Scale up of Cell Culture Bioreactors from Bench to Pilot to Commercial Scale ... 103
William Tran, Garima Chaudhary, Jan-Erik Schaefer, Erik Hasenfus, Thomas Wucherpfennig, Daniel Bock, Henry Lin

(233a) Chemical Engineering Principles in Oral Solid Dosage Process Development-Where Are We? ... 104
Weixian Shi

(233o) Characterization of Mixing and Heat Transfer in Continuous Lab Scale Reactors ... 105
Jerry S. Salan, David am Ende

(233as) Screening a Library of Commercially Available Compounds for Drugs with Novel Anti-Biofilm Activity 106
Anand Ramasubramanian, Jose Lopez-Ribot, Kai Leung, Anand Srinivasan, Johnathan Abercrombie, Nelson Torres

(233av) A Small Volume Continuous Facility for GMP API Manufacture 107
Rick Spencer

(233ad) Controlling Amorphization Versus Crystallization of Carbamazepine Nanoparticles Using a Supercritical-assisted Spray Drying Process 108
Luis Padrela, Kevin M. Ryan, Barry Long

(233an) Evaluation of the Therapeutic Potential of FOXC1 Knockdown for Treatment of Basal-Like Breast Cancers 109
Landon A. Mott

(284a) Mass Transfer Characterization of a Biphasic Continuous Reaction System for Improved Process Understanding and Control 110
Joseph Reckamp

(284b) The Development of Integrated Continuous Drug Substance Processes 111
Jessica Whelan, Mark Barrett, Brian Glennon

(284d) Process Intensification through Continuous Spherical Crystallization Using an Oscillatory Baffled Crystallizer 112
Ramon Pena, Zoltan K. Nagy, Christopher L. Burcham, Daniel Jarmer

(284e) Designing for Continuous: A Step Change in Performance 113
Philip Donnellan, Roderick Jones, Brian Glennon

(284f) Continuous Drying of Powders and Slurries in an Extruder without Changing Product Particle Size Distribution 114
Manuel Kreimer, Isabella Aigner, Gerold Koscher, Daniel Lepek, Johannes G. Khinast

(284g) Continuous Co-Crystallization of Novel Theophylline and 4-Aminobenzoic Acid Co-Crystals from Twin Screw Extrusion 115
Rahamatullah Shaikh, Shaza Darwish, Manuel Kuhs, Ahmad B. Albadarin, Denise Croker, Gavin Walker

(342a) Streamlined Design for Continuous Production of a Pharmaceutical Drug Product 116
Stephen L. Conway, Robert F. Meyer, Frank D. Witulski, Manoharan Ramasamy, Oliver Clarke, Palmerly Tom, Mason Gibbs, Mark Purcell, Lalloo Anita, Julie Pentz, Ken Thompson, Brendon G. Ricart, Jon Hoza, Donghui Hong, Gina Thompson, Philip Cryan, Andrew Owen

(342b) Flowsheet Modeling of a Continuous Direct Compression Process at Production Scale 117
Shaun Galbraith, Bumjoon Cha, Zhuangrong Huang, Seongkyu Yoon, Robert F. Meyer, Samantha Hurley, Matthew H. Flamm

(342c) Characterizing a Drug Product Continuous Manufacturing Platform's Residence Time Distribution Using Major and Minor Unit Formula Excipient Step Changes and Soft Sensors 118
Leo Manley

(342d) Establishing a Procedure for Transferring HME Processes Using Physical Similarities 119
Ândreas Witschnigg, Theresa R. Hörmann, Michaela Zagler, Gerold Koscher, Stephan Laske, Johannes G. Khinast

(342e) Investigation of the Influence of Continuous Blending Process Parameters on the Lubrication of Pharmaceutical Powder Blends 122
Pongpumin Bunchatheeravate, Anni Zhang, Joseph Bullard

(342f) "Correlating Raw Material Properties to Critical Processing Parameters to Critical Quality Attributes in a Direct Compaction Continuous Line." 123
Gerardo Callegari, Thomas O'Connor, Fernando Muzzio

(342g) A Novel Analysis of Residence Time Distributions for Continuous Powder Mixing 124
Geng Tian, Xiaochuan Yang, Sau Lee, Robert Fisher, Shuaili Li, Moo Sun Hong, Zongyu Gu, Thomas O'Connor

(342h) Model Based Loss-in-Weight Feeder Design and Optimization 125
Eric Jayjock, M. Sebastian Escotet-Espinoza

(384a) Use of Real-Time Analytics and Apoptosis Assays for Enhanced Understanding of Cell Growth and Viability in a CHO-Based Process for Monoclonal Antibody Production 126
Yongqi Wu, Nobel Vale, Vivian Deng, Mina Chaudhry, Dimuthu A. Jayawickrama, Timothy Stevens, Frank Ritacco

(384b) Providing a Molecular Basis to Modeling Protein Adsorption to Multimodal Carbohydrate-Based Ligands 127
Shishir PS Chundawat, Sonia K Brady, Vibha Narayanan, Matthew J. Lang

(384c) Operational Control Strategy Development Via First Principles Mapping 128
Darrell Staggs, Matthew Foster

(384d) Nanoextrusion: A Platform Enabling Comparative Assessment of Nanocomposites Vs. Amorphous Solid Dispersions for Drug Dissolution Enhancement 129
Ecevit Bilgili, Meng Li, Nicolas Ioannidis, Indiana Suriel, Costas G. Gogos

(384e) Evaporation of Binary Mixtures and Shell Formation in Spray Dried Droplets 130
Pedro Valente, Íris Duarte, Tiago Porfirio, Xu Liu, Feng Zhang, Márcio Temtem

(384f) Impact of Tooling Design During Development and Risk Evaluation of Tri-Layer Tablet Formulations 132
Sophia Robertson-Lavalle, Jing Tao, Preetanshu Pandey, Sherif Badawy

(384g) End-Point Determination and Scale-up/Scale-Down of High Shear Wet Granulation Process Using an in-Line Di-Electric Probe 133
Mehrdad Kheiripour Langroudi, Jennifer Sun, W. Mark Eickhoff

(400a) Selection Support Methodology Between Batch and Continuous Technologies for Pharmaceutical Tablet Manufacturing Processes 134
Kensaku Matsunami, Takuya Miyano, Hiroaki Arai, Hiroshi Nakagawa, Masahiko Hirao, Hirokazu Sugiyama

(400b) Development of a Continuous Production Line for Solid Dosage Forms Via Dry Granulation 135
Isabella Aigner, Stephan Sacher, Julia Kruisz, Michael Martinetz, Patrick R. Wahl, Johannes G. Khinast

(400c) Considerations in Developing a Control Strategy for Continuous Direct Compression 136
Donald Kientzler, Aditya Vanarase, Anthony Tantuccio, Kevin Macias, Timothy Stevens, Admassu Abebe, Alexander Schaepman, James Holman
(400d) Applying Supervisory Process Control Practices to Continuous Pharmaceutical Manufacturing 137
Niels Nicolaï, Ingmar Nopens, Krist V. Gernaey, Thomas De Beer
(400e) Control of a System of Loss-in-Weight Feeders in a Continuous Manufacturing Drug Product Application 138
Joshua Hanson, Ian Leavesley
(400f) Dynamic Optimization of Continuous Manufacturing of Pharmaceuticals 139
Michael Shoham Patrascu, Paul I. Barton
(400g) Extended Iterative Optimzation Technology for Mixture Composition Estimation in Moving Powders 140
Salvador García-Muñoz, Zhenqi Shi
(424a) Enhancing Process Understanding and Control with Real-Time Analytics for a Challenging Solid Form System of a Pharmaceutical Compound 141
Joe Gao, Bing-Shiou Yang, Michelle Raikes, Francesco Ricci, Ankit Gargava, Soojin Kim, Xiaorong He, Keith Horspool
(424b) Simultaneous Fast Measurements of Tablet Porosity/Moisture: Real-Time Release Testing for QC Dissolution Prediction 142
Mehrdad Kheiripour Langroudi, Morgan Crawford, Lee Klein, Craig B. Ikeda, W. Mark Eickhoff
(424c) Real Time Prediction of Tablet Tensile Strength By Ultrasound in Continuous Manufacturing 143
Golshid Keyvan, Sonia Razavi, German Drazer, Alberto Cuitiño, Fernando Muzzio
(424d) Toward Predicting Tensile Strength and Characterizing the Acoustic Properties of Pharmaceutical Tablets on Continuous Manufacturing Platforms 144
Sonia M. Razavi, Alberto M. Cuitino
(424e) Overcoming the Challenges of Noisy Data during the Calibration of NIR As a PAT Technique to Measure Roller Compacted Ribbon Density 145
Mary Ellen Crowley, Avril Hegarty, Michael McAuliffe, Graham O'Mahony, Luke Kiernan, Kevin Hayes, Abina Crean
(424f) PAT Based Batch Statistical Process Control for Pharmaceutical SEMI-SOLID and Liquid Manufacturing Processes 150
Nils Bostijn
(424g) Real Time PAT Monitoring of Bioreactor Mab IgG3 Cell Culture Process Dynamics Via Fourier Transform Infrared Spectroscopy: Implications and Challenges 153
Huiquan Wu
(428a) Food, Pharmaceutical & Bioengineering Division Award: Control, Sensing, and Chemical Analysis Using Microfluidic Devices 154
Mark A. Burns
(428b) Area 15A, Food - Plenary: Bioelectronic Nose: Integration of Biotechnology and Nanotechnology 155
Tai Hyun Park
(428c) Area 15B/PD2M, Pharmaceuticals - Plenary: Human Multi-Organ-on-a-Chip Platform for Pharmaceutics 156
Ali Khademhosseini
(428d) Area 15C, Bioengineering - Plenary: Status of Biocatalysis: Protein and Process Engineering 157
Andreas S. Bommarius
(428e) Area 15D/E, Engineering Fundamentals in Life Science - Plenary: Cellular Bioengineering: Past, Present, and Future 158
Rena Bizios
(487a) Evaluating Mixing Performance in a Stirred Tank Reactor with Different Impeller Geometries and Reactor Size By CFD 159
Rui C. Silva, Filipe A.P. Ataíde, José L.C. Santos
(487b) CFD Simulation on Assessment of Biologic Product Mixing Performance in Single Use Technology 160
Wei Chen, Chris Lee, Rao Mantri, Howard Stamato
(487c) Characterization of Lab and Plant Reactors Using Fast Competitive Reactions and a Predictive Model 161
Joe Hannon, Andrew Bird, Andrew Derrick
(487d) Computational and Experimental Determination of the Hydrodynamics in a Scaled-Down Pharmaceutical Vessel Provided with Angle-Mounted Impellers 162
Ji Ma, Piero M. Armenante
(487e) Experimental Study of Wet Milling in a Rotor-Stator Mixer 163
Richard V. Calabrese, Kanan Ghaderzadeh
(487f) Practical Large Eddy Simulation 164
John A. Thomas, Kevin Smith
(487g) Scale-up Rules for the Generation of Solid-Stabilized Emulsions 165
Alexandre Al-Haiek, Emir Tsabet, Louis Fradette
(494d) Developing Therapeutic Medicines through Chemical Engineering Innovations 166
Ann L. Lee, Judy F. Hsii
(522b) A Regulatory Perspective for Manufacture of Complex Drug Substances 167
Jane Chang
(522c) CQA Control Strategy for a Synthetic Peptide Therapeutic Candidate Etelcalcetide 168
Asher Lower
(522d) Development and Implementation of a Mathematical Model As a Key Component of a Palladium Control Strategy for a Semi-Continuous Process to Produce a Small Molecule Drug Substance 169
Kevin Seibert, Carla Luciani, Colm O'Mahony

(522e) Monitoring the Trace Levels of Free Sulfate Release to Terminate a Depolymerization Reaction Using ATR-FTIR 170
Satyanarayana Thirunahari
(522f) FDA Review Chemist's Participation in Pre-Approval Inspections: Case Study from Drug Substance Facility 171
Kshitij A. Patkar, Delaram Moshkelani
(522g) Panel Discussion: Current Issues and Challenges for CQA Monitoring and Control for Drug Substance Manufacturing 172
Naiqi Ya
(551a) A Rapid, Materials-Sparing Approach to the Exploration of a Roller Compaction Process Design Space 173
Catherine Metzler, Pongpumin Bunchatheeravate, Gregory Buonocore, Hanlin Li, Hailey MacDonald, Ryan Peterson, Noreen Zaman, Joseph Bullard
(551b) Advances in Discrete Element Modeling of High-Shear Wet Granulation Process Using Rocky-DEM 174
Preetanshu Pandey, Brenda Remy, Rahul Bharadwaj, Lucilla Almeida
(551c) Scale-up of HME Process from a Trilobed 16mm to a Bilobed 27mm Extruder; SPH Simulations and 1D Mechanistic Modelling 175
Josip Matic, Stephan Laske, Amrit Paudel, François Pfefferlé, Jessica Lovey Martinetti, Sophie Martel, Johannes G. Khinast
(551d) Scale-up and Blender Change Model for the Pharmaceutical Lubricated Mixing Process 176
Yasuhiro Suzuki
(551e) Scaling Intra-Tablet Coating Variability 177
Siddhartha Agarwal, Carl Wassgren
(551f) Improving Workflow for Modeling of High-Shear Rotor-Stator Wet Milling for Pharmaceutical Applications 178
Derek Starkey, Carla Luciani, Salvador García-Muñoz
(551g) Characterizing Mixing Time Scales in an Exothermic Mixing-Limited Reaction 179
Anuj A. Verma, Niharika Chauhan, Samrat Mukherjee
(579a) Pharmaceutical Crystallization: A Manufacturing Process Perspective 180
Huiquan Wu
(579b) Cooling Crystallization Within Micromixers: Mathematical Modeling, Theoretical Analysis, and Experimental Validation 181
Mo Jiang, Richard D. Braatz
(579c) Real Time PAT Monitoring of the Solvent-Mediated Transformation of Ciclesonide Acetone Solvate into Methanol or Cyclohexane Solvate 182
Lina Zhou, Qiuxiang Yin, Ting Wang, Jingkang Wang
(579d) Crystal Morphology and Size Control in Pharmaceutical Industry 183
Junbo Gong, Yan Wang, Jingkang Wang
(579e) Liquid -Liquid Phase Separation during the Cooling Crystallization of Pyraclostrobin 184
Junbo Gong, Kangli Li, Jingkang Wang, Xiaowen Zhu
(579f) Discriminate the Differences Between Surface Water and Channel Water in the Formation of Solvates 185
Yumin Liu, Dandan Han, Jingkang Wang, Junbo Gong
(579g) Influence of Ultrasonic Wave on the Crystallization of Lysozyme 186
Yanan Zhou, Hongxun Hao, Jingkang Wang
(596a) Site-Specific Conjugation of a Drug to an Antibody Via Nucleotide Binding Site 187
Nur Mustafaoglu, Franklin Mejia, Jonathan D. Ashley, Tanyel Kiziltepe, Basar Bilgicer
(596b) Acid-Sensitive Silylated or Acetalated Polysaccharides for Vaccine Applications 188
Kristy M. Ainslie
(596c) The Effect of Particle Physico-Chemical Properties on Phagocytosis 189
Anusha Garapaty, Julie A. Champion
(596d) Process Development and Scale-up of an ADC Conjugation Process 190
Selin Aytar, Shawn Brueggemeier, Michael Hay, Annie Tam, Michael Smith
(596e) In Situ crosslinked Endosomolytic Polymeric Vesicles for Immunotherapeutic Targeting of the Sting Pathway 191
Daniel Shae, Sema Sevimli, Alyssa Merkel, Denise Buenrostro, Anna Caldwell, Julie A. Sterling, John Wilson
(596f) Impact of Dose, Route, and Composition on the Immunogenicity of Polyelectrolyte Multilayers Assembled on Gold Templates 192
Peipei Zhang, Christopher M. Jewell
(596g) Diminution of Macrophage Attachment on CD47-Functionalized and Polydopamine-Coated Surface 193
Jun Zhang, Ching-An Peng
(642a) De Novo Design of Skin-Penetrating Peptides for Enhanced Transdermal Delivery of Peptide Drugs 194
Stefano Menegatti, Michael Zakrewsky, Sunny Kumar, Joshua De Oliveira, Samir Mitragotri
(642b) Ligand-Targeted Conjugate Systems for Delivery of siRNA to Tumors 195
Hok Hei Tam, Omid Veiseh, Robert Langer, Daniel G. Anderson
(642c) Design,Development and Optimization of pH-Responsive Hydrogels for the Oral Delivery of Human Growth Hormone 196
Stephanie Steichen, Nicholas Peppas, Colleen O'Connor
(642d) Design of Biodegradable Layer-By-Layer Films Capable of Localized, Efficient, and Sequential Gene Delivery 197
Lingxiao Xie, Xiong Ding, Guangzhao Mao
(642e) Aminoglycoside-Derived Liposomes for Drug Delivery 198
Sudhakar Godeshala, Cathy S. Madsen, Sandra J. Gendler, Kaushal Rege

(642f) Water Soluble Enteric Polymer 199
Jin Zhao, Oliver Petermann, Matthias Knarr, Josh Imperial, Debora Holbrook, Harold Bernthal

(642g) Preparation of a Drug Delivery System (DDS) of Hydropylic and Hydrophobic Drugs: In Vitro Validation in Human Lung Cancer Cells 201
Fernando Mesías-Recamán, Jesus Rodriguez-Rodriguez, Miguel A. Galán, Eva Martín del Valle

(661a) A QbD Approach for Identifying Critical Process Paramerters in Drug Substance Manufacturing 202
Michael Ryan, Geny Doss, Thomas Gandek, Patricia Hurter, Stephanie Krogmeier, Drew Kuzmission, Adam Looker, William Rowe, Derek Vrieze, David Willcox

(661h) QbD and Continuous: A Step Change in Performance 203
Philip Donnellan

(661c) A QbD Approach to Experimental Designs with Kinetic Parameters 204
Michael Ryan, David Willcox

(661d) New Trends in Materials-Sparing and Modelling Approaches Towards Pharmaceutical Development 205
Jasmine M. Rowe, Xiaodong Chen, Johnathan M. Goldman, Faranak Nikfar

(661e) The Journey of Mechanistic Models from Design to Optimization/ Launch… and Beyond 206
Carla Luciani, Salvador García-Muñoz, Kevin Seibert, Eric Crick, Thomas Wilson, Michael Kobierski, Colm O'Mahony

(661f) Scale-up of Reaction and Crystallization Steps of a Commercial Manufacturing Process in a Multi-Purpose Plant Using Mechanistic Modeling 207
Filipe A.P. Ataíde, Rui C. Silva, José L.C. Santos

(661g) Case Study: QbD Approach to Design Space Verification 208
David Willcox, Brian Phenix, Billie Kline

(683a) Development of Injectable Microgels for Galns Enzyme Replacement Therapy 209
Era Jain, Yasaman Chehreghanianzabi, Shiragi Patel, Michael Flanagan, Qi Gan, Adriana M. Montaño, Scott A. Sell, Silviya P. Zustiak

(683b) Computationally-Driven Design of a Heterobivalent and Gate to Improve Targeting Agent Selectivity 210
Sadie M. Johnson, Aakash Deshpande, Benjamin J. Hackel

(683c) Supramolecular Pegylation of Biopharmaceuticals 211
Matthew Webber

(683d) Non-Invasive Screening of Diseases Using an Orally Available Near-Infrared Imaging Agent 212
Sumit Bhatnagar, Greg Thurber

(683e) Investigation of the Cellular Uptake and Cytotoxicity of Cell-Penetrating Peptides in Candida Fungal Pathogens 213
Zifan Gong, Mackenzie T. Walls, Amy J. Karlsson

(683f) Catalase-Laden Microdevices for Cell-Mediated Enzyme Delivery 214
Junfei Xia, Zhibin Wang, Yuanwei Yan, Zhijian Cheng, Li Sun, Yan Li, Yi Ren, Jingjiao Guan

(683g) Ragweed Pollen-Based Oral Vaccine Delivery 215
Md Jasim Uddin, Harvinder Gill

(683h) The Roles of Competing Dissolution, Diffusion and Transient Marangoni Convection Fluxes in Surfactant-Enhanced Spreading of Aerosols for Pulmonary Delivery 216
Todd M. Przybycien, Robert D. Tilton, Stephen Garoff, Ramankur Sharma, Steven Iasella, Timothy Corcoran

(692a) Improving Single Use Bioreactor Design and Process Development, New Research Towards Intensifying Seed-Train and Scale-up Methods Using 5:1 Turn-Down 217
Nephi Jones, Christopher Brau, Ben Madsen, Tony Hsiao, Scott Collins

(692b) Characterization and Design of a 3L Single-Use Bioreactor to Support Process Development and Tech Transfer 218
Ankit Mehta, T. Craig Seamans, Zbynek Kalal, John Bowers

(692c) Foam Control in the Mammalian Cell Culture Process through Extrinsic and Intrinsic Methods 219
Kevin Chang, Douglas Osborne, Rachel Ferguson, Richard Gilmore, Dane A. Grismer, Orlin D. Velev

(692d) Morphological and Life Cycle Changes Exhibited at the up-Scaling of a Saccharomyces Cerevisiae High-Cell Density Fed-Batch Biopharmaceutical Process 220
Juan Aon, Ricardo Tucson, Vani Patel, Vakhtang Loladze

(692e) Accelerating Manufacturing in a Clinical Drug Substance Plant through Application of Plant Simulation 221
John Robinson

(692f) Model Predictive Control: A Process Development Tool for Mammalian Cell Culture Processes 222
Jessica Whelan, Mark Barrett, Brian Glennon, Stephen Craven

(692g) Accurate Determination of B22 Data – Multivariate Analysis to Improve B22 As a Predictive Tool 223
Sarah Hedberg, Daryl Williams, Jerry Heng

(692h) Multivariate Data Analysis (MVDA) for Cell Culture Process Development 224
John Bowers, John Higgins, Maria Khoiuzam, Louis Obando, Balrina Gupta

(701a) Gold-Layersome Nanoshells: Preparation and Characterization for Photothermal Therapy 225
Akram Abbasi, Keunhan Park, Arijit Bose, Geoffrey D. Bothun

(701b) Chiral Nanoparticles for Enantioslective Adsorption and Separations 226
Nisha Shukla, Andrew J. Gellman, Darwin Yang

(701c) Triggered Ligand Clustering on Lipid Nanoparticles Enables Selective Targeting and Killing of Untargetable Cancer Cells: The Case for 'Sticky Patches' 227
Michelle Sempkowski, Stavroula Sofou

(701d) Superparamagnetic Nanoparticles for Triggered Drug Release from Alginate Hydrogels 228
Georgios A. Sotiriou, Alexandra Teleki

(701e) Dual Drug-Loaded Janus Particles for Co-Delivery of Hydrophobic and Hydrophilic Compounds 229
Silvina Tomassone, Jennifer Winkler

(701f) Preparation of Colloidal Microcapsules for Controlled Drug Delivery Systems 230
Gokce Dicle Kalaycioglu, Nihal Aydogan
(701g) Stimuli Responsive Lipid Nanotubes for Drug and Protein Delivery 231
Hande Unsal, Nihal Aydogan
(708a) Uncertainty Analysis As Essential Step in the Establishment of the Dynamic Design Space of Primary Drying During Freeze-Drying 233
Thomas De Beer, Séverine T.F.C. Mortier, Krist V. Gernaey, Ingmar Nopens
(708b) Tunable Atomization of Viscous Liquids in the Production of Amorphous Solid Dispersions 234
Mark Kastantin, John Baumann
(708c) Secondary Drying Scale-up Methodology: Eliminating a Bottleneck with a Lean Development Approach 235
Tiago Porfirio, Pedro Valente, Ines Matos, Jorge Moreira, João Vicente, Márcio Temtem, Viriato Semião
(708d) Scientific and Regulatory Considerations on Particle Engineering for Dry Powder Inhaler (DPI) 237
Zhigang Sun
(708e) Production of Pharmaceutical Tablets By FDM 3D Printing 238
Matej Novak, Ales Zadrazil, Frantisek Stepanek, Pavel Kovacik
(708h) Experimental Analysis of Pharmaceutical Twin-Screw Wet Granulation Using Hydrophilic and Hydrophobic Formulations 239
Maxim Verstraeten, Michael Ghijs, Daan Van Hauwermeiren, Ingmar Nopens, Thomas De Beer, Kai Lee, Neil Turnbull, Mary T. am Ende, Pankaj Doshi
(732a) Alginate/Chitosan Microparticles for Gastric Passage and Intestinal Release of Therapeutic Protein Nanoparticles 242
Kevin Ling, Julie A. Champion
(732b) Photodynamic Enhancement of Drug Delivery to Tumors 243
Huang-Chiao Huang, Imran Rizvi, Joyce Liu, Ashish Kalra, Helen Lee, Jaeyeon Kim, Jonathan Fitzgerald, Tayyaba Hasan
(732g) Highly Efficient Encapsulation of Small-Molecule N-Acetylcysteine within PLGA Nanoparticles to Restore Redox Balance in Oxidant-Stressed Environments 244
Nick P. Murphy, Kyle Lampe
(732d) Fabrication and Characterization of PLGA Nanoparticle-Bacteria Conjugate Nanobeads for Increased Tumor Penetration 245
Richey M. Davis, SeungBeum Suh, Ami Jo, Bahareh Behkam
(732e) Interstitial Release of Cisplatin from Triggerable Liposomes Enhances Efficacy Against Triple Negative Breast Cancer Solid Tumor Analogues 246
Sally Stras, Tim Holleran, Alaina Howe, Stavroula Sofou
(732f) Bioresponsive Polymer Coating on Targeted Drug Nanorods 247
Tunyaboon Laemthong, Caitlin Brocker, Dipak Barua, Daniel Forciniti, Sutapa Barua
(732c) Internalization of Nanoparticles Functionalized with Low Molecular Weight Protamine into Erythrocytes 248
Rangika Hikkaduwa Koralege, Kaustuv Sahoo, Nicholas Flynn, Steven Hartson, Jing Pope, Ashish Ranjan, Carey Pope, Joshua Ramsey
(732h) Enzyme Replacement Therapy Extended to the Brain through Nano-Polymersomes 249
Jessica Kelly, Doug Martin, Mark Byrne
(737a) Regulatory Considerations for Introducing New Technologies during Drug Development 250
Rapti Madurawe
(737b) Viscoelasticity and Molecular Mobility of Celecoxib/Pvp/Tpgs Melt: A New Approach for Amorphous Drug Dispersion Assessment during Hot Melt Extrusion Product Design 251
Elba Garcia, Adlin Mendoza, Zuleyka Morales, Darlene Santiago
(737d) Miniaturized Purification Platform for Automated Screening of Flow Synthesis Using Extraction and Crystallization 255
Andy Koswara, Edward Barks, Claire Yiqing Liu, Joseph Oliva, Bradley P. Loren, Michael Wleklinski, Zinia Jaman, Ryan T. Hilger, James R. Zimmerman, Conor Parks, David H. Thompson, Robert G. Cooks, Zoltan K. Nagy
(737e) Novel Validation Method Development for a Drug Delivery Device Used During Breastfeeding 256
Rebekah Scheuerle, Stephen Gerrard, Krishnaa Mahbubani, Richard Kendall, Sylvaine Bruggraber, Catherine Tuleu, Nigel Slater
(737f) Photothermal Therapy of Bladder Cancer and Melanoma Via Targeted Single-Walled Carbon Nanotubes 257
Needa Virani, Patrick McKernan, Barry Lavine, Rajagopal Ramesh, Robert Hurst, Joel Slaton, Roger Harrison
(737g) Bioproduct Design through Market Assessment: NSF Innovation Corps (I-CorpsTM) Participation Guides the Development of New Escherichia coli Strains 258
Bob Beitle, Ellen M. Brune, McKinzie Fruchtl, Rudra Palash Mukherjee
(737h) Recombinant Snake Antivenom Peptide 259
Claire F. Komives, Michael Balderrama, John Li, Elda E. Sanchez
(742a) CFD MODEL Based Near-Infrared Spectroscopy Implementation for in-LINE ASSAY Monitoring of a Pharmaceutical Suspension 260
Maxim Verstraeten, Thomas De Beer
(742b) Using Optical Coherence Tomography (OCT) to Analyze in Real-Time the Tablet Coating Variability of a Continuous Drum Coating Process 261
Patrick R. Wahl, Manuel Zettl, Stephan Sacher, Rainer Bastian, Thomas Fränkel, Johannes G. Khinast, Matthias Wolfgang
(742c) Advanced Flexible Control System Implementation into Direct Compaction Continuous Pharmaceutical Manufacturing Pilot-Plant 262
Ravendra Singh, Fernando J. Muzzio, Rohit Ramachandran, Marianthi Ierapertritou

(742d) Raman Spectroscopy- Towards the Prediction of Quality Attributes and Application in Cell Culture Process Development 263
Fabian Feidl, Michael Sokolov, Ines Da Silva Costa, Jean-Marc Bielser, Jonathan Souquet, Hervé Broly, Carsten Uerpmann, Massimo Morbidelli, Alessandro Butté
(742e) Hydrodynamics and Mass Transfer Effects in Mini Vessel Dissolution Testing Systems 264
Bing Wang, Piero M. Armenante, Gerard Bredael
(742f) 3D Raman Spectroscopy – A Method for Mapping Ingredient Distribution Throughout a Tablet Volume 265
Savitha Panikar, Wei Meng, Pallavi Pawar, Sarang Oka, Rudy Hofmeister, Fernando Muzzio
(742g) Real-Time Monitoring and Modeling of Milling in a Continuous Tableting Line Via Dry Granulation 266
Mariana Moreno, Qinglin Su, Zoltan K. Nagy, Gintaras Reklaitis
(749a) Reducing Cut Size of Uniflow Cyclones with Optimized Swirl Vane Design 267
Tobias Kofler, Martin Pillei, Michael Kraxner
(749b) Development of Framework for Undesired Agglomeration in an Agitated Filter Drying Process 268
Hong Lee Lim, Karen Hapgood, Brian Haig
(749c) Modeling of the Pressure Drop Across Polydisperse Packed Beds in Depth Filtration 276
Siying Zhang, Joseph J. McCarthy
(749d) Measurement of the Axial Dispersion of Cohesive Powders in Rotary Kilns 277
Ingrid J. Paredes, Benjamin Glasser, William G. Borghard, Bereket Yohannes, Alberto Cuitiño, Fernando J. Muzzio
(749e) Residence Time Distribution of Fine to Coarse Particles in Rotary Kilns 278
Maryanna De Melo Soledade, Alex Stéphane Bongo Njeng, Stéphane Vitu, Jean-Louis Dirion, Marie Debacq
(749f) Evaluation of Small-Scale Powder Flow Characterization Tests 279
Aaron Zettler
(749h) Assessment and Predictive Modeling of Pharmaceutical Powder Flow Behavior in Small-Scale Hoppers 280
Mark Polizzi, Jon Hilden, Justin Franchville
(766a) Application of PAT-Based Feedback Control for the Crystallization of Pharmaceuticals in Porous Media 281
David Acevedo, Jing Ling, Keith Chadwick, Zoltan K. Nagy
(766b) Optimization of the Bottom-up Precipitation of Nano-Particles of an Ultra-Fast Crystallizer Through the Qualitative Characterization of API-Polymer Interactions 282
Athanas Koynov, Luke Schenck, Busolo Wabuyele, Amy L. Quach
(766c) Core-Shell Nanocomposites for Fast Release of Hydrophobic Active Pharmaceutical Ingredients: Impact of Carrier Size 283
Mohammad Azad, Jacqueline Moreno, Ecevit Bilgili, Rajesh N. Dave
(766d) 'Designer' Hydrogel Particles for Controlled Crystallization and Release of Hydrophobic Active Pharmaceutical Ingredients 284
Abu Zayed Md Badruddoza, P. Douglas Godfrin, Allan S. Myerson, Bernhardt L. Trout, Patrick S. Doyle
(766e) Dissolution of Carbamazepine Crystallized Directly Onto Excipients 285
Vivek Verma, Clare M. Crowley, Peter Davern, Sarah Hudson, B. Kieran Hodnett
(766f) Modeling Spray Dried Dispersion: From Droplet to Particle 289
Jaime Curtis-Fisk, Priti Jain, Jodi Mecca, Bart Rijksen, Shrikant Khot, William Porter
(766g) Modeling Dissolution of Polymer Strip Films Loaded with Poorly Water-Soluble Drug Nanoparticles 293
Scott M. Krull, Ecevit Bilgili, Rajesh N. Davé
(766h) A Novel Method for Grindability Assessment of Pharmaceutical Powders 294
Tina Bonakdar, Muzammil Ali, Mojtaba Ghadiri, Arjen Tinke
(767a) Feedforward Control of Continuous Pharmaceutical Manufacturing Process 295
Ravendra Singh, Glinka Cathy Pereira, Nikita Soni, Andrés D. Román-Ospino, Marianthi Ierapertritou, Rohit Ramachandran
(767b) Simultaneous Measurement of Granule Disintegration and API Dissolution Profiles As a Tool for Maintaining Product Quality during Process Scale-up 296
David Smrcka, Jiri Dohnal, Veronika Lesakova, Zdenek Grof, Frantisek Stepanek
(767c) Particle Characterization Technique to Enable Rapid Pharmaceutical Formulation and Process Optimization 297
Mark Menning
(767d) HPMC Cloud Point: Exploring Hydroxypropylmethyl Cellulose Behavior in Pharmaceutical Formulations 298
Maryam Mohammadpour, Brian Krins, Mary Ellen Crowley, Gavin Walker, Abina Crean
(767e) QbD for Generic Modified Release Drug Products 302
Yue Teng
(767f) Utilization of Quality By Design Principles to Define Formulation Best Practices for a Direct Compression Excipient 303
Kathryn Hewlett, True Rogers, Karen Balwinski, Jaime Curtis-Fisk, Robert Schmitt, Shrikant Khot
(767g) Microcrystalline Cellulose – Does Wood Pulp Source Impact the Direct Compression Performace of This Excipient? 306
Mary Ellen Crowley, Abina Crean
(774a) Understanding the Role of Hpmcas in Phenytoin Solid Dispersion Formulation Using Coarse Grained Molecular Dynamics Simulations 311
Wenjun Huang, Ronald G. Larson
(774b) Assessing the Efficacy of Thermo-Responsive Poly(N-isopropylacrylamide) in Phenytoin Solid Dispersion Formulation Using Atomistic Molecular Dynamics Simulations 312
Soroush Moghadam, Ronald G. Larson
(774c) Solid State Form Understanding Applied to the Design of a Spray Dried Dispersion Process: A Quality By Design Approach to Optimization of Feed Solution Loading 313
Kyle Blakely, Christopher L. Burcham

(774d) Ternary Dispersions of Itraconazole: A Comparison of Hot Melt Extrusion and Spray Drying 314
Mark Davis, Catherine Kelly, Gavin Walker
(774f) Tunable Drug Loaded Stable and Fast Release Amorphous Composites Production Using Mesoporous Neusilin®UFL2 315
Mohammad Azad, Jacqueline Moreno, Meng Li, Rajesh Dave
(774g) Hot Melt Extrusion of Ternary Solid Dispersions of Itraconazole: Amorphous Vs Liquid Crystal Intermediate 316
Ahmad B. Albadarin, Catherine Kelly, Mark Davis, Maryam Mohammadpour, Sachin Korde, Sudhir Pagire, Anant Paradkar, Gavin Walker
(780a) Investigation of the Dfferences in Material Properties Between Amorphous Dispersions Created By Conventional Methods and High-Shear Co-Precipitation 317
Athanas Koynov, Luke Schenck, Narayan Variankaval, Andrew Brunskill, Amanda Mann, Yongjun Li, Dina Zhang, Dirk Stueber, Paul Walsh
(780b) Melt Rheology of Cellulose Ethers Designed for Hot Melt Extrusion 318
Tirtha Chatterjee, Kevin O'Donnell, Yongfu Li, Uma Shrestha, Robert Sammler, Shrikant Khot
(780c) The Water Sorption Behaviour of Amorphous Solids: Experimental and Theoretical Insights Using a Model Protein BSA 319
Daryl Williams, Niklas Henning Jahn
(780d) A Kinetic Approach to Polymorph Prediction: Polymorph Specific Alpha and Beta Glycine Nucleation Rates and Solubility Curves from Molecular Dynamics 320
Conor Parks, Andy Koswara, Hsien-Hsin Tung, Nandkishor Nere, Shailendra Bordawekar, Zoltan K. Nagy, Doraiswami Ramkrishna
(780e) Silica Nanoparticle-Seeded Microfluidic Emulsion Crystalisation - Crystalline Drug Composite Microparticles 321
Saif A. Khan, Denise Zi Ling Ng, Eunice Wan Qi Yeap
(780f) The Effects of Crystal Size and Habit on the Dehydration Induced Polymorphism; The Case of Carbamazepine Dihydrate 322
Eftychios Hadjittofis, Geoff G. Z. Zhang, Jerry Heng
(780g) Raman Mapping for Analyzing Multi-Component Solid Dosage Formulations 325
Zhen Liu, Haichen Nie, Patrick Marsac, Brian Marks, Wei Xu, Anthony Leone
(787a) Optimizing Process Yields of a Legacy 25,000L Mammalian Cell Culture Process Using Integrated Upstream and Downstream Process Yield Models 326
Simpson Gregoire, Jessica Woodman, Sivaprakash Agastin, Sivashankar Sivakollundu, Jeffrey Savard
(787b) Intelligent Control Strategies Towards the Intensification of Monoclonal Antibody Production Via Continuous Operation 327
Maria M. Papathanasiou, A. Mantalaris, Efstratios N. Pistikopoulos
(787c) Development of a Scalable Process for Production of a Broadly Protective Influenza Antigen 328
Julie Fogarty, Yuan Lu, James Swartz
(787d) Application of Process Monitoring Data Management Systems to Make Multiple Incremental Improvements to Cell Cultures Processes at the Manufacturing Scale 329
Christopher C. VanLang, Erin K. Abbott, June Axelson, Hugh Graham
(787e) Scalable Production of Glioblastoma Cancer Stem Cells in 3 Dimension Thermoreversible Hydrogels for Drug Discovery 330
Qiang Li, Yuguo Lei, Haishuang Lin, Ou W Ang
(787f) Levers to Modulate Product Quality Attributes of Product-a in CHO Cell Culture 331
Jayashree Subramanian, Abigail Pynn, Lauren Tang
(787g) Data-Driven Optimization Using an Evolutionary Design of Dynamic Experiments for Biopharmaceutical Processes 332
Zhenyu Wang, Christos Georgakis
(788a) In Silico Formulation Development of Oral Solid Dosage Forms Using a Material Property Database 333
Maxx Capece, Ping Gao, Raimundo Ho, David O'Brien, Ahmad Sheikh, John C. Strong, Divya Sunkara
(788b) Prediction of Tablet Dissolution from API and Excipient Polymer Properties 334
Jaime Curtis-Fisk, True Rogers, Karen Balwinski, Kathryn Hewlett, Robert Schmitt, Shrikant Khot
(788c) Connecting Processed Material's Properties to Process Development Strategy 337
Ajit Narang
(788d) Decoupling of a Low-Dose Dosator Capsule Filling Process in Dynamic and Static Mode Tests to Understand the Extend of the Effect of Powder and Process Parameters on Capsule Quality Attributes 338
Eva Faulhammer, Sandra Stranzinger, Peter Loidolt, Vittorio Calzolari, Stefano Biserni, Stanko Srcic, Rok Dreu, Amrit Paudel, Johannes G. Khinast
(788e) Quantitative Investigation on Segregation Phenomenon Using Discrete Element Method in Large Scale Pharmaceutical Compression Process 342
Yijie Gao
(788f) Two-Phase Flow (Solids: Interstitial Air) Modeling of a Press Feed System 343
Thomas Baxter, Roger Barnum, James K. Prescott, John Carson
(788g) Use and Misuse of the Diametrical Compression Test 344
Jon Hilden, Mark Polizzi, Aaron Zettler
Author Index

462053 Maximizing Poorly Water-Soluble API Nanoparticle Loading in Polymer Strip Films

Abstract:

One of the most prevalent misconceptions of the polymer strip film format for pharmaceutical applications is its limited active pharmaceutical ingredient (API) loading. While this is partially due to the size restriction of a typical orodispersible dosage (~6 cm^2 and ~40 μm thick), this limitation also stems from API particle stability issues in traditional pharmaceutical film manufacturing processes. For instance, solvent casting imposes an inherent limit on the API loading in the resulting film, as exceeding a certain API loading leads to uncontrollable precipitation of API particles upon removal of organic solvent during drying. The objective of this work was to challenge the previously conceived limits of API loading in pharmaceutical strip films using a combination of API nanomilling and slurry casting, as well as evaluate the practical limitations of strip films with high loadings of API nanoparticles. Hydroxypropyl methylcellulose (HPMC) was used as the film-forming polymer and griseofulvin (GF) was used as a model BCS class II API. HPMC-E15 films were prepared with API loadings between 9â€“49 wt% GF and HPMC-E4M films were prepared with API loadings between 30â€“73 wt% GF. All HPMC-E15 films exhibited good content uniformity and API nanoparticle redispersibility, whereas HPMC-E4M films had slightly worse content uniformity and poor API nanoparticle redispersibility above 50 wt% GF loading. Although film strength remained strong with increasing GF loading, a significant decrease in percent elongation was observed, resulting in film brittleness. Results demonstrate that, with proper particle engineering and stabilization protocols, high loading of poorly water-soluble API nanoparticles is indeed achievable in polymer films.

451652 Computational Approaches for Modeling Fluid Bed Granulation Process

Abstract:

Understanding the fundamental multi-physics behind the fluid bed granulation process and its projections in macroscopic scale provide the essentials to fabricate granulate with desired material properties. A proper prediction of physicochemical properties and morphological characteristics of a continuous wet granulation process product, particularly in response to operation conditions, is contingent upon the knowledge of the contact pattern evolution and the microstructure formation under the effect of the fluidizing medium. In this regard, current study looks at unveiling correlation between the process conditions and the main mechanisms that determine the material properties of granulate during fluid bed granulation process.

A multi-component modeling approach was explored that enables us to understand the fundamentals of fluid bed granulation process through both statistical modeling and first-principle modeling tools. A supervised machine learning algorithm that can simulate the evolution of process variables during fluid bed granulation process was explored, including: particle diameter, particle size distribution and loss on drying. Model predictions agree well with experimental analysis. Currently the focus is on a numerical framework that integrates design of experiment and the findings of statistical models with multi-scale mechanistic models such as discrete element model and computational fluid dynamics models. Using such models in early and late stage development phases will enhance the development timelines and increase the process product understanding.

461274 Comparison of the Electrostatic Properties of Amorphous and Crystalline Lactose

Abstract:

Background

Conversion from a crystalline to amorphous state in solid materials may be observed as a result of various processes applied to pharmaceutical materials during manufacturing. Examples include rapid crystallisation from solution to mechanical activation of the crystalline phase upon milling.[1-2] The resulting products usually display different thermodynamic properties to pure crystalline forms, hence affecting their subsequent performance.[2-3] Furthermore, powders are susceptible to a process called tribocharging whenever movement of particles across the surface is involved. If surface amorphization is observed during processes involving the movement of powder, such as during milling, generation and accumulation of electrostatic charge could be affected significantly, hence affecting the overall powder flow and sticking.

Aims

The aim of this work is to determine how the presence of an amorphous state influences the electrostatic performance of lactose. The main objectives involve determining whether the material electrostatic characteristics such as the maximum charge accumulated and charge decay behaviour are directly dictated by the materialâ€™s state.

Methods

Amorphous lactose was prepared by spray drying from water and storing at dry conditions to limit its re-crystallisation. The electrostatic properties of both pure crystalline and pure amorphous lactose with known particle size were studied using JCI 155v6 Charge Decay Analyser[4] under controlled environmental conditions (25 Â°C and 25% RH). Subsequently, four blends containing 15%, 30%, 50% and 75% amorphous content by mass were prepared and characterised using the same method. All samples studied were further characterised using PXRD, RST and SEM. The analysis for further applied to anhydrous lactose prepared by dehydration of crystalline lactose by exposing to elevated temperatures.

Results

Electrostatic characterisation of pure amorphous and crystalline lactose revealed that the materials showed significantly different behaviour in terms of maximum charge accumulated as a result of corona charging and decay characteristics. The maximum charge of -684.65 ± 45.04 V/g acquired by the pure amorphous lactose was twice as large as the charge accumulated by the pure crystalline sample (-337.64 ± 5.56 V/g). Although the smaller particle size of amorphous lactose could potentially account for its greater overall charge accumulation, the decay profiles are significantly different for the two forms with the amorphous lactose showing significantly slower charge decay. The time of 113.19 ± 8.30 min is required for the charge on the amorphous lactose to decay to 10% of its initial value, whereas only 10.03 ± 0.16 min is needed for the crystalline sample to reach the same level. Results obtained for four blends with varied amorphous content show that the material performance is directly dictated by its composition and a linear correlation was obtained between charge accumulated and amorphous content of the blends. The same correlation was observed for the charge decay profiles.

Conclusion

The results obtained show that amorphous lactose is able to accumulate a larger amount of electrostatic charge, which is retained for a prolonged time compared to its crystalline form. This behaviour is then directly translated into lactose blends having varied amorphous content, where both components exist in a form of physical mixture. Even though amorphous lactose is unstable and tends to convert to crystalline form quite rapidly, its generation during processing could lead to significant charge generation and accumulation that otherwise would not be expected for a crystalline material.

Acknowledgments

The authors acknowledge Dr Jin Wang Kwek for his contribution to this work.

References:

(1) Descamps, M.; Willart, J. F.; Dudognon, E.; Caron, V., Transformation of pharmaceutical compounds upon milling and comilling: the role of T(g). *J. Pharm. Sci.* **2007,** *96* (5), 1398-1407.

(2) Hancock, B. C.; Zografi, G., Characteristics and significance of amorphous states in pharmaceutical systems. *J. Pharm. Sci.* **1997,** *86* (1), 1-12.

(3) Fitzpatrick, J. J.; Hodnett, M.; Twomey, M.; Cerqueira, P. S. M.; O'Flynn, J.; Roos, Y. H., Glass transition and the flowability and caking of powders containing amorphous lactose. *Powder Technol.* **2007,** *178* (2), 119-128.

(4) Rowland, M. G. Electrostatic Properties of Particles for Inhalation. University of Bath, 2014.

458202 Particle Engineering Research from the NSF Erc Research Experience for Teachers (RET) Program at Njit

Abstract:

We aim to give highlights of particle engineering research from the New Jersey Institute of Technology (NJIT)'s Research Experiences for Teachers (RET) Program, which is a collaboration of the NSF Engineering Research Center for Structured Organic Particulate Systems (ERC-SOPS) and the Center for Pre-College Programs. This NSF-funded program engaged pairs of 14 high school teachers in 2015 summer research program, where they performed research on various particulate materials and processes relevant to pharmaceutical products. In addition, various faculty and educational experts delivered workshops on technical writing, best research practice, effective collaboration in a team, pharmaceutical industry and process–formulation development, educational module development, etc. The teachers acquired the skills and knowledge of research and the subject of particle engineering and pharmaceutical engineering that they incorporated into their teaching practice. In the research component of the program, the teachers collaborated with graduate student mentors under the guidance of professors and worked on various particle–pharmaceutical engineering projects. In the educational component, the teachers came up with a professional development plan and prepared "educational modules", which were delivered to high school students. While this presentation will expose few educational highlights including sample modules developed by the teachers, it will mainly focus on two summer research projects about the bioavailability enhancement of poorly water-soluble drugs via drug nanoparticle composites and drug nanoparticle-laden polymer strip films. Through application of particle engineering approaches such as nanoparticle formation–stabilization and drug encapsulation within hydrophilic polymeric matrices, we have ensured fast redispersibility of drug nanoparticles from solid dosages and demonstrated significant improvement of the dissolution rate of poorly water-soluble drugs.

463026 Study of DPI Carriers Design Parameters for Dry Powder Inhalation through Artificial Neural Network Modeling

Abstract:

Carriers are commonly used in dry powder inhalation to enhance the flowability, aerosolization and drug delivery efficiency of fine drug particles. The properties of carriers such as, size, shape, density, surface roughness, as well as carrier-to-drug ratio, inhalation flow rate, type of dry powder inhalers have important impacts on the drug delivery efficiency. However, there are contradictory reports in the literature on the effects of individual parameters on drug delivery efficiency. In this study, the effect of carrier properties, carrier-to-drug ratio, and inhalation flow rate on drug delivery efficiency in terms of emitted dose (ED) and finc particle fraction (FPF) are examined using artificial neural network (ANN) modeling. Image analysis is performed on scanning electron microscopy (SEM) images of carriers to determine the surface properties. Sensitivity analysis is carried out to determine important parameters that have the most impact on ED and FPF. Genetic algorithm (GA) is applied to determine the ANN modeling parameters. The effects and interactions of various factors on ED and FPF are revealed.

453857 The Surface Properties of Organic Crystalline Solids and Their Interactions with Polymeric Excipients and Binders

Abstract:

The physiochemical properties of solid state pharmaceuticals are crucial in understanding drug manufacturing, processing, and biological absorbtion. They govern the flowability, dispersibility, bulk density, dustiness, compressibility, and particle aggregation of active pharmaceutical ingredients (APIs)[1]. Binders and excipients are added to the formulation so as to achieve greater uniformity in the final product, and to enlarge the drug particles by agglomeration. Thus, particle design and characterozation are crucial for the understanding of the product being delivered to the consumer[2,3]. Many advances have been made in this research but given there are many mechanisms that affect the final product, itâ€™s difficult to model and predict what will occur accurately, such as adhesion, attrition, coalescence etc[4].

In pharmaceutical processes micron and submicron sized particles are used. In this scale, surface properties such as surface free energy heavily influence particle behaviour[5]. Surface free energy, or more conveniently surface energy, is defined as the energy required to form a surface of a particular material per unit area, where various components make up the total surface energy for one specific facet of a crystal. By having a better understanding of these components and the surface energies, researchers can better predict the properties of drugs. This includes how particles form, grow, break, stick, flow, and more. This then offers more data for researchers to understand how the APIs interact with polymeric excipients and binders.

Current research and models have too often made assumptions disregarding the surface heterogeneity of the particles[6]. Due to the anisotropy of the crystal, the properties are vastly more complex than the simple assumptions of spherical homogeneous particles[6]. Measuring these differences is crucial in the further development of this field and drug formulation as a whole. The most appropriate way to measure the facet specific surface energies is by sessile drop contact angle measurements or inverse gas chromatography[5]. This can be expanded further to compare the energetic interactions of pharmaceuticals with solvated excipients and see how we may better predict the physiochemical properties of the resulting particles[4].

Our study chooses to focus on the formulation of pharmaceuticals alongside polymers so as to offer a better understanding of their properties during such processes. This will offer better control in their manufacturing and later ties in to the drugs bio-availability. Our investigation is divided into two parts, the first involves property analyses of the crystals and powders of the API, and the second the interactions between the drugs and excipients. Here we chose to use both three polymorphs of anhydrous Carbamazepine and Copovidone as our model API and excipient respectively. Together, we offer further data to facilitate the modelling in the particle design of pharmaceuticals.

We performed studies on the polymorphs of anhydrous Carbamazepine using: sessile drop contact angle measurements (CA), finite dilution inverse gas chromatography analysis (FD-iGC), and X-ray photoelectron spectroscopy (XPS). The interactions between Copovidone and the main facets of P-Monoclinic Carbamazepine were analysed with two experiments. In the first, we measured the contact angle of Copovidone dissolved in water at various amounts, on the crystal facets of Carbamazepine. In the other we measured the contact angle of water on the crystals facets with spun cast Copovidone on them. The combination of these instruments helps elucidate and predict some of the surface properties that may affect drug formulation, offering more comprehensive data on their anisotropy.

P-Monoclinic seeds (with dimensions about 0.5 x 0.5 x 0.1 cm) were prepared by two step cooling crystallization of a supersaturated methanol solution. These seeds were suspended to grow in a, slowly evaporating, supersaturated methanol solution6[1]. The crystals grew to sizes up to 5x3x3 cm, with three main facets exposed; (100), (002) and (010). CA measurements, using diiodomethane, water, formamide and ethylene glycol, were performed on these facets to determine their surface energy, both the Lifshitz-van der Waals and the acid-base components (using VOCG approach), and their affinity to water. The dispersive surface energy ranges from about 48 mJ/m^2 to 40 mJ/m^2 depending on the facet. The acid-base component has a smaller contribution, with its value going up to 10 mJ/m^2. However, there is large variation between the acid and the base component. The acidic component can be more than one order of magnitude larger than the basic one.

XPS analysis was conducted on the facets showing that strong amine presence on the surface of two of them. Contact angle measurements suggests that these two facets are the more hydrophilic. From the other hand the carbonyl component of the facets, less abundant than the amine one, does not seem to correlate with the water affinity.

P-Monoclinic Carbamazepine was recrystallized by stirring a supersaturated ethanol solution and the Triclinic polymorph was prepared by heating the P-Monoclinic polymorph at 160°C overnight[7]. FD-IGC measurements were performed on the two polymorphs to determine their surface energy heterogeneity. It was found that the most stable Monoclinic polymorph is lower energetically than the Triclinic. Computational analysis of the surface energy map of the Monoclinic polymorph shows some agreement with the measured surface energies[8].

We measured the contact angles of 0.1%, 0.25%, 0.5%, 1%, 1.5%, 2%, 2.5%, 3%, 4%, 5%, 10%, and 20% polymeric solutions by weight on three facets of P-Monoclinic Carbamazepine: (002), (010) and (100), to better understand how Copovidone interacts with facets of Carbamazepine. The amount of polymer in the solution was positively correlated to the solutions wettability on all three facets; the contact angle became smaller. Between the three facets the more apolar they were according to the FD-iGC, the more wettable they became as polymeric content increased. These increases were not identical, but were more pronounced as the dispersive component on the contact angle solution increased, i.e. as the polymeric content increased.

The data above was corroborated with water contact angle measurements of crystalline facets spun cast with Copovidone on top of them. This led to a mark increase in wettability, therefore the surface became more hydrophilic, in agreement with the data supplied in the previous paragraph. Della Volpe proposes that the acidic component of the surface energy of water is 6.5 times stronger than the basic one, from the other hand FD-IGC measurements on Copovidone reveal that it has a very strong basic component and a Lifsitz-Van der Waals component similar to that of Carbamazepine. Thus, it can be concluded that the increase in hydrophilicity is attributed to the acid-base interactions.

Previous studies have established the concept of anisotropy in pharmaceutical crystals. These anisotropic properties were proved to affect dry properties of pharmaceutical processes such as cohesion. However, this study goes beyond dry interactions revealing the importance of crystal anisotropy in solid-liquid interactions, particularly important in a wide range of pharmaceutical processes such as wet granulation, dissolution etc. It highlights the need for in depth understanding of crystal anisotropy in order to develop robust model predictive tools describing solid-liquid processes. Furthermore, it highlights the need for the application of complementary techniques to accurately describe crystal anisotropy and it also indicates the importance of computational models in these studies

References:

1. Ho, Raimundo, Majid Naderi, Jerry Y. Y. Heng, Daryl R. Williams, Frank Thielmann, Peter Bouza, Adam R. Keith, Greg Thiele, and Daniel J. Burnett. 2012. "Effect Of Milling On Particle Shape And Surface Energy Heterogeneity Of Needle-Shaped Crystals". *Pharm Res* 29 (10): 2806-2816.
2. Krycer, Ian, David G. Pope, and John A. Hersey. 1983. "An Evaluation Of Tablet Binding Agents Part I. Solution Binders". *Powder Technology* 34 (1): 39-51.
3. Simons, S.J.R., D. Rossetti, P. Pagliai, R. Ward, and S. Fitzpatrick. 2005. "The Relationship Between Surface Properties And Binder Performance In Granulation". *Chemical Engineering Science* 60 (14): 4055-4060.
4. Ho, Raimundo, Sarah E. Dilworth, Daryl R. Williams, and Jerry Y. Y. Heng. 2011. "Role Of Surface Chemistry And Energetics In High Shear Wet Granulation". *Industrial & Engineering Chemistry Research* 50 (16): 9642-9649.
5. Ho, Raimundo, Steven J. Hinder, John F. Watts, Sarah E. Dilworth, Daryl R. Williams, and Jerry Y.Y. Heng. 2010. "Determination Of Surface Heterogeneity Of D-Mannitol By Sessile Drop Contact Angle And Finite Concentration Inverse Gas Chromatography". *International Journal Of Pharmaceutics* 387 (1-2): 79-86.
6. Heng JYY, Bismarck A, Lee AF, Wilson K, Williams DR. Anisotropic surface energetics and wettability of macroscopic form I paracetamol crystals. *Langmuir*. 2006;22:2760-2769.
7. Grzesiak AL, Lang M, Kibum K, Matzger AJ. Comparison of the four anhydrous polymorphs of carbamazepine and the crystal structure of form I. *Journal of Pharmaceutical Sciences*. 2003;92(11):2260-2271.
8. Smith RR, Williams DR, Burnett DJ, Heng JYY. A new method to determine dispersive surface energy site distributions by inverse gas chromatography. *Lagmuir*. 2014;30(27):8029-8035.

NARROW TUBE SPRAY DRYING WITH PRECISE DRYING HISTORY CONTROL FOR PARTICLE ENGINEERING

May Ginn (Jaclyn) Lee[1], Shahnaz Mansouri[1], Karen Hapgood[1] and Meng Wai Woo[1],
(1) Chemical Engineering, Monash University, Clayton, Melbourne, Australia
Email: meng.woo@monash.edu

Introduction

The use of spray drying technology has become more prevalent and popular over the years in the drying industry. It is a one step process, using hot air to dry atomised sample solution, resulting dry powdered product in a matter of seconds. Spray drying is particularly suitable to be used in the pharmaceutical industry as the spray dried products offers consistency and meet the standards on particle size, density and morphology.[1] Amorphous and crystalline structures are most commonly discovered in spray dried particles. Amorphous structure provides high dissolution rate which is an advantage.[2] However, in the amorphous state, products are generally unstable and undergo phase transition during shelf life. This alters the functionality and properties of product which could lead to degradation.[3] In addition, amorphous particles are prone to develop into sticky particles causing wall deposition and reduce the overall efficiency.[4] In contrast, crystalline particle exhibit properties that are opposite to that of amorphous. Partial or complete crystallisation is capable of reducing stickiness and prevents wall deposition. They are flowable thereby making them simple and easy in handling. Most importantly, crystalline particle are stable and does not undergo phase transition upon storage, unlike amorphous particles which is only stable in low humidity.[5] Thus, many researchers are focusing on increasing the degree of particle crystallinity in spray dried products in order to overcome the issue cause by amorphous particles.

Many factors and parameters are involved in determining the degree of particle crystallinity with temperature being the highest impact followed by atomiser rate, pump rate, feed concentration and aspirator rate.[6] Previously, Islam and Langrish, 2009 used lactose as the study material and reported with a higher inlet drying temperature from 134°C to 210°C, the degree of crystallinity increases from 55% to 76% using a Buchi-290 laboratory scale spray dryer.[4] On the contrary, when it is scaled-up to a pilot plant, results shows the opposite as an increase in inlet drying temperature from 170°c to 210°C yields a decrease in particle crystallinity from 72% to 22%.[7] These contradicting results were due to inlet temperature being used as the parameter to describe the condition suitable to encourage particle crystallinity. Experiments carried out thus far were focusing on controlling the inlet or outlet drying temperature without any in-situ study on the effect of particle drying history along the chamber on particle crystallinity. Hence, this paper would be addressing this issue by creating various drying history using temperature as the primary factor in controlling particle crystallinity.

Current laboratory spray dryers have limitations in which only the inlets or outlet temperature is explicitly controlled. Thus, a new spray dryer capable of controlling the drying history is required for this study. Mansouri et al, 2015 proposed using a narrow tube spray dryer with internal diameter of 1.27-4.8 cm by introducing hot air at the inlet similar to current spray dryers.[8] With such narrow tube, heat loss through the walls was significant, therefore insulation was essential to meet the targeted drying temperature. A new concept was discovered based on the narrow tube dryer by utilising the rapid heat transfer through the tube walls as a mean to control the particle drying history. Instead of providing heat at the inlet of the drying chamber, heat can be provided through the walls of the tubes. The amount of heat transferred to the drying chamber can be easily manipulated by varying the wall tube temperature. As a result, in depth study on the effect of various drying history on particle crystallinity can be undertaken.

Modified Narrow Tube Dryer

In order to address the current limitations of spray dryers in controlling the drying history of particles throughout the drying chamber, a new modified spray dryer called *'narrow tube dryer'* was designed. The narrow tube dryer was designed based on the idea of controlling the temperature profile of air in the drying chamber through the heated tube walls. This is of paramount importance as the air would ultimately act as a drying medium for the particles through convective heat transfer. Thus, by controlling the temperature profile of air in the drying chamber, the effect of drying history on particles can be studied.

The narrow tube dryer consists of twelve 0.5 m length of copper tubes with inner and outer diameters of 1.7 and 1.9 cm respectively. Resistors connected in series are used as heating elements and are directly mounted on the tube walls which are controlled by a PID controller. Two thermocouples are placed on each tube to ensure uniform heating is achieved for each tube. In addition, a thermocouple is placed in between two connecting copper tubes at 0.5 m interval to measure the bulk temperature in the drying chamber. Since the wall temperature of each tube needs to be controlled independently, all tube are isolated from one another by placing gasket between them. By doing this, different wall temperature can be applied and varied accordingly for each independent tube. Particles are collected at the end of the drying chamber using a high efficiency particulate arrestance (HEPA) filter with the aid of a vacuum suction.

Material and Method

Lactose solution with 10 wt% solute was used as the study material in this experiment. Alpha lactose monohydrate was purchased from Sigma-Aldrich while milli-Q water was used to prepare the solution. The lactose solution was left to be stirred for 4 hours before it is spray dried in the narrow tube dryer. The average duration of spray drying for each run is approximately 1 minute. Final samples were collected on the HEPA filter paper before they are transferred to a Scanning Electron Microscope (SEM) stubs for viewing. All image analysis on particle morphologies were carried out using FEI Nova Nano SEM 450 FEG and a bench top SEM (Phenom World). Prior to viewing on SEM, samples are prepared on aluminium stub using double sided carbon tape and sputter coated with approximately 1nm Iridium. The accelerating voltage used was 3 kV with magnification ranging from 1000 to 40000 x. All images were taken using secondary electron imaging.

Manipulating the Particle Drying History

As the objective of this paper is to study the effect of drying history on particle morphology, 1-Dimension (1D) modelling of the temperature profile along the tube is paramount as it is capable of predicting important parameters of both the drying medium and the particle itself such as bulk temperature, particle temperature and moisture content. With 1D modelling, it helps in experimental planning since the results can be predicted before the actual run.

Heat, mass and momentum balance equations were used to determine the bulk temperature profile along the tube. Equation 1, 2 and 3 below capture the temperature profile of a particle, the change in absolute humidity during the drying process and the velocity of particle travelling horizontally respectively. These three equations were previously used by Patel for a vertical plug flow spray dryer. With the narrow tube dryer, equations for heat and mass balance will still be valid while some modifications were done for momentum balance since particles are travelling horizontally. [9]

$$\frac{dT_p}{dt} = \frac{hA_p(T_b - T_p) + \Delta H_L m_s \frac{d\bar{X}}{dt}}{m_w C_{p,w} + m_s C_{p,s}} \qquad \text{Equation 1}$$

$$\frac{dY}{dl} = -\frac{\theta}{\dot{V}\rho_b v_p}\frac{dm_w}{dt} \qquad \text{Equation 2}$$

$$\frac{dv_p}{dt} = \frac{-0.75 C_D \rho_b (v_p - v_b)|v_p - v_b|}{d_p \rho_p} \qquad \text{Equation 3}$$

Another important parameter to be considered is the rate of evaporation of the droplet which accounts for the changes in the particle moisture content. Equation 4 was used to predict the rate of evaporation.[10]

$$-\frac{dm_w}{dt} = -m_s\frac{d\bar{X}}{dt} = h_m A_p\left[\rho_{v,sat} exp\left(-\frac{\Delta E_v}{R_g T_p}\right) - \rho_{v,b}\right] \qquad \text{Equation 4}$$

The apparent activation energy, ΔE_v in Equation 4 is unique for each type of material and its initial concentration. Though Equation 5 below is used for lactose solution with initial concentration of 20%, it is concluded that Equation 5 is suitable to be used for initial concentration of 10% as well since the difference is negligible. [11]

$$\frac{\Delta E_v}{\Delta E_{v,b}} \approx 1.017 exp[-1.678(\bar{X} - X_b)^{1.018}] \qquad \text{Equation 5}$$

By incorporating Equation 1-5, the bulk temperature profile along the tube can be predicted using Equation 6 and this was used to compare with the experimental results.

$$\dot{V}\rho_b C_{p,b}\frac{dT_b}{dl} = \frac{\theta}{v_p}\left[\frac{dm_w}{dt}\{\Delta H_L + C_{p,v}(T_b - T_p)\} - hA_p(T_b - T_p)\right] - \dot{V}\rho_b(\Delta H_L + C_{p,v}T_b)\frac{dY}{dl} + [U(\pi D_e)(T_w - T_b)] \qquad \text{Equation 6}$$

The overall heat transfer coefficient, U in equation 6 was calculated by taking into consideration the thermal conductivity of the copper tube, k which is 400 W/(m.K) and the internal convective heat transfer coefficient. The external heat transfer coefficient can be eliminated since a constant wall temperature was maintained for all tubes. The overall heat transfer coefficient, U is calculated using equation 7.

$$\frac{1}{U_o} = \frac{d_o ln\left(\frac{d_o}{d_i}\right)}{2k} + \frac{d_o}{d_i} \times \frac{1}{h_i} \qquad \text{Equation 7}$$

All the parameters in Equation 6 were calculated using the equations mentioned above. The only manipulated parameter was the wall temperature of the tube, T_w . Thus, by varying the wall temperatures of each tube, the bulk temperature, T_b was calculated along the tube creating unique drying history specific to the wall temperatures. Equation 6 was used as the basis in comparing and validating the experimental data. With this verification, other crucial parameters such as T_p and X can be predicted and aid in analysing particle morphology. Overall, 1D modelling was used as a tool to predict particle drying history and study its effect on particle morphology.

Results and Discussion

Control Model

Prior to using 1D model as a tool in predicting crucial parameters throughout the drying process, it was important to first validate its applicability on the modified narrow tube dryer. In the first section, two runs were carried out and for each run, theoretical bulk temperatures were calculated with and without the presence of atomised droplets by equating θ which refers to the mass flow rate of reactant to 1and 0 respectively in Equation 6.

Table 1. Wall temperatures for Run 1 and 2

Tube No.	1	2	3	4	5	6	7	8	9	10
Run 1 (°C)	29	42	49	61	34	79	84	91	92	99
Run 2 (°C)	68	76	82	74	34	60	69	67	67	86

Table 1 shows for both Run 1 and 2, in which the wall temperatures were set in an increasing manner with a sudden decrease in the middle section. Run 1 was set for a gradual increase while Run 2 was set for a rapid increase in bulk temperature. Based on Figure 1 and 2, experimental results follow closely to the theoretical results, displaying similar trends. Theoretical results shows that with the presence of atomised droplets, the bulk temperature was approximately 2-3°C lower than the run with air alone. This agrees with the experimental results whereby bulk temperature with atomised droplet is lower than that of air alone. The small difference is contributed by the low mass loading of atomised droplets, leading to a negligible effect on bulk temperature. [12]

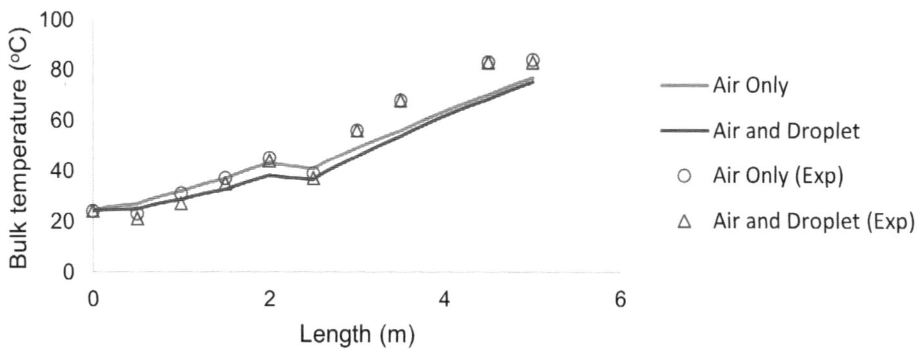

Figure 1. Run 1-Bulk temperature with and without the presence of atomised droplet

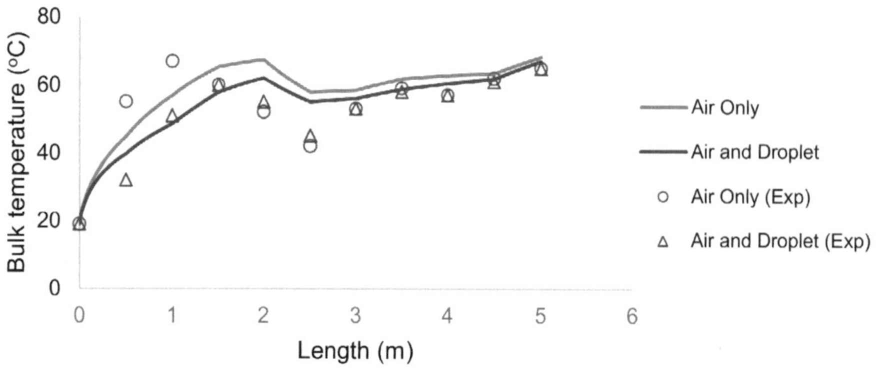

Figure 2. Run 2-Bulk temperature with and without the presence of atomised droplet

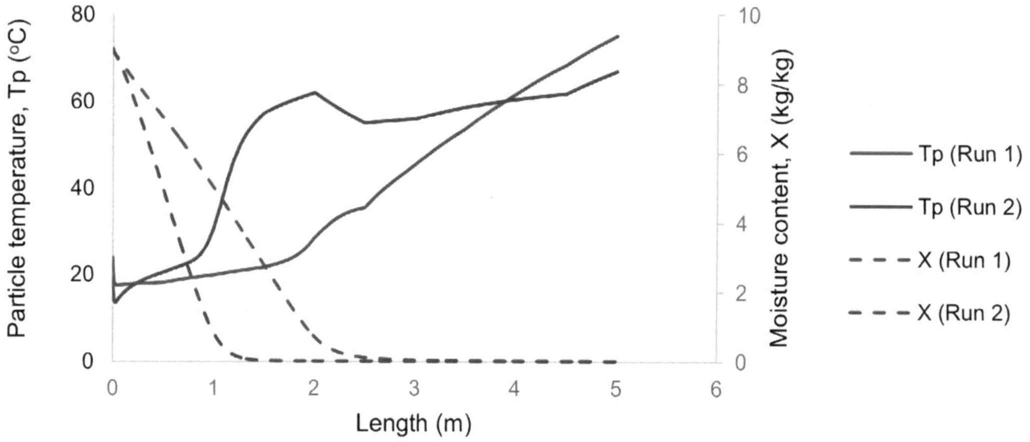

Figure 3. Particle temperature and moisture content of Run 1 and 2

The validation of control model is pivotal for the prediction of particle temperature and moisture content. Figure 3 shows the different drying rates are for Run 1 and 2 as the bulk temperature for both runs are notably different. Both runs exhibit a kink in the middle section due to the sudden drop in wall temperature before it increases again. This demonstrates the sensitivity of bulk temperature to wall temperature and substantiates the idea of precisely controlling the drying history of particle using the narrow tube dryer.

Effect of Drying History on Particle Morphology

In this section, the effect of drying history on particle morphology will be analysed with the aid of the predicted particle temperature and moisture content as shown in Figure 4. Three runs with different wall temperatures applied were carried out by mainly manipulating the first six tubes while maintaining the remaining tubes at a constant temperature as shown in Table 2. This was aimed at investigating the significance of initial drying rate on spray dried particles. The last two tubes were intentionally left at room temperature to induce a cooling effect at the end right before particle collection to maintain a T_p-T_g<0, avoiding any sticky particle.

Table 2. Wall temperatures for Run 3,4, and 5

Tube No.	**1**	**2**	**3**	**4**	**5**	**6**	**7**	**8**	**9**	**10**
Run 3 (°C)	90	90	60	80	90	90	90	90	90	90
Run 4 (°C)	70	30	30	30	90	75	90	90	90	90
Run 5 (°C)	30	35	44	47	57	80	90	90	90	90

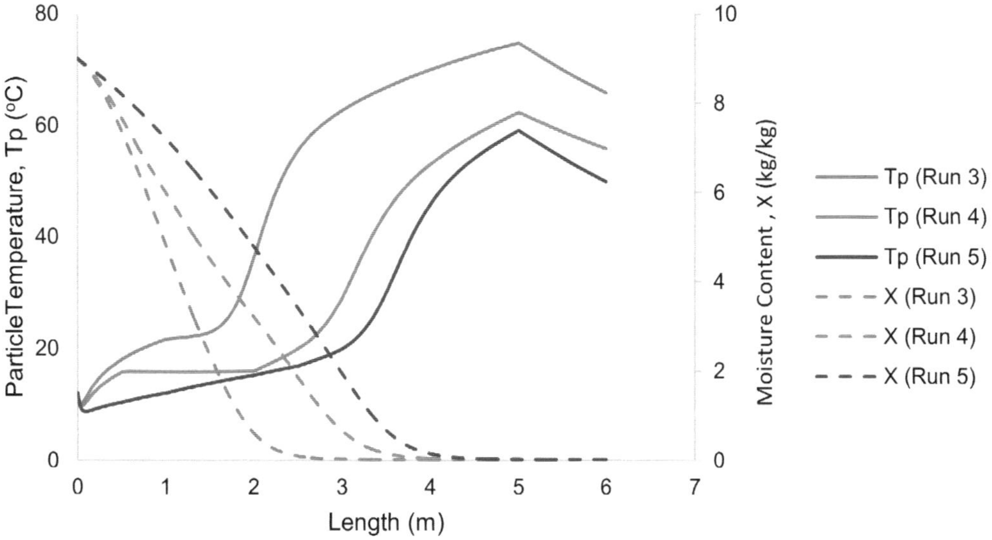

Figure 4. Particle temperature and moisture content of Run 3,4 and 5

Wall temperature in Run 3 was set to be the highest followed by Run 4 and Run 5. Based on the predicted moisture curve in Figure 4, this agrees with the wall condition as Run 3 and 5 display the highest and lowest gradient signifying the evaporation rate as well. These 3 different runs resulted in various particle morphology as shown in the SEM images tabulated in Table 3. It is notable that successful spray drying was achieved for all three runs despite the low drying temperature in comparison to typical spray drying conditions which are above 100°C. Mixture of amorphous and crystalline structures were observed for all runs.

Run 3 with the highest particle evaporation rate produces bigger crystalline particles, approximately 2-3 μm while Run 5 with the lowest particle evaporation rate yields finer crystalline structure, less than 1 μm. This could be due to a lower particle evaporation rate allowing sufficient time for molecules to rearrange themselves as the drying process continues. For the condition in Run 3 with high particle evaporation rate, limited time was provided for molecules rearrangement during the evaporation stage. Once evaporation slows down, a sudden increase in particle temperature can be observed. Based on Figure 2, this situation happened at length 1.8, 2.8 and 3.4 m for Run 3, 4 and 5 respectively which is in accordance to the time frame for particle evaporation. In addition, flat piece shaped crystals were found in both Run 3 and 4. This suggests that particle morphology is not subjected to a specific drying history but an evaporation time frame may exist which results in a unique particle morphology.

Table 3. SEM results for Run 3, 4 and 5

Run 3		
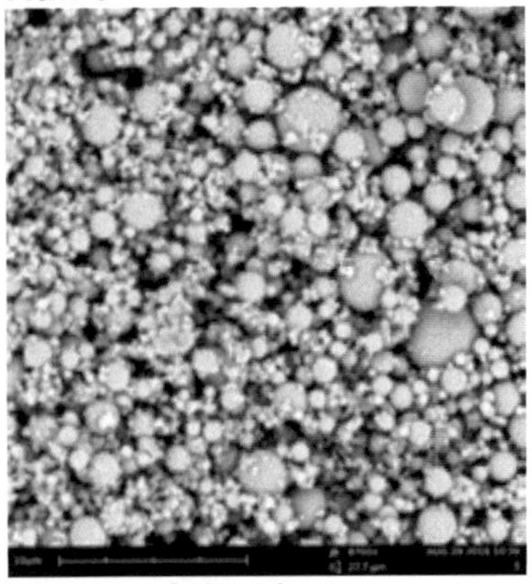	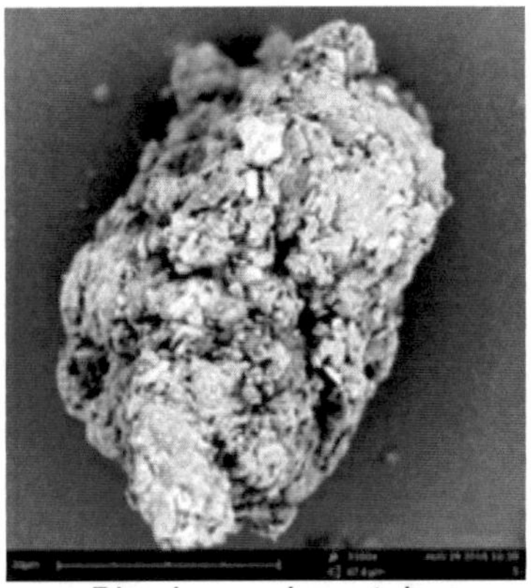	
Amorphous	Big clumped crystals	Flat piece shaped crystals
Run 4		
	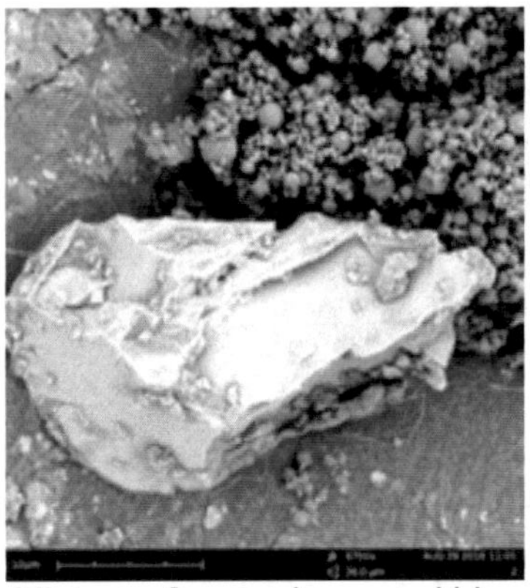	
Amorphous	Mixture of amorphous and big crystals	Flat piece shaped crystals
Run 5		
	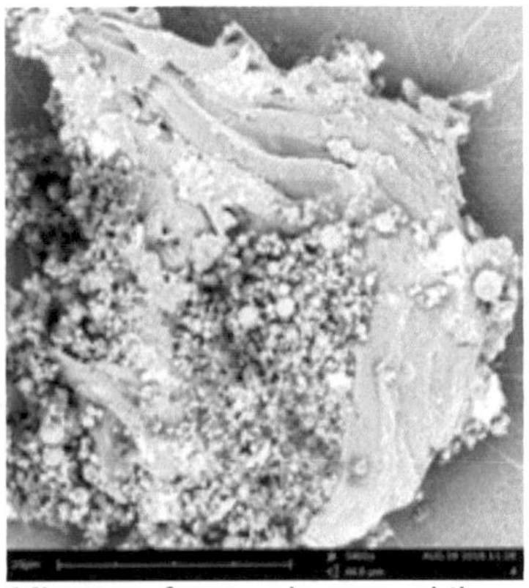	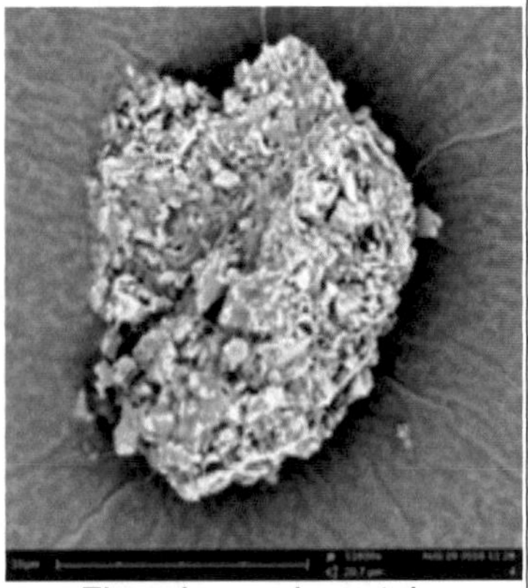
Amorphous	Mixture of amorphous and tiny crystals	Fine clumped crystals

Conclusion

The newly modified narrow tube dryer is capable of precisely manipulating the drying temperature along the tube through convective heat transfer from the heated tube walls. 1D modelling has proven to be a useful tool in predicting temperature profile and providing valuable insights along the dryer. By incorporating 1D modelling, the drying history can be controlled to study its effect on particle morphology. Results shows a low drying temperature, below 100°C is sufficient to yield successful spray dried product which is in fact beneficial for thermal labile materials. Mixture of amorphous and crystalline structures were produced for all runs. However, different initial drying rate affects the size and shape of crystalline particles. Further research is currently underway to unveil the underlying phenomena on crystallinity.

References

1. Sollohub, K. and K. Cal, *Spray drying technique: II. Current applications in pharmaceutical technology.* Journal of Pharmaceutical Sciences, 2010. **99**(2): p. 587-597.
2. Chen, J., B. Sarma, J.M.B. Evans, and A.S. Myerson, *Pharmaceutical Crystallization.* Crystal Growth & Design, 2011. **11**(4): p. 887-895.
3. Shekunov, B.Y. and P. York, *Crystallization processes in pharmaceutical technology and drug delivery design.* Journal of Crystal Growth, 2000. **211**(1–4): p. 122-136.
4. Chiou, D., T.A.G. Langrish, and R. Braham, *The effect of temperature on the crystallinity of lactose powders produced by spray drying.* Journal of Food Engineering, 2008. **86**(2): p. 288-293.
5. Woo, M.W., N. Fu, F.T. Moo, and X.D. Chen, *Unveiling the Mechanism of in situ Crystallization in the Spray Drying of Sugars.* Industrial & Engineering Chemistry Research, 2012. **51**(36): p. 11791-11802.
6. Chiou, D. and T.A.G. Langrish, *A comparison of crystallisation approaches in spray drying.* Journal of Food Engineering, 2008. **88**(2): p. 177-185.
7. Das, D. and T.A.G. Langrish, *An activated-state model for the prediction of solid-phase crystallization growth kinetics in dried lactose particles.* Journal of Food Engineering, 2012. **109**(4): p. 691-700.
8. Mansouri, S., V.S. Hena, and M.W. Woo, *Narrow Tube Spray Drying.* Drying Technology, 2015.
9. Patel, K., X. Chen, R. Jeantet, and P. Schuck, *One-dimensional simulation of co-current, dairy spray drying systems — pros and cons.* Dairy Science & Technology, 2010. **90**(2-3): p. 181-210.
10. Chen, X.D. and S.X.Q. Lin, *Air drying of milk droplet under constant and time-dependent conditions.* AIChE Journal, 2005. **51**(6): p. 1790-1799.
11. Lin, S.X.Q. and X.D. Chen, *A Model for Drying of an Aqueous Lactose Droplet Using the Reaction Engineering Approach.* Drying Technology, 2006. **24**(11): p. 1329-1334.
12. Mostafa, A.A. and H.C. Mongia, *On the interaction of particles and turbulent fluid flow.* International Journal of Heat and Mass Transfer, 1988. **31**(10): p. 13.

465150 Soluble Tag-Assisted Liquid-Phase Peptide Synthesis Using Hydrophobic Benzyl Alcohols

Abstract:

[Introduction]
Peptides have attracted much attention as promising medicinal candidates. As compared with traditional small molecules and antibody medicines, peptides have medium-sized molecular structures. Therefore, they are expected to have both advantages of small molecules and antibody medicines, for example, chemical synthetic accessibility and highly selective pharmacological activity. Current chemical synthesis of peptides is mainly carried out through solid-phase methods, which use insoluble resins as supports to realize easy separation processes and also enable fully automated synthesis. However, soluble tags assisted liquid-phase approach has also been studied intensively since large scale preparations are still challenging in solid-phase method. In this context, polyethylene glycols(PEGs) are the first option as soluble tag and they are central to many applications in the field. Since PEGs are soluble in polar solvents and insoluble in less-polar solvents, a reaction of polymer-attached reactant or catalyst can be performed under homogeneous conditions in a polar solvent. After the reaction completion, the polymer is precipitated by the addition of a less-polar solvent. However, as for peptide synthesis, because both amino acids and coupling reagents are polar, the opposite solubility is preferable for soluble tag to rinse away excess chemicals by washing with polar solvents.
We have been working on the use of hydrophobic benzyl alcohols as soluble tags in liquid-phase method. In this method, reactions take place in homogeneous conditions in less-polar solvents and after the reaction completion, polar solvents are added to precipitate the tag-attached product. Since excess chemicals can be rinsed away efficiently by washing with polar solvents, this approach allows a production of peptide with high quality in relatively larger scale. For this strategy to be successful, high control of the tag-attached product solubility and the physical property must be addressed. In this study, the liquid-phase chemical synthesis of several therapeutic peptides and the investigation on poor-solvents-addition-induced precipitation mechanism will be presented.

[Experimental]
The hydrophobic benzyl alcohols synthesized from the corresponding hydroxyl benzaldehyde in 2 steps, including alkylation by respective bromide and reduction by sodium borohydride. At each step, the desired products were washed with polar solvents repeatedly to give pure forms, which were characterized by NMR. The chemical synthesis of peptides was carried out in less-polar solvents using N-protected amino acids and COMU as a coupling reagent. After the reaction completion, polar solvents were used as poor solvents to induce precipitations of the tag-attached products. The physical property of the recovered precipitates were characterized by scanning electron microscopy, particle size distribution measurement, and X-ray diffraction.

[Results and discussion]
High control of the solubility property was achieved for the hydrophobic benzyl alcohols. The tag-attached products were able to be precipitated quantitatively at most steps. Although dichloromethane was a good solvents for all chemicals used, including the tags, N-protected amino acids, and a coupling reagent, we found that tetrahydrofuran was also effectives as a reaction solvent. This is preferable for the large scale production of peptides from the viewpoint of environmental aspect. As for the poor solvent, methanol and acetonitrile were found to be effective to induce precipitation of the tag-attached products. By using this methodology, both couplings and deprotections were carried out in homogeneous reaction conditions to give several therapeutic peptides in gram-scales. In addition, the physical property of the precipitation was found to be dramatically different when cyclohexane was used as a reaction solvent instead of tetrahydrofuran. Especially, the biphasic system composed of cyclohexane and acetonitrile provided an ideal reaction condition, which solubilize the all chemicals used for the synthesis. The precipitations obtained from this reaction condition were fine powders that were able to be recovered readily through paper filtration.

[Reference]
(a) Okada, Y. et. al, Org. Lett. 2015, 17, 4264-4267. (b) Okada, Y. et. al. J. Org. Chem. 2013, 78, 320-327. (c) Tana, G. et. al. Chem. Commun. 2010, 46, 8219-8221

473183 Biomanufacturing Science and Technology Consortium to Advance U.S. Manufacturing of Biopharmaceuticals

Abstract:

Biomanufacturing has become an emerging industry of strategic importance worldwide and is one of the key technology areas identified by the Advanced Manufacturing Partnership (AMP) as an appropriate target for public-private investment to support advances in manufacturing and U.S. competitiveness. The development of advanced biomanufacturing technologies will foster and support the fast growing biomanufacturing industry and for establishing sustained leadership of the US in the area (PCAST, 2012) (AMP 2, 2014).

The Biomanufacturing Science and Technology Consortium (BSTC) is hereby proposed, which will be led by academic and industry leaders across the nation. The proposed consortium is founded on the operational experiences of the Biopharmaceutical Process and Quality Consortium (BPQC) at University of Massachusetts Lowell (MA), which has been building collaborative efforts in advancing research and practice outcomes in the biopharmaceutical industry for four years. The proposed consortium represents a comprehensive team of actively participating stakeholders covering the entire spectrum of the industryâ€™s value chain to identify challenges that the biomanufacturing industry is currently facing and to develop a dynamic roadmap to accelerate solutions. The consortium will be comprised of biomanufacturers including small and medium size businesses, research universities and colleges, government institutes and non-profits focusing on achieving the sustained U.S. leadership in biomanufacturing. The consortium will encompass multiple geological locations of strategic importance and disciplinary expertise areas, e.g., bio-engineering, regulatory and measurement sciences, education and workforce development. (BPQC, 2012).

The consortiumâ€™s roadmapping exercise will take place at several strategic locations mainly through focused workshops. Each regional workshop will have an open brainstorming session to identify challenges and to encourage innovative ideas. The last concluding workshop will bring an active group of stakeholders and subject matter experts to synthesize and prioritize the outcomes from the multiple national and regional workshops. This consortium will also attempt to leverage the concept of *crowdsourcing* to inform additional ideas for solutions to key questions and to identify research projects. The proposed planning efforts will lead to a clear roadmap for a next generation biomanufacturing paradigm, reliable industry standard systems, and substantive knowledge spanning the industry challenges, as well as a list of potential pre-competitive projects covering the entire spectrum of topics toward the identified new paradigm. Through the roadmapping effort, this proposed consortium aims to lay the foundation for a sustained partnership among the members that will foster the U.S. competitiveness in the global market.

The consortium seeks to make all-encompassing plans to promote the development of advanced biomanufacturing science and technology, and also actively looks for the upgrade and transformation of existing biomanufacturing, as well as sharing resources and promoting innovations. Upon successful completion of this proposed planning effort, the consortium aims to conduct the following tasks:

- Prototyping a system of innovation for advanced biomanufacturing
- Developing breakthroughs in a number of key technologies
- Establishing technological standards and sustained global leadership of the U.S. biomanufacturing industry
- Building a robust and integrated supply chain network of an advanced biomanufacturing

463406 Defining Challenges and Creating Solutions in Biopharmaceutical Manufacturing

Abstract:

In this presentation, we will discuss a number of technology challenges associated with the manufacturing of biopharmaceuticals. The opportunity to address pre-competitive challenges through public-private partnerships is a potential solution to address these challenges. However, careful consideration of the role of relevant federal agencies, health authorities, along with the global nature of the biopharmaceutical industry is required to identify the most effective path forward to advance biopharmaceutical manufacturing.

460344 Technology Advances That Will Enhance Continuous Biomanufacturing

Abstract:

The development of advanced biomanufacturing technologies involves consideration of converting batch manufacturing to continuous operation. Continuous manufacturing has a significant number of advantages that include operation of a manufacturing method that will produce consistent, high quality products at lower costs. This is achieved by using quality by design (QbD) principles which have been historically demonstrated to work in the continuous production of small molecules. It is done by reducing overall variability in the process which enables enhanced control within an optimized operating range. To implement this approach effectively for continuous biopharmaceutical processing, technology is needed in improved sampling, process analytical technology (PAT), and data handling. This will then allow for the determination of the critical quality attributes (CQA) and relate them to the critical process parameters (CPP) and show how these can be used to control process operation within the design space to enable real-time release testing (RTRT).

467812 Energy-Based Medium Design for Mammalian Cell Fed-Batch Cultures

Abstract:

Mammalian cell cultures are the main industrial platform for the production of high-value biologics, such as monoclonal antibodies (mAb) [1]. One way to reduce operational costs is by maximising mAb yield and cell growth whilst minimising cell culture medium expenditure. Current optimization strategies are typically empirical requiring a multitude of expensive, intensive and time consuming experiments. Model-based optimization approaches seek to reduce unnecessary experimentation. Previously, we have applied model-based optimization methods to identify feeding regimes to maximize mAb titter in GS-NS0 cultures limited only on glucose and glutamate supply while neglecting the exhaustion of other essential amino acids [2]. In addition, culture medium design has been based on stoichiometric balances considering only cell composition [3] and metabolite profiling where the medium was designed based on the exhaustion of a limited number of nutrients [4]. Alas, the above-mentioned studies neglect the energy requirements of the cell not only for maintenance and proliferation but also for mAb production.

Herein, a novel predictive model for GS-NS0 cells producing cB72.3 mAb has been developed considering cell and product composition as well as cellular energy requirements for maintenance, proliferation, and mAb production. The development of this model follows the framework described in [5] and describes growth kinetics, metabolism, mAb secretion and the production/consumption of ATP based on glucose and 15 amino acid energy metabolic pathways. The full range of the model parameters was subjected to Global Sensitivity Analysis (GSA) to identify the significant parameters that were then estimated from specifically-designed batch experiments. Subsequently, the validated model was utilised to design a low cost supplemental medium for the implementation of an optimized fed-batch strategy which increased production by 60% and extended viability and culture longevity by 24 hrs. As a result of this feeding strategy, nutrient and ATP concentrations were controlled in a tight range without accumulation. The successful coupling of cell and product composition with energy requirements has enabled the development of a tailored medium composition reducing operational costs whilst maximising production. This integrated model-based platform, which is experimentally validated, presents a cost-effective and efficient approach for media development and upstream optimisation of a wide range of biopharmaceutical products.

References

1. Zhu, J., Mammalian cell protein expression for biopharmaceutical production. Biotechnology advances. 2012. 30(5): p.1158-1170.
2. Kiparissides,A., Kountinas,M., Kontoradvi,C., Mantalaris, A., Pistikopoulos, E.N. 'Closing the loop' in biological systems modeling — From the in silico to the in vitro. Automatica. 2011. 47(6): p.1147-1155.
3. Xie, L, Wang, D.I.C. High Cell Density and High Monoclonal Antibody Production Through Medium Design and Rational Control in a Bioreactor. Biotechnology and Bioengineering.1996. 51: 725-729.
4. Sellick, C.A., Croxford,A.S., Maqsood,A.R., Stephens, G., Westerhoff,H.V., Goodacre,R., Dickson, A.J. Metabolite Profiling of Recombinant CHO Cells: Designing Tailored Feeding Regimes That Enhance Recombinant Antibody Production. Biotechnology and Bioengineering. 2011. 108 (12): 3025–3031.
5. Kiparissides,A., Pistikopoulos,E.N., Mantalaris,A. On the model-based optimization of secreting mammalian cell (GS-NS0) cultures. Biotechnology and Bioengineering. 2015. 112: 536–548.

461381 Evaluating the Potential of Miniature Scale Chromatography Process Characterization to Support Biomanufacturing Strategies and Regulatory Filings

Abstract:

High throughput downstream chromatography techniques such as resin slurry plates and miniature columns have been well described and implemented for biologics research, discovery, and early process development applications. These techniques have seen relatively little application in late stage development areas, though, mainly due to the necessity of qualified scale-down models and increased precision in this space that is required for regulatory acceptance. However, an automated parallelized scale-down approach could potentially provide much needed reductions in time and resources for activities like formal process characterization where representative data across many conditions is needed to support regulatory filings.

This talk will highlight some of the differences and challenges for utilizing high throughput techniques in late stage development compared to earlier stages of process development and will suggest potential applications in manufacturing space. For miniature columns, theoretical limitations such as transport phenomena and wall effects will be discussed. Practical considerations like pulse testing, pooling strategies, and resin reuse will also be presented. Case study comparisons of formal process characterization strategies for miniature scale and traditional lab scale as well as direct scale-down comparisons across four orders of magnitude will be shown. Finally, recommendations for the ultimate limitations of miniature columns will be provided to suggest future state directions and applications in late stage development spaces.

470644 Ecoprime Twin - Continuous Two-Column Chromatography at the Process Scale

Abstract:

The call for implementing continuous processing into the manufacturing of bio-pharmaceuticals requires re-thinking of process chromatography and its equipment designs. Proof-of-concepts of innovative multi-column continuous processes were introduced by the industry, which verified improvements in resin utilization and increased productivity. However, questions remain remains how to scale up the standard chromatographic batch processes to continuous production scale. Our focus is here on the process optimization of the twin-column technology that has a less overall complexity in process and system design and performs comparable if not better than other multi-column setups. The successful scale-up from bench-top to twin-column pilot scale will be supported by performance data. The requirements for the equipment design, process automation and regulatory aspects due to the implementation of the continuous processing are highlighted. Furthermore, future trends how the EcoPrime Twin technology will enable the integrated continuous downstream processing are pointed out.

452951 Enzymatic Synthesis of Complex Natural Products

Abstract:

On account of their structural and biosynthetic complexity, most natural products have traditionally been discovered and produced from fermentable microorganisms. Within the past decade however, there has been an explosive growth of "orphan" biosynthetic pathways whose products are unknown. Many of these compounds cannot be accessed from their natural sources. In vitro reconstitution of these multi-enzyme systems has enabled efficient synthesis of relatively small quantities of some of these products for discovery applications. The benefits of scaling up these preparative methods are enormous, but significant challenges remain to be addressed. Using glycolipids and polyketides as examples, I will illustrate some of these opportunities and challenges.

466988 Quality-by-Control (QbC) Technologies for Pharmaceutical Process Development

Abstract:

The advent of process analytical technologies (PAT) more than a decade ago has brought the applications of advanced control of pharmaceutical manufacturing processes in the realm of possibility. The role of advanced feedback control concepts in improving the performance of batch processes, as well as enabling technologies in the paradigm shift from batch to continuous manufacturing will be corroborated. Examples how advanced control can be used for the improved production of pharmaceutical particulates will be provided, with application to both continuous drug substance and drug product manufacturing. Novel control approaches for integrated continuous crystallization and wet milling systems will be presented that can produce crystalline materials with tailored maid properties with excellent consistency. The role of rigorous mathematical modelling in the systematic optimal design, start-up and control of continuous crystallization processes to achieve and maintain the desired controlled stage of operation will be illustrated through examples for both cascade of mixed suspensions mixed product removal (MSMPR) as well as plug flow crystallization systems. Anti-fouling control (AFC) will be presented as an illustrative example demonstrating how active feedback control can be the key enabling technology for the implementation and smooth operation of continuous crystallization processes. The application of a systematic framework for the design of hierarchical control systems for continuous tablet production will also be described. A novel view for implementing continuous drug manufacturing processes that are based on spatially distributed control of various particle formation processes within a single continuous plug flow system, rather than the integration of individual unit operations, will be presented. The system uses spatially distributed antisolvent addition, heating/cooling cycles and/or binder/excipients addition to control the individual mechanisms of nucleation, growth and spherical agglomeration to produce particles with controlled internal and external size and micromeritic properties. The talk intends to provide motivating examples of the next stage of innovation in pharmaceutical manufacturing, illustrating the potential benefits of the new quality-by-control (QbC) framework in improving product quality and process efficiency while reducing costs and time-to-market.

470571 Drug Product Continuous Manufacturing Control Strategy Considerations

Abstract:

This presentation will cover the key components of a fully automated drug product continuous manufacturing control strategy. In-process control, real time decision making, sampling plans and sampling plan justification, and real time release testing will be discussed. Experience with filings and inspections along with a few learnings and considerations for continuous manufacturing will be shared.

470871 Regulatory Approach for Supporting Emerging Pharmaceutical Technology

Abstract:

Office of Pharmaceutical Quality (OPQ) supports modernization of pharmaceutical development and manufacturing. Therefore, OPQ encourages the adoption emerging pharmaceutical technology that has potential to enhance the process capability and therefore product quality. Emerging technology may include novel process and product designs, as well as innovative analytical technology. Examples of emerging technology are continuous manufacturing and 3-D printing technologies. Because emerging technologies, by definition, will generally be unfamiliar, in both industrial and regulatory contexts, OPQ established the Emerging Technology Team (ETT) to facilitate the identification and implementation of emerging technology in at least three ways. 1) by enhancing communication among sponsors, manufacturers, and regulators, both early in the engagement of emerging technology design and development and at appropriate later time points (including the course of regulatory review); 2) by leading or otherwise supporting the quality assessment team, in partnership with relevant OPQ offices, in regulatory reviews; and 3) by documenting ETT/OPQ resolutions to scientific and policy issues that may inform FDA approaches to the review and implementation of emerging technologies. Furthermore, OPQ have initiated a regulatory science program to enhance the agencyâ€™s knowledge in the emerging technology area. For example, OPQ initiated several research projects in continuous pharmaceutical manufacturing, including the development of integrated process models similar to the approaches utilized by the chemical industry; the development of a database of excipient properties to capture the behavior of common raw materials within different processing environments, the standardization of processing equipment integration components to increase flexibility; and the advancement of automatic process control systems.

470944 The Integrated Quality Assessment: Office of Process and Facilities' Perspective on the Holistic Evaluation of Pharmaceutical Quality

Abstract:

The Office of Pharmaceutical Quality (OPQ) was formed for the purpose of ensuring a consistent and unified approach to FDAâ€™s assessment of pharmaceutical quality for pharmaceutical products. This restructuring involved a dramatic modification in the structure of the organization, changes in roles and responsibilities for review and inspection functions, and an elevated stature for the quality assessment function within CDER. The reorganization and resulting formation of OPQ has been in effect since January 2015. Since that time there has been continued implementation and improvement upon the integrated quality assessment within OPQ. This assessment now includes a greater level of communication and interdependency within the assessment team, simultaneous with the specialization of the assessment disciplines for products, processes and facilities. In particular, there has been an evolution of the facility assessment: from the initial risk-based for the assessment of the facility; to the composition of the inspection team and focus of the pre-approval inspection; concluding with the final integration of the inspection results into the quality assessment. The changes have positioned OPQ to provide most appropriate level of engagement with sponsors and the development of expertise necessary to perform the technical assessment of increasingly complex technologies, such as continuous manufacturing of pharmaceutical products. The basis for the changes with OPQ will be provided, along with a description of important initiatives within the office, concluding with a description of the programmatic successes resulting from the formation of OPQ.

469279 A Phenomenological Model for the Prediction of Granule Drying Times in a Fluid Bed Dryer

Abstract:

A computational model designed to predict the drying times of pharmaceutical granules in a fluid bed dryer is described. The model is comprised of two major components, the first of which utilizes a well-established two-stage phenomenological drying model, and assumes that granule drying consists of evaporation of liquid 'domains' from the granules surfaces followed by liquid evaporation from within the intra-granular porous network. The transient evolution of the granule moisture content is solved by numerically integrating the one-dimensional heat equation. The moving liquid/air interface is solved in a stationary computational domain via a Landau-style mapping. The second component numerically simulates the stochastic nature of the fluidization behavior of granules inside the fluid bed dryer. The combination of this stochastic simulation with the existing phenomenological approach allows for the study of mechanistic correlations between the granule physical characteristics (size, porosity, permeability, and initial moisture content) and the process conditions (e.g., drying air flow rate and temperature, and fluidization "efficiency"). We demonstrate that center-point conditions from an extensive series of process development runs can be used to derive the empirical inputs to the model, and that these inputs can be scaled according to the aforementioned process conditions—ultimately allowing for the prediction of drying times across a broad range of processing conditions.

458414 Practical Application of a Mechanistic Model to Scale up Fluid Bed Drying in Drug Product Development

Abstract:

Fluid bed drying is one of the core processes in pharmaceutical manufacturing of oral dosage forms. For several decades models based on mechanistic understanding of the fluid bed drying process have been used. In practice however, in the pharmaceutical industry extensive experimentation is still carried out during drug product development, to investigate fluid bed drying of wet granules at different operating conditions and equipment scales. Typically experimentation is guided by Design of Experiments and results are studied by multivariate analysis. This procedure does not take advantage of the existing fluidized bed drying theory that is already available, and that would allow to reduce the number of experiments performed, at the same time providing better process understanding and more reliable scaling to different operating conditions [1].

In this work a case study is presented, the goal of which was to show that it is possible to use a mechanistic model to scale up fluid bed drying of granules containing active pharmaceutical ingredients (API) from the laboratory to the pilot scale. To this end, two different formulations containing two model APIs were chosen and granulated via high shear wet granulation. The obtained wet granules were then dried in a fluid bed dryer at the laboratory scale. The fluid bed drying model available in the software gSolids v.4.1 (PSE, London, UK) [2, 3], was calibrated for the two formulations with the laboratory scale experimental results. The model was then used to predict the fluid bed drying process of the two formulations at pilot scale. Experiments were run at the pilot scale and model predictions could be confirmed without further adjustments.

It was so demonstrated that it is possible to scale up the fluid bed drying process of API containing granules from laboratory to pilot scale by using a mechanistic model. A workflow was established based on the experience gathered in the case study that details experimental data needed for model set up and calibration and possible solutions to overcome challenges in using the model to predict the process at pilot scale.

It can be concluded that using a mechanistic model constitutes a viable alternative to empirical modelling allowing the saving of API during drug product development by reducing experimentation at large scale.

References

[1] Kemp, I. (2011) Drying models, myths and misconceptions. *Chem. Eng. Technol.,* **34** (7), 1057-1066.

[2] Burgschweiger, J., Groenwold, H., Hirschmann, C. And Tsotsas, E. (1999) From hygroscopic single particle to batch fluidized bed drying kinetics. *The Canadian Journal of Chemical Engineering* **77,** 333-341.

[3] Burgschweiger J, Tsotsas E (2002) Experimental investigation and modelling of continuous fluidized bed drying under steady-state and dynamic conditions. *Chem. Eng. Sci.*, **57**, 5021-5038.

461578 Modelling of Co-Rotating Twin-Screw Extruders in the Pharmaceutical Industry I: Single Component Model

Abstract:

Modelling of co-rotating twin-screw extruders in the pharmaceutical industry I: single component model

Pavol Rajniak[1], Maria Fuentes-Gari[1], Niall Mitchell[1], Tariq Islam[2], John Robertson[2] and Sean K. Bermingham[1], (1)Process Systems Enterprise, London, United Kingdom, (2)CMAC, Strathclyde University, Glasgow, United Kingdom

Twin-screw extruders are being increasingly applied in the pharmaceutical industry for the manufacture of solid dispersions. In particular, Hot Melt Extrusion (HME) is a viable manufacturing alternative for poorly soluble drugs that are difficult to process. This is due to the high shear stress applied in the process, which enhances mixing of the base polymer with the dispersed API. Delivery systems that can be obtained using this technology include pellets, granules, sustained release tablets and implants. Experiments for HME are typically very labour-intensive, involving the use of highly viscous polymers, high pressures, and require proper cleaning between runs. Identifying the critical experiments to perform based on model simulations would thus be highly desirable. Single-component modelling of twin-screw extrusion processes could be beneficial in the following areas: identifying optimal screw configurations, tracking the degree of melting and when identifying the area within the extruder where the polymer becomes completely melted.

An axially distributed mechanistic model of HME was previously developed by Eitzlmayr et al.[i], wherein the extruder elements were described using physically-meaningful parameters estimated from experimental data and any given extruder could be built into the model in a modular way. Their model was implemented into gPROMS FormulatedProducts®, and the simulation results obtained for different HME configurations using Soluplus® were reproduced. Important process variables such as filling ratio or element pressure, which are often difficult to measure, were given as simulation outputs; measurable variables, including the temperature of the screws and the melt and residence time distribution, were validated against experimental results obtained in a 40:1 ThermoFisher Eurolab co-rotating twin-screw extruder at CMAC. In addition, experiments with a chosen extruder configuration operating under fixed conditions were carried out with Soluplus®, Plasdone™ and Affinisol™. The effect of polymer nature on screw element parameters (A1, A2 model parameters) was investigated by comparing the results obtained across the three polymers.

Overall, the existing model was proven to perform well for the single component polymer case. Further work focuses on extending this model to account for multi-component mixtures and capturing API solubility.

i. Eitzlmayr, A., Koscher, G., Reynolds, G., Huang, Z., Booth, J., Shering, P. and Khinast, J. (2014) "Mechanistic modeling of modular co-rotating twin-screw extruders", *International Journal of Pharmaceutics,* 474(1), 157-176.

**Corresponding Author:* m.fuentesgari@psenterprise.com

464894 Modeling and Optimization of Agglomeration in Suspension through a Coupled Population Balance Framework

Abstract:

Since its introduction by Randolph and Larson[1] (1971) and the solution methods described by Ramkrishna[2] (2000), the population balance model (PBM) has been widely used and accepted as model formulation method for simulation and prediction of the size distribution and other properties of particulate systems. PBMs allow for systems that include any or all of the following mechanisms: nucleation, growth, breakage and agglomeration. Following the initial work by Smoluchowski[3] (1917) on the rate of aggregation for spherical particles, there have many contributions for systems that exhibit agglomeration including dispersion (bubble) coalescence[4,5], granulation[6,7] and particle aggregation during crystallization[8,9,10]. The commonality in the limitations of many of the previous studies were physically irrelevant and/or empirically based agglomeration kernels, difficulty in assessing the influence of process conditions (*e.g.* hydrodynamics, particulate physical properties), solution method efficiency for optimization and control applications, and loss of information of constituent particles. These limitations present obstacles for the simulation, optimization, and control of the increasingly popular agglomeration technique of spherical crystallization.

Peña and Nagy[11] studied and showed the benefits of spherical crystallization as a process intensification technique whereby both internal (primary crystals) and external (agglomerates) properties can be controlled experimentally through a decoupled continuous spherical crystallization (CSC) approach; providing the means by which both biopharmaceutical (bioavailability, dissolution) and manufacturing (flowability, filtration, drying) properties can be simultaneously adapted to meet desired quality specifications. This technique opens the door for combined experimental and modeling approaches for the optimization and control of both the primary crystal and agglomerate properties in spherical crystallization processes. However, many of the PBMs currently in literature would fail to accomplish this because of the aforementioned limitations and loss of information.

To overcome the issues of loss information a coupled PBM formulation is required. A coupled PBM formulation could simultaneously track the evolution in the primary crystals and the evolution of the agglomerates. The relationship between primary crystal properties and their effect on final agglomerate properties would thereby be more evident and more efficient than traditional approaches. To the best of our knowledge, the only previous work that presented this approach is that of Ochsenbein et al.[12]. In their study, Ochsenbein et al.[12] focused on the agglomeration of needle-like crystals in suspension. Through a coupled PBM framework Ochsenbein et al.[12] were able to develop a population balance equation (PBE) to describe the evolution of the primary crystals by a two-dimensional growth rate to represent the needle like structure of the crystal. They then used another PBE to describe the evolution of the agglomerates as a function of the primary crystals. For the agglomeration kernel, they derived their own modified kernel that include both characteristic lengths of the primary crystals participating in the agglomeration. The new PBM formulation also allowed for the development of new parameters that add value to the simulations due to their experimental relevance. However, the work of Ochsenbein et al.[12] neglected nucleation something common to previously developed agglomeration models. The coupled population balance model framework will be extended herein.

The contribution of this work is the extension of the coupled PBM framework for application in the simulation and optimization of a spherical crystallization system. A coupled PBM framework has been developed for a semi-batch, reverse addition, anti-solvent crystallization system with agglomeration. The system includes nucleation and growth the primary crystals and subsequent agglomeration. The purpose of the work is to exploit the advantages presented by a coupled PBM framework; for example, the ability to optimize for specific primary and agglomerate sizes. This presents an opportunity to find optimal operating conditions that meet both bioavailability and manufacturability demands. It also allows for the ability to develop first principles based parameters for agglomeration efficiency and porosity. Additionally, through the retention of the information of the primary particles the interplay between the effects of operating conditions on the properties of the primary crystals versus the agglomerates will be clear.

1. Randolph, A.; Larson, M. Theory of particulate processes; analysis and techniques of continuous crystallization. 1971. New York: Academic Press.
2. Ramkrishna, D. Population balances theory and applications to particulate systems in engineering. 2000. San Diego, CA: Academic Press.
3. Smoluchowski, M. V. Z. *Physical Chemistry*, 1917, 19, 129–168.
4. C. a. Coulaloglou and L. L. Tavlarides. Description of interaction processes in agitated liquid-liquid dispersions. *Chemical Engineering Science*, 1977, 32(11), 1289–1297.
5. M. J. Prince and H. Blanch. W. Bubble coalescence and break-up in air-sparged bubble columns. *AIChE Journal.* 1990, 36(10), 1485–1499.
6. Simon M. Iveson. Limitations of one-dimensional population balance models of wet granulation processes. *Powder Technology*, 2002, 124(3), 219-229.
7. L. X. Liu and J. D. Litster. Population balance modelling of granulation with a physically based coalescence kernel. *Chemical Engineering Science*, 2002, 57(12), 2183-2191.
8. P. Marchal, R. David, J.P. Klein, and J. Villermaux. Crystallization and precipitation engineering—I. An efficient method for solving population balance in crystallization with agglomeration. *Chemical Engineering Science*, 1988, 43, 59–67.
9. R. David, P. Marchal, J.P. Klein, and J.V. Klein. Crystallization and Precipitation Engineering III. A Discrete Formulation of the Agglomeration Rate of Crystals in a Crystallization Process. *Chemical Engineering Science*, 1990, 46(1), 205–213.
10. Sanjeev Kumar, D. Ramkrishna, On the solution of population balance equations by discretization—III. Nucleation, growth and aggregation of particles. *Chemical Engineering Science*, 1997, 52(24), 4659-4679.
11. R. Peña and Z. K. Nagy. Process Intensification through Continuous Spherical Crystallization Using a Two-Stage Mixed Suspension Mixed Product Removal (MSMPR) System. *Crystal Growth & Design*, 2015, 15(9), 4225-4236.
12. D.R. Ochsenbein, T. Vetter, S. Schorsch, M. Morari, and M. Mazzotti. Agglomeration of Needle-like Crystals in Suspension: I. Measurements. *Crystal Growth & Design*, 2015, 15(4), 1923–1933.

457293 Modelling of Water Absorption By a Pharmaceutical Tablet

Abstract:

Aqueous film coating is a crucial step in the manufacture of solid-dosage drugs in pharmaceutical industries. It is well understood that the shelf life of pharmaceutical tablets depends on the amount of humidity to which they are exposed during the coating process, the handling of the intermediate coated product and the packaging. Understanding, and being able to predict the mechanisms of water absorption onto and into tablets, is therefore important, to avoid accelerating the degradation mechanisms caused by high water content.

Mathematical models that efficiently simulate the overall coating process have already been developed. Recent models permit calculating the size of the spray droplets before impact, simulating the tablet movement inside the coater and predicting coating uniformity. However, there is little, if any, fundamental understanding of the sorption phenomena at the point of contact between the coating droplet and the porous tablet. This information is essential to perform a full-system global optimisation and determine the optimal design of the tablet.

This work aims to define and describe the driving forces that play a role in water-based coating sorption after impingement on a porous substrate. We aim to demonstrate two mathematical models. The first takes into account the spreading and sorption phenomena, whereas the second predicts the evaporation rate of the water absorbed in the porous matrix of the tablet. The ambient conditions; such as temperature, pressure and humidity, are set to be similar to the conditions present inside a pan-coater.

The spreading sorption model is divided into two separate phases: the kinematic phase and the capillary phase. During the first, inertial forces govern the spreading, while during the second, capillary forces become dominant. To acquire results for the kinematic phase, we developed a 1D model that solves the energy balance equation for the first milliseconds after impact. The results for the capillary phase come from the solution of the Navier Stokes equation, obtained using the lubrication approximation theory.

For the second model, of the evaporation process, we adopted a model for the second drying stage of a slurry droplet inside a spray dryer. During this stage, we can no longer describe the droplet as a liquid system containing solids, having to regard it as a wet particle with a dry crust and a wet core. The dry core represents the surface of the tablet, while the wet core represents the wetted areas inside the tablet.

The information provided by these mathematical models, is important to understand the spreading, absorption and evaporation phenomena and predict the moisture content inside the tablet after the impingement of a liquid coating droplet. We implemented the models in gPROMS, employing the Modelbuilder modelling platform. The results (droplet height, spreading and wetting front profile) are in good agreement with recent experimental data and CFD simulations.

455995 Design Problem of Pharmaceutical Manufacturing Processes

Abstract:

The modern pharmaceutical industry is requested to realize superior manufacturing processes regarding multiple aspects such as quality, economy, supply reliability, and many others. To this end, the setup of design problem is critical because the choice of design variables, constraints and objective functions determines the capability of the process to be realized. The aim of this presentation is to explore the characteristics of design problem of pharmaceutical manufacturing processes based on the two research examples from the presenterâ€™s group.

The first example is on the design of pharmaceutical tablet manufacturing processes considering the choice between batch and continuous technologies. While pharmaceutical tablets are usually manufactured in batch processes, continuous manufacturing technology is in intensive development in both industry and academia. In this work, a decision-support method was proposed for the choice between two technologies with four steps of: (i) process modeling, (ii) evaluation, (iii) sensitivity analysis, and (iv) interpretation. The method was applied to the following design situation, i.e., â€œa company launches a new product which is supposed to be marketed for 30 years. The tablet product can be manufactured by either technology which needs to be specified based on the economic evaluation.â€ After performing the four steps of method, a better process regarding net present cost could be suggested depending on the API price as well as the maximum demand quantity over the production period of 30 years.

The second example is on the design of solvent recovery processes in API manufacturing. Generally in API manufacturing, various types and large amount of organic solvents are used, which raises concerns regarding environment, health and safety (EHS) risks. In this work, an actual API manufacturing was investigated, and a process was designed for recovering tetrahydrofuran (THF) from the azeotropic mixture with methanol and water. The objective functions were set as the net present value as well as EHS risk evaluation scores. As the process, four alternatives were generated regarding the core separation technology, i.e., (a) zeolite membrane (b) pressure-swing distillation (c) azeotropic distillation, and (d) entrainer. The target THF concentration was set as 99.95 wt% in the initial evaluation, which was in the secondary evaluation relaxed to 95 wt%, a sufficient concentration in the actual API manufacturing. The initial multiobjective evaluation identified the alternatives (a), (b) and (d) as Pareto optima, among which, the alternative (b) was the best in all aspects in the secondary evaluation.

The outcomes in these two research examples were sensitive to the setup of the design problem, i.e., design variable, constraint and objective function. In the first example, treating the choice of batch and continuous as a design variable was meaningful because the superior technology found was depending on the API price and the product demand. In the second example, different results were obtained by regarding the target concentration changeable, which would be considered fix in reality due to the required effort of change management in GMP. The design problem of pharmaceutical manufacturing process has large degree of freedom in nature, however in practice, many variables are considered as hard constraints or determined on an empirical basis. In this regard, MSO can be regarded as the supportive technology for investigating wide range of items in a speedy manner in the design phases of pharmaceutical manufacturing processes.

467775 Multi-Scale Modeling in Pharmaceutical Drug Substance Manufacturing

Abstract:

The role of modeling in pharmaceutical drug substance manufacturing is becoming increasingly more important as the regulators push to the use of more sophisticated methods and tools to address process development and manufacturing. Those who adopt the best modeling approaches benefit from such decision in terms of right-first-time and confidence resulting from enhanced process understanding.
This work presents a multi-scale modeling paradigm applied to drug substance manufacturing. Starting at the molecular level, it is shown how Ab initio modeling (Density Functional Theory) can bc uscful for determining the IR spectra of the molecules that in specific cases can facilitate analytical work. As the modeling scale is increased, CFD simulations are shown to be key to study in detail the flow patterns inside a reactor. At the unit operation level, Dynochem is used to model the reaction kinetics and crystallization steps, as well as the work-up part of the process. Process trains are then addressed resorting to SuperPro Designer flowsheet software that can be useful for determining bottleneck and COGs. In the next modeling scale scheduling of a multi-process train facility is simulated with SchedulePro, that includes details also on operators and stocks. Finally, company-wide capacity (multi-site) is assessed by making use of Monte-Carlo stochastic simulations.

468026 A Simultaneous Approach for Kinetic Parameter Estimation and Curve Resolution Based on Spectroscopic Measurements

Abstract:

Identification of reaction kinetics plays an important role in the development of the commercial product and process development for chemical and bio-chemical systems. Precise knowledge of kinetic parameters is critical in the reactor scale-up from laboratory to industrial levels and the design of robust, controllable and safe processes. Spectroscopic techniques are commonly used for chemical reaction monitoring especially within ultraviolet-visible, infrared and near-infrared wavelength regions; these measurements provide vast amounts of data related to the concentration of the species in the reaction medium. Ample research has been conducted in developing techniques to resolve the information in the spectral data, and identify the kinetic parameters of the underlying chemical reaction (MCR-ALS).

This work describes an efficient, unified framework for kinetic parameter estimation and curve resolution that is based on maximum likelihood principles and a simultaneous collocation approach that leads to improved estimates of parameters.

This study considers the dynamic equation model for the reaction kinetics as a stochastic differential-algebraic equation system,Â $dc = f(c, y, q) + k\, dW(t)$, $g(c, y, t) = 0$, where $c(t)$ is the concentration vector at time t, $y(t)$ are additional algebraic states, q is the kinetic parameter vector and $W(t)$ is the reaction system noise described by standard Brownian motion or Wiener processes. The spectroscopic measurements $d(t, l)$ are governed by Beer-Lambertâ€™s law, $d(t, l) = S_k\, c_k\, (t)\, s_k\, (l) + z(t, l)$, with absorbance $s_k\, (l)$ for component k and wavelength l, and measurement noise $z(t, l)$.

Determination of q by parameter estimation requires consideration of a dynamic optimization problem derived from maximum likelihood principles, and this leads to two important challenges. First, an efficient and reliable optimization strategy is required, even for kinetic systems that exhibit open loop instability and dependent kinetic equations. We successfully address this task by applying a direct transcription nonlinear programming (NLP) approach that determines the optimal kinetic parameters and the concentration profiles simultaneously. Second, since only spectroscopic measurements are available, covariances and other distributional information for $W(t)$ and $z(t, l)$ cannot be determined directly. Instead, we apply maximum likelihood principles in order to derive the covariances needed for parameter estimation. Coupled solution of these two tasks leads to an iterative optimization-based approach that provides covariance estimates and initializes the kinetic parameters. These results allow the subsequent solution of the parameter estimation problem with fixed covariance terms. Confidence regions for the kinetic parameter estimates are then obtained through the application of NLP sensitivity, a byproduct of the direct transcription approach.

We demonstrate our proposed approach on six case studies, including a comprehensive treatment of aspirin kinetics. These cases include simulated as well as actual experimental datasets. Detailed numerical results for these cases are presented and compared with the MCR-ALS GUI 2.0 toolbox. These show significant improvements in performance and computational cost over state-of-the-art approaches.

465303 Application of Computational Chemistry in Pharmaceutical Process Development

Abstract:

Increasing speed of drug discovery and chemical development often requires complementing experimental data with a variety of modeling tools to provide fundamental knowledge at a molecular level. With advances in computational power and theoretical techniques (e.g. density functional theory), implementation of quantum mechanical (QM) calculations for systems relevant to small molecule pharmaceutical process development has become more readily available. Process design and understanding in several areas of chemical development are significantly enhanced by utilization of QM modeling. In this work, several industrially relevant case studies will be presented to provide an overview of QM modeling and its applications in both early and late phase development. The case studies will discuss QM calculations in three key areas related to reaction and crystallization development: a) evaluation of relative stabilities and activation energies (pKa estimation, conformational analysis, transition structures); b) investigation of parallel reaction pathways (chemo- and stereo-selectivities); c) spectral predictions for reactive intermediates (NMR, FTIR, Raman). Impact to process understanding and design gained from computational chemistry modeling will be highlighted.

466350 Systematic Optimisation of Continuous Separation for Artemisinin Recovery in Continuous Pharmaceutical Manufacturing

Abstract:

An environment of growing R&D expenditure and increasing competition from generic manufacturers has led to a sustained research interest in Continuous Pharmaceutical Manufacturing (CPM), which has emerged as a promising alternative to the current paradigm of batch production. The latter, while possessing benefits including equipment flexibility and the extensive industry know-how of a mature technology, is hampered by disadvantages including low efficiency (of both material and energy use), problematic process scale-up and poor heat and mass transfer. Moreover, its status as a mature technology means improvements are incremental at best. Novel CPM methodologies provide enticing options toward fostering the concept of Quality by Design (QbD) which is enthusiastically pursued by the US pharmaceutical industry, and the closely related development of Process Analytical Technology (PAT), which focuses on efficient, online pharmaceutical process monitoring and control. The concepts of CPM, QbD and PAT have accordingly received attention and encouragement by influential, world-leading regulatory authorities (Lee et al., 2015).

Artemisinin is one of the most important antimalarial substances available today. First identified and isolated from the plant *Artemisia annua* in the late 1970s (Tu, 2011), work for which the *2015 Nobel Prize in Physiology/Medicine* was awarded, artemisinin is currently produced via batch extraction from the cultivated plant. However, long product lead times, and fluctuating demand due to unpredictable burden of malaria, lead to highly variable prices and production levels (Jolliffe and Gerogiorgis, 2015a). Recent research has demonstrated the continuous synthesis of artemisinin (Kopetzki et al., 2013; Seeberger et al., 2014). This novel synthesis uses a waste product from the current batch production process as a feedstock, and produces artemisinin via two sequential plug flow reactors, the first of which employs photo-oxidation. Remarkable product yields can be achieved, and the process has been further adapted to produce several artemisinin derivatives (Gilmore et al., 2014).

Critical to harnessing the full potential of CPM is the design and employment of continuous product separation systems. A recent published study of ours has explored the potential performance of eight solvents for continuous crystallisation of artemisinin via both computationally predicted (UNIFAC) and experimentally reported solubilities; results indicate promise for the use of ethanol or ethyl acetate (Jolliffe and Gerogiorgis, 2015a, 2016). We have also previously employed nonlinear optimization toward finding ideal continuous separation scenarios for another API: promising temperatures, solvent and liquid-liquid extraction (LLE) configurations have been determined toward minimising the total CPM process cost, by means of comprehensively considering mass transfer, thermodynamics and detailed economic cost modelling (Jolliffe and Gerogiorgis, 2015b).

The present paper expands upon our previous contributions in the field of CPM: we investigate further the continuous separation of artemisinin by formulating a nonlinear optimization problem comprehensively considering mass transfer modelling via empirical correlations, solubility prediction via group contribution methods (UNIFAC, NRTL), and environmental and sustainability metrics (E-factor, Mass Intensity), for several potential separation unit operations and solvents. The CPM process studied here, based on the chemistry developed by Kopetzki et al. (2013), has been shown to be compatible with a range of separation methods (Horváth et al., 2015). The design space illustrates clear nonconvexities for several key process variables, and potential local minima in the final objective surface arise from regions of applicability of empirical relations. The solved problem and our novel results illustrate how the crucial consideration of less easily quantifiable characteristics (e.g. environmental impact, safety, toxicity) can qualify optimised solutions of superior technical performance.

LITERATURE REFERENCES

1. Gilmore, K., Kopetzki, D., Lee, J.W., Horváth, Z., McQuade, D.T., Seidel-Morgenstern, A., Seeberger, P.H., 2014. Continuous synthesis of artemisinin-derived medicines. *Chem. Commun.* **50**: 12652–12655.

2. Horváth, Z., Horosanskaia, E., Lee, J.W., Lorenz, H., Gilmore, K., Seeberger, P.H., Seidel-Morgenstern, A., 2015. Recovery of artemisinin from a complex reaction mixture using continuous chromatography and crystallization. *Org. Process Res. Dev.* **19**: 624–634.

3. Jolliffe, H.G., Gerogiorgis, D.I., 2015a. Process modelling and simulation for continuous pharmaceutical manufacturing of artemisinin. *Chem. Eng. Res. Des.* (in press, available online).

4. Jolliffe, H.G., Gerogiorgis, D.I., 2015b. Technoeconomic optmisation for green Continuous Pharmaceutical Manufacturing (CPM) of analgesics, *Proceedings of the AIChE Annual Meeting*, Salt Lake City, UT, USA.

5. Jolliffe, H.G., Gerogiorgis, D.I., 2016. Systematic solvent evaluation for artemisinin recovery in continuous pharmaceutical manufacturing, *Comput.-Aided Chem. Eng.*(in press, available online).

6. Kopetzki, D., Lévesque, F., Seeberger, P.H., 2013. A continuous-flow process for the synthesis of artemisinin. *Chem.–Eur. J.* **19**: 5450–5456.

7. Lee, S.L., O'Connor, T.F., Yang, X., Cruz, C.N., Chatterjee, S., Madurawe, R.D., Moore, C.M.V., Yu, L.X., Woodcock, J., 2015. Modernizing pharmaceutical manufacturing: from batch to continuous production. *J. Pharm. Innov.* **10**(3): 1–9.

8. Seeberger, P.H., Kopetzki, D., Levesque, F., 2014. *Method and device for the synthesis of artemisinin.* US20140364630 A1.

9. Tu, Y., 2011. The discovery of artemisinin (qinghaosu) and gifts from Chinese medicine. *Nat. Med.* **17**: 1217–1220.

468494 Prediction of Liquid-Liquid Partitioning and Solubility of APIs with the SAFT-γ Mie Approach

Abstract:

Lipophilicity is a fundamental physicochemical property in drug discovery and is usually quantified from the partitioning equilibrium of solute molecules between water and an immiscible organic solvent. The most popular system in this context is the 1-octanol + water mixture. The partitioning of solutes in aqueous alkane systems (*e.g.*, hexane + water and hexadecane + water) is also of interest for drug discovery and the difference between the partition coefficients serves as a measure of the hydrogen bonding potential of a molecule. The solubility of APIs in both aqueous and organic solvents is an essential property for drug development and manufacturing. It serves as essential information in the design of the downstream processes of APIs.

A general approach to predict the solubility and partition coefficients in octanol+water (K_{OW}) and alkane+water (K_{alk}) systems is developed. The core of our predictive framework lies in the use of the recently developed group-contribution SAFT-γ Mie approach [1-4]. SAFT-γ Mie is a predictive equation of state which allows one to determine the thermo-physical properties of molecules in terms of the constituent functional groups that represent their unique molecular identity. These functional groups are then modelled as fused spherical segments that interact via Mie (generalized Lennard-Jones) potentials of variable repulsive and attractive ranges. The parameters for each functional group are developed from fluid-phase equilibrium data for simple compounds and, once estimated, they are applied to the study of more complex multifunctional molecules in a predictive manner. This key feature of the SAFT-γ Mie approach enables the prediction of the properties of organic molecules in various solvents and solvent mixtures over wide thermodynamic conditions without the need for experimental solubility data for a specific molecule.

The necessary alkane-water and octanol-water group interactions are established providing an excellent description of the partition coefficients of several families of organic compounds which are used to test the reliability of the models. The SAFT-γ Mie models are then applied to predict the K_{OW} and solubility of various APIs in different organic solvents; the results are found to be in excellent agreement with the experimental data. The APIs employed in this study range from a relatively simple molecule, azelaic acid, to multifunctional molecules, such as ibuprofen, ketoprofen, lovastatin, and simvastatin. No experimental data related to the API are used in the K_{OW} prediction. For the solubility calculation, in addition to the molecular structure, a few properties of the pure API are required, namely the melting temperature, heat of fusion, and the heat capacity difference between the liquid and the solid state of API. This work highlights the extension of the SAFT-γ Mie equation of state in predicting important thermodynamic properties of complex multifunctional molecules in different solvents and demonstrates that the appraoch can be employed as a novel predictive platform for the prediction of important API properties from molecular structure alone.

References

[1] T. Lafitte, A. Apostolakou, C. Avendaño, A. Galindo, C. S. Adjiman, E.A. Müller, and G. Jackson "Accurate statistical associating fluid theory for chain molecules formed from Mie segments" The Journal of Chemical Physics, 2013, 139: 154504

[2] V. Papaioannou, T. Lafitte, C. Avendaño, C. S. Adjiman, G. Jackson, E. A. Müller, and A. Galindo "Group contribution methodology based on the statistical associating fluid theory for heteronuclear molecules formed from Mie segments" The Journal of Chemical Physics, 2014, 140: 054107

[3] S. Dufal, V. Papaioannou, M. Sadeqzadeh, T. Pogiatzis, A. Chremos, C. S. Adjiman, G. Jackson, and A. Galindo "Prediction of thermodynamic properties and phase behavior of fluids and mixtures with the SAFT-γ Mie group-contribution equation of state" Journal of Chemical & Engineering Data, 2014, 59: 3272-3288

[4] S. Dufal, T. Lafitte, A. J. Haslam, A. Galindo, G. N. Clark, C. Vega, and G. Jackson, "The A in SAFT: Developing the contribution of association to the Helmholtz free energy within a Wertheim TPT1 treatment of generic Mie fluids" Molecular Physics, 2015, 113: 948-984

Acknowledgements

We gratefully acknowledge financial support from Pfizer Inc.

461660 Use of a First Principles Model during QbD Studies to Assess Impurity Formation for a Coupling Reaction

Abstract:

The overall strategy that was used in QbD studies for the evaluation of reactions in the synthesis of an API will be assessed by reviewing the specific case for the coupling reaction of an amine with an acid chloride to form the corresponding amide. For each reaction in the experimental design set, a model was developed based on kinetic data that was acquired. Kinetic parameters were then obtained for each reaction and were assessed as a set to determine if the studied variables impacted impurity formation over their desired manufacturing range. In the reaction presented, two main impurities are tracked: 1) an imide impurity is formed from the product and excess acid chloride and 2) a carboxylic acid impurity is formed by hydrolysis of the acid chloride. The formation of impurities was investigated varying the amount of acid chloride, reagent level, catalyst loading, solvent amount and temperature. The resulting model was used to calculate impurity levels at reaction completion for a variety of conditions. The development and use of the kinetic model during QbD studies allowed for a deeper understanding of the reaction over its entire time course and enhanced confidence in how to control the formation of impurities.

461275 Verification and Use of CFD Modeling in Stirred Tank Reactors for Pharmaceutical Applications

Abstract:

Mixing often determines the success or failure of pharmaceutical process scale-up. Many unit operations, including heterogeneous reactions, fed-batch processes, liquid-liquid extractions, and crystallizations are susceptible to changes in performance due to variable mixing. To predict mixing performance between scales, we have adopted a computational fluid dynamics (CFD) modeling approach using MixIT, a CFD package specialized for mixing in stirred tanks. In this talk, we present a comparison of CFD modeling with experimental measurements.Â To assess the model's accuracy, we measured agitator power consumption, a representative mixing attribute in stirred tanks, up to kilo-scale and compared the measurements with the model's predictions. The measurements were taken across a variety of common agitator types and baffle configurations, with a wide range of power per unit volume values. The results demonstrated good agreement between experiments and simulations, showing the modelâ€™s accuracy in predicting mixing attributes. We then demonstrated the use of CFD to predict performance of a mixing sensitive liquid-liquid reaction as it was scaled up to 20L. In this example, reaction monitoring with process analytical technology (PAT) was combined with CFD predictions to develop correlations between mixing attributes and the mass transfer coefficient.

469808 Product-Agnostic Computational Fluid Dynamic (CFD) Model for Single-Use Mixing System

Abstract:

As the interest in single-use systems continues to grow in the biopharmaceutical community, mixing is a phenomenon that needs to be understood and characterized routinely, taking effort and time. Insufficient mixing can lead to content inhomogeneity, while excessive mixing can lead to protein denaturation as well as an increase in cycle time. To help reduce experimentation for future process transfer and scale-up, a product-agnostic multiphase computational fluid dynamic (CFD) model was developed. The CFD model establishes a process design space for 10L and 50L single-use mixing bags. The design space can be used by product teams to define impeller speeds in order to achieve the dcsired blend times based on working volumes of fluid, and fluid properties. These studies will accelerate the design and use of single-use stirred bags to formulate drug solutions.

447399 Continuous Biocompatible Reactive Extraction Concept for the Purification of Di-Carboxylic Acids from Dilute Aqueous Solutions

Abstract:

Product concentrations for the biocatalytic production of dicarboxylic acids achievable by whole-cell biocatalysis have recently been shown to reach industrial levels. Caused by their high reactivity, dicarboxylic acids are an attractive substance class for the chemical industry and are even described to serve as future platform chemicals. The development of downstream processing (DSP) concepts still is the bottleneck towards an industrial application of dicarboxylic acids from biocatalytic origin.

Reactive extraction (RE) has shown to be applicable as selective and efficient purification step for the direct recovery of dicarboxylic acids from a dilute aqueous solution, with the mechanism of the RE recently described in our previous work [1]. A hydrophobic complex is formed between hydrophilic dicarboxylic acid (*cis,cis* muconic acid and itaconic acid) and the extractant (tri-n-octylamine) and is extracted into the organic phase. For the recovery of the dicarboxylic acid from the organic RE phase two different strategies have been developed recently: *ph*-shift and addition of a water soluble amine lead to a full recovery of the acid and allow for a recycle of the organic RE phase containing the extractant.

Within this work batch extraction experiments published previously [1] were transferred into a continuous DSP concept using a 15 stage lab-scale Mixer-Settler setup. The transfer to continuous operation and the scale up (factor 20) in the Mixer-Settler experiments marks a first step to fulfill the requirements for an industrial application. Different process parameters (e.g. mass flow, number of stages for RE and REEX) were investigated with respect to the overall DSP yield achievable. Additionally a recycle of the organic phase (including amines for RE) and a combined RE and REEX was investigated in order to prove the applicability of the developed DSP concept. Stability investigations over the course of up to 12 h at equilibrium state within the Mixer-Settler set up of the system

demonstrate the robustness of the presented DSP concept. Aqueous phase after REEX was shown to enable final purification and thus allow for a further development of an in situ DSP concept for dicarboxylic acids from biocatalytic origin.

[1] J. Gorden, T. Zeiner, C. Brandenbusch, Fluid Phase Equilibria, 393 (2015) 78-84

462006 Combining PAT with Data Visualization and Machine Learning to Understand and Control Batch Cooling Crystallization

Abstract:

With the rapid development of PAT, online crystallization monitoring tools have become commonplace. Spectroscopic monitoring tools, such as ATR-FTIR, and particle-detecting tools, such as FBRM, are especially popular in that regard [1].

Using these tools, large data sets can be collected, which presumably contain detailed information on crystallization dynamics. But this is where the promise of big data often falls short. Simply collecting raw data does not automatically lead to understanding. Instead, the data most be processed, analyzed, and presented in a coherent way.

In this talk, we will describe how data from ATR-FTIR, FBRM, and temperature measurements can be brought together to visualize batch cooling crystallization as movement through a two-dimensional space [2]. The data visualization, we will show, fosters an intuitive understanding of the crystallization dynamics. It also sets the stage for a low-dimensional mathematical model of crystallization and dissolution dynamics that can be learned directly from data.

Combined, the data visualization and application of machine learning creates a framework for establishing control. In particular, we demonstrateâ€"on two experimental case systemsâ€"the use of the framework to accurately control the mean size of crystals produced by unseeded batch cooling crystallization [2, 3].

[1] L. L. Simon, H. Pataki, G. Marosi, F. Meemken, K. HungerbÃ¼hler, A. Baiker, *et al.Â* 2015. Assessment of Recent Process Analytical Technology (PAT) Trends: A Multiauthor Review. *Organic Process Research & Development* 19, 3-62.

[2] Griffin, D. J., Grover, M. A., Kawajiri, Y. and Rousseau, R. W. 2015. Mass-Count Plots for Crystal Size Control. *Chemical Engineering Science* 137, 338-351.

[3] Griffin, D. J., Grover, M. A., Kawajiri, Y. and Rousseau, R. W. 2016. Data-Driven Modeling and Dynamic Programming Applied to Batch Cooling Crystallization. *Industrial & Engineering Chemistry Research* 55, 1361-1372.

467965 Stability Analysis of Solar Silicon Wafer Manufacturing By the Horizontal Ribbon Growth Process

Abstract:

The increasing demand for energy and the prospect of global warming urges the world community to seek cleaner and cheaper energy resources to replace non-renewable fossil fuels. The silicon based photo-voltaic system is one of the most promising renewable energy resources. Its application has grown rapidly during the past 2-3 decades as production cost has been reduced and the potential for application is virtually without limit.

This promotes the project of cost reduction for silicon wafer via different technology development. Traditional silicon growth processes are either expensive or compromise the material efficiency for faster production speed. The horizontal ribbon growth (HRG) process has the potential to overcome these limitations. In the HRG process, heat is removed through the top surface of the molten silicon pool, meanwhile, a thin silicon solid sheet is produced and extracted continuously. However, despite the theoretical and experimental efforts to develop and commercialize this technology for more than 60 years, many technical challenges have been reported to achieve a stable and fast ribbon production.

One of the principal challenges encountered is the instability at the wafer-melt interface due to various coupled effects. Theoretical studies have been developed for the instabilities during unidirectional solidification. In 1964, Mullins and Sekerka developed the foundation for the thermal and solutal conditions to ensure the stable crystallization front under an arbitrary sinusoidal perturbation of infinitesimal amplitude to the interface [1]. However, the flow pattern in the melt during the crystallization also plays an important role on the impurity distribution, which alters the micro-structure and macro material property of the grown crystal. This flow pattern can be induced by the rotation of the crucible or crystal, withdrawal of the material, density change during crystallization, etc. Coriell and Sekerka performed extensive studies in the coupled convective and morphological instabilities. [2][3]

In this work, we developed a linear stability theory to study the crystallization interface of the Horizontal Ribbon Growth (HRG) process. The conditions for the onset of stability are investigated theoretically and numerically on the basis of diffusion-convection equations for the thermal, solutal fields coupled to the Navier-Stokes equation describing the flow field in the molten pool near the interface. In order to understand the unexplained "facet lines" observed in the experiments, the interface stability conditions of the system under different operating conditions are tested.

1. **Mullins, W. W., & Sekerka, R. F,** "Stability of a planar interface during solidification of a dilute binary alloy," *Journal of applied physics*, **35**(2), 444-451(1964).
2. **Coriell, S. R., Cordes, M. R., Boettinger, W. J., & Sekerka, R. F,** "Convective and interfacial instabilities during unidirectional solidification of a binary alloy," *Journal of Crystal Growth*, **49**(1), 13-28 (1980).
3. **Coriell, S. R., McFadden, G. B., Boisvert, R. F., & Sekerka, R. F,** "Effect of a forced Couette flow on coupled convective and morphological instabilities during unidirectional solidification," *Journal of crystal growth*,**69**(1), 15-22 (1984).

458953 Protein Enzymatic Hydrolysis Integrated with Ultrafiltration

Abstract:

Thermolysin was used to hydrolyse serum bovine albumin. The molecular mass and concentration of hydrolysates were identified by SE-HPLC. The substrate pre-incubation at high temperatures (>46°C) contributed to the increase of the hydrolysis rate and final efficiency, but at 70°C a negative effect caused by the aggregation of albumin molecules was noticed. Pre-incubation time is significant up to 0.5 hour.

Strong product inhibition was detected, thus the values of hydrolysis degree obtained in classical reactors were low. The kinetic equation was elaborated and its constant values were estimated.

The data yielded in an integrated bioreactor-ultrafiltration process provides strong evidence that considered enzymatic process could be significantly intensified. The ultrafiltration membrane with a small pores size was selected to take off the products (peptides), including enzyme inhibitors.

The utility of a previously elaborated kinetic model of biopolymer hydrolysis was checked and a qualitative and quantitative description of the proteolysis products was performed. The data gathered suggest that process efficiency and permeate composition depend on the proper selection of dilution rate and the separation properties of the membrane.

469298 The Dynamics of Biodiesel Profile and Fungal Diversity of Activated Sludge during Lipid Accumulation on Acetic Acid

Abstract:

Recent studies indicated that the microbial consortia of wastewater treatment plant (WWTP) activated sludge feeding on acetic acid can accumulate lipids, which can be transesterified into biodiesel. The composition of the raw activated sludge, however changes significantly with the fluctuations of operational and environmental conditions in a WWTP. The variations in raw activated sludge composition may pose additional processing cost after extraction and transesterification of the lipids if there is significant variation of the biodiesel profile (measured as fatty acid methyl esters or FAMEs) among raw activated sludges. This study describes the changes on biodiesel profile during lipid accumulation on acetic acid. The changes on the diversity of fungal community are also correlated to the changes in biodiesel profile. The findings suggest that even with variations in the biodiesel profiles from lipids of raw activated sludge, the enhancement of lipid content by feeding on acetic acid induces speciation of microbial community, specifically fungi, resulting in the homogenization of biodiesel profiles.

460002 Computational Solid State Chemistry Applications in Pharmaceutical Industry

Abstract:

Given the complexity of the pharmaceutical solid state landscape and challenges facing the pharmaceutical industry, an accelerated Drug Development greatly benefits from guidance provided by computational methods. The emerging field of Computational Pharmaceutical Solid State Chemistry covers the whole spectrum of state of the art computational approaches, which are used to support all steps related to the development of solid state pharmaceuticals.[1] This work will consider virtual screening approaches allowing solvent selection for impurity purge, coformer selection for cocrystallization and solid solvate/hydrate formation screening.

1. Abramov, Y.A., Ed. *Computational Pharmaceutical Solid State Chemistry*, John Wiley & Sons, 2016.

461286 Process Optimization Using a Distillation Model Coupled with a Solubility Model for Increased Process Yield Following Multiple Distillations and Solvent Swap

Abstract:

Yield variability and extended distillation times lead to the need for process modeling. The combined distillation and solubility model predicts both the acetonitrile concentration and resulting yield loss in the mother liquors. The model was used to reduce yield loss and cycle time while providing additional potential for solvent recovery of acetonitrile. The distillation model was generated using Aspen properties with the system at 0.1 bar absolute pressure and Euler integration for a stepwise mass balance approach while the solubility model was generated using data regression from mother liquors. The solubility data was necessary to be used from mother liquor rather than in a pure system due to increased solubility of the intermediate while in the presence of the byproducts of the desired reaction. The model has predicted the acetonitrile composition to be within 0.03 mass fraction of the experimental data. The yield loss using the mother liquors model with modeled acetonitrile mass fraction resulted in a difference of no more than 1.30% yield over the 0.001 to 0.114 acetonitrile mass fraction range. The model was used to maximize yield and was verified on pilot plant scale.

464090 Large-Scale Crystallization of a Pure Metastable Polymorph By Reaction Coupling

Abstract:

A maximum of more than 40 g of acetaminophen Form II crystals with a yield of 65 mol % is made reproducibly in a 500 mL reactor by coupling the acetylation of *p*-aminophenol with the neutralization of acetic acid before the crystallization of acetaminophen Form II crystals. This novel working principle involves a sudden drop of the solubility of acetaminophen from the acetic acid–water environment to the acetate–water system in addition to temperature cooling but without agitation. This particular processing pathway is capable of maintaining a large supersaturation for quite some time for the system to enter the metastable zone of Form II with respect to the acetate–water system. The large amount of Form II crystals so produced are then isolated by filtration and oven drying to prevent Form II crystals in the mother liquor from transforming to the thermodynamically stable Form I crystals by the Ostwald's rule of stages.

461590 PAT Enabled Drying Platform for API Process Development â€“ Investigation of Solvent Trapping Associated with Form Transition and Crystal Collapse during Drying

Abstract:

An active pharmaceutical ingredient was isolated as a crystalline hydrate which converts to an amorphous solid during drying. Challenges associated with material handling and incomplete drying were identified at the factory scale. It was suspected that differences in processing equipment and conditions were leading to rapid loss of water, collapse of the crystalline lattice, and trapping of the process solvent at levels above specification.

A PAT enabled drying platform was used and advanced laboratory methodologies were developed to capture detailed process data and gain a fundamental understanding of the drying process, test various theories for solvent trapping, and evaluate drying from an alternative solvent system. Experiments performed led to increased process understanding, identification of conditions needed for successful drying, and ultimate process selection to improve process efficiency, robustness, and portability. The newly developed process was successfully demonstrated at pilot plant scale with performance in line with model projections. Results of this investigation and the methods applied will be described in detail during this presentation.

454184 Modeling and Simulation Efforts in Scale up of Sustained Release Microspheres

Abstract:

Abstract

Generic pharmaceutical industry has to operate in tight timelines of product development and scale up from lab to plant. To meet such needs, Dr Reddys Laboratories use approaches like modeling and simulation for better process understanding and to reduce scale up risks. This presentation provides example cases of such efforts in a microspheres product.

Microencapsulation process development demands understanding of the impacts of various material properties and process parameters to achieve the right product characteristics. Employing appropriate modeling approaches at different scales can aid in developing this understanding. In this work, a first principle based diffusion model which uses Monte Carlo simulation helps link the impact of microsphere product attributes like size distribution, porosity, and polymer properties on the sustained release of drug from microspheres.**[1]** Results from a numerical simulation of this model gauge the risk of each of the product characteristics on the release. This understanding identifies the key product attributes to focus on and manipulate with the process conditions.

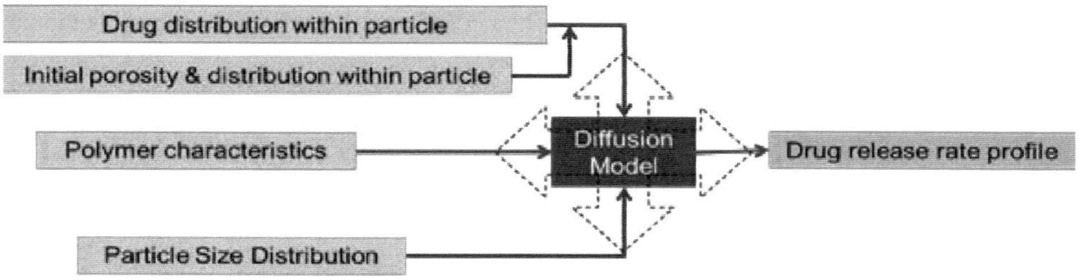

Manufacture of such micro-encapsulated products consists of multiple unit operation steps like phase separation, extraction and drying to get the desired product characteristics. A combination of first principle, and CFD based approaches for the unit operations links how each of the process steps impacts the product attributes. Some of the important linkages made include scale dependent variations of: mixing conditions on particle size using vessel averaged Reynolds and Weber number obtained from CFD simulations; cycle times for extractions by single particle spherical diffusion model; and drying cycle predictions by first principle discrete element model with mixing simulated by randomization of the discrete elements. Custom lab scale experimentation provides the necessary model parameters and enables extension of the understanding towards the plant scale. This provides the necessary framework for a successful scale-up.

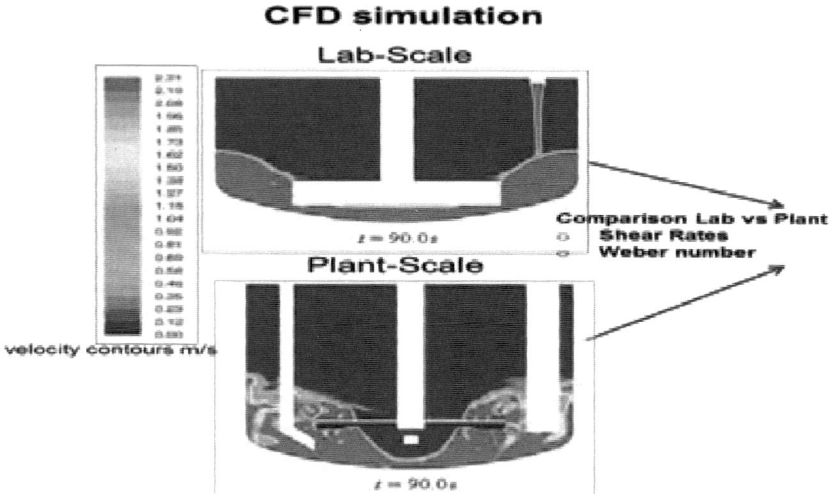

[1] J. Siepmann, N. Faisant, J.P. Benoit, A new mathematical model quantifying drug release from bioerodible microparticles using Monte Carlo simulations, Pharm.Res. 19 (2002) 1885–1893.

455552 Combined Task Bayesian Optimization for Efficient Scale-up and Technology Transfer

Abstract:

Operating condition optimization is crucial to assure product quality and reduce operation cost in any industry. In particular, one of the most challenging problems is to optimize operating conditions of commercial-scale equipment when only a small number of data are available just after scale-up from pilot-scale equipment. Such a situation is quite common because the experimental cost is significantly higher with commercial-scale equipment than pilot-scale equipment. In the present work, we aim to solve this problem and propose a new operating condition optimization method, which is referred to as combined task Bayesian optimization (CTBO). The proposed method is applicable not only to the scale-up problem but also to the problem of technology transfer from a mother plant to a copy plant; that is, when two plants are the same but operating conditions need to be optimized at the copy plant due to various uncertainties.

A conventional method builds a model that relates product qualities with operating conditions, and then optimizes the operating conditions by using the model so that the target product qualities are realized and operation cost is minimized under various constraints. Unfortunately, this method does not work well when only little data is available at a commercial-scale plant, because it is difficult to build an accurate model. Such a situation always occurs just after scale-up from a pilot-scale plant. The optimization performance would be improved if data of the pilot-scale plant could be used jointly with data of the commercial-scale plant. However, combining data of different scales is not straightforward because pilot-scale and commercial-scale plants have different numbers and types of sensors, which are operated under different conditions.

The goal of this research is to develop a new method that can efficiently and accurately derive optimal operating conditions by using data obtained from a commercial-scale plant and a pilot-scale plant even when only little data is available at the commercial-scale plant. The proposed method uses Bayesian optimization (BO).

BO can systematically determine a plan for new operating conditions to be evaluated for further optimization. Thus, without experimental design, BO can find a better solution through fewer experiments than conventional methods. BO uses two important tools: Gaussian process regression (GPR) and an acquisition function. GPR is used as a nonlinear regression method, and the acquisition function is used to determine new operating conditions to be evaluated by taking account of the trade-off between exploration and exploitation. By combining them, BO can efficiently optimize operating conditions.

Transfer learning aims to exploit knowledge from one or more source tasks and to apply the knowledge to the target task. A key idea of the present work is that transfer learning is useful for solving the scale-up problem by regarding the source task and the target task as the pilot-scale plant and the commercial-scale plant, respectively. Among transfer learning algorithms, adaptive transfer learning and frustratingly easy domain adaptation are used in this work. The former is an algorithm established for GPR, and the latter is a famous, simple algorithm.

Conventional transfer learning algorithms assume that the number of input variables in the source task is the same as that in the target task. In practice, however, the number and types of sensors of a pilot-scale plant are different from those of a commercial-scale plant. Thus, to apply transfer learning to the scale-up problem, we introduce a transformation matrix, which transforms data of the pilot-scale plant into data of the commercial-scale plant so that data from both can be used for modeling the commercial-scale plant. To derive the transformation matrix automatically, we propose to use Markov Chain Monte Carlo (MCMC), which is also used to determine GPR hyperparameters.

In the present work, combined task Bayesian optimization (CTBO) is proposed by integrating BO, transfer learning, and the transformation matrix. To validate the effectiveness of the proposed method, CTBO was compared with BO and LW-PLS + jDE (locally weighted partial least squares + self-adaptive differential evolution) through their applications to numerical examples and a pharmaceutical granulation process. Two types of CTBO were tested: CTBO(ATGP), i.e., CTBO with adaptive transfer learning, and CTBO(DAGP), i.e., CTBO with frustratingly easy domain adaptation.

The results of the numerical examples showed that the optimization performance of BO was superior to that of LW-PLS + jDE since BO appropriately determined operating conditions one by one. CTBO further improved the optimization performance by combining data of both the source task and the target task. Similarly, in the pharmaceutical granulation process, CTBO outperformed LW-PLS + jDE and BO. The improvement rate of CTBO(DAGP) compared with LW-PLS + jDE and BO were 48.6% and 15.2%, respectively, when the number of samples on the commercial-scale plant was 10. The results demonstrated that CTBO used data of the pilot-scale plant effectively to optimize the operating conditions of the commercial-scale plant.

456730 Development and Optimization of a Mixed Suspension Mixed Product Removal (MSMPR) Crystallization Process Incorporating Wet-Milling

Abstract:

In recent years, there has been an increased interest in moving pharmaceutical manufacturing from batch to continuous processing. The potential to reduce cost, simplify production and improve product quality have driven significant investment in continuous manufacturing. While crystallization is an important separation and purification unit operation, continuous crystallization processes have not been widely studied within the pharmaceutical industry primarily because of the design complexity and difficulty of robust operation. The MSMPR crystallizer is arguably one of the most promising technologies as its operation is conceptually simpler than for e.g. tubular crystallizers and it is easy to convert batch crystallizers to continuous operation. However, issues such as fouling and encrustation, transfer line blockage, and classification due to limited nucleation are frequently experienced with numerous reports in the literature.

The work presented herein solves the fouling and classification issues encountered for one of Takedaâ€™s APIs by adding wet-milling to the first stage of an MSMPR crystallizer. The API studied was found to be characterized by slow nucleation kinetics resulting in a prolonged wash out period. During this period, large (>1mm) particles were formed that were unable to be transferred leading to classification issues, build up of supersaturation and severe fouling.

The high shear imparted by the wet-mill was found to control particle size and increase both primary and secondary nucleation rates. This reduced supersaturation and therefore fouling and encrustation. The wet-mill was incorporated in the first stage of the cascade which was then experimentally optimized investigating factors such as solvent composition, types and materials of construction of the impeller, wet-mill rotational speed, wet-mill turnover numbers, residence time and start-up methods. Finally, a multi-stage system was developed to optimize productivity.

457164 Developing Process Models of Hydrogen Peroxide Decontamination in Pharmaceutical Manufacturing

Abstract:

Decontamination is a critical technology for providing sterile environment in the production of sterile drug products such as injectables, intravenous bags and eye-drops. In the modern production facilities, these drug products are filled in the equipment such as isolators, which are decontaminated before the operation using vapor hydrogen peroxide. This industry-standard decontamination agent is diffused in the equipment as vapor and mist, kills the microorganisms on the surface, and removed by aeration before the filling can start. The success of this decontamination process is a prerequisite for assuring the quality of the above mentioned sterile drug products. Furthermore, decontamination is a relevant process regarding productivity, because the entire process including conditioning and aeration dominates the production time. Despite of such relevance in quality and productivity, scientific elucidation of decontamination is still in infancy, which is the obstacle to more rational design of decontamination processes.

The objective of our research is to develop a model of decontamination using hydrogen peroxide, which can serve as the basis for process design in pharmaceutical manufacturing. We aim to create a model with linking the design parameters to so-called â€œD-valueâ€, and further to the objective functions such as time required for decontamination or energy consumption. D-value is the representative parameter of decontamination efficacy, and is defined as the time required to kill 90% of surviving microorganisms. As the design parameters, concentration of hydrogen peroxide as well as humidity are considered, and the influence of condensation is also taken into account in the model. The relationship between the design parameters and D-value is to be investigated by experiments using biological indicators with following the standard application procedure in the pharmaceutical industry. In addition to decontamination efficacy, our investigation will cover time and energy consumption required for the decontamination process including conditioning and aeration.

So far we set up experimental apparatus for the decontamination process in the steady-state condition, and measured D-values with varying humidity and hydrogen peroxide concentration. The preliminary results of the experiments suggested that the influence of humidity and concentration on decontamination efficacy is strong as known in the previous studies. A mathematical expression has been firstly created considering the relationship between the varied parameters and the observed D-values. This initial mathematical model will incorporate the influence of condensation on decontamination efficacy by using vapor-liquid equilibrium, diffusion, and mass transfer. Additionally, the model will cover the effect of the concentration of residual hydrogen peroxide on the product. After consideration of the condensation and the residual concentration, the time required for decontamination process will be evaluated, and the time will be linked to the objective functions. In future, we intend to create a design method for the decontamination processes using the process model to be developed in this work.

457244 Comparison of Raman and UV Spectroscopy for the in-Line Quantification of Active Ingredient in Pharmaceutical Semi-Solid and Liquid Formulations

Abstract:

Comparison of Raman and UV spectroscopy for the in-line quantification of active ingredient in pharmaceutical semi-solid and liquid formulations

N. Bostijn[1,*], C. Vervaet[2], T. De Beer[1]

[1]Laboratory of Pharmaceutical Process Analytical Technology, Ghent University, Ottergemsesteenweg 460, 9000 Ghent, Belgium

[2]Laboratory of Pharmaceutical Technology, Ghent University, Ottergemsesteenweg 460, 9000 Ghent, Belgium

Introduction

In-line spectroscopic techniques are increasingly proposed as alternative methods for the quantification of active ingredients in pharmaceuticals. This is due to their advantages over the traditional methods, such as fast and non-destructive measurements without the need of sample pretreatment. One of these spectroscopic techniques is Raman, that is until now, mostly applied for solid dosage forms. UV spectroscopy is a widely used quantification technique, although less frequently implemented inside a process environment for in-line monitoring.

This study compares the feasibility of in-line Raman and UV spectroscopy for the quantification of active ingredients in pharmaceutical formulations. A suspension and gel containing 0.09 % (w/w) and 2 % (w/w) active ingredient, respectively, were used as model formulations. The performance of both spectroscopic techniques was assessed via accuracy profiles [1-3].

MATERIALS AND METHODS

Materials

Two commercial available pharmaceutical formulations were selected from a pharmaceutical company. The suspension contained 0.09 % (w/w) active ingredient and the gel 2 % (w/w). Both formulations were prepared based on confidential information.

Methods

Calibration (80, 90, 95, 100, 105, 110 and 120 % of target) and validation (85, 95, 100, 105 and 115 % of target) samples were created by producing a batch for the lowest concentration. After batch completion, a specific amount of active ingredient was added, corresponding to the subsequent concentration level, followed by the collection of spectra. This procedure was repeated until the highest concentration level was reached and spectra were recorded from each concentration. In total, one calibration batch and three validation batches (three days) were produced for each formulation.

The suspension and gel were prepared using a customized IKA LR2000 modular mini plant reactor system (IKA, Staufen, Germany). The mixing vessel was equipped with a double wall for controlling the temperature of the process. A peristaltic pump (Watson-Marlow, Falmouth, Cornwall, UK) was used for pumping water from a water bath (Type 1032, GFL, Burgwedel, Germany) to the mixing vessel.

In-line Raman spectra were recorded using a Raman Rxn2 spectrometer (Kaiser Optical Systems, Ann Arbor, MI, USA), equipped with a fiber-optic PhAT probe. The laser wavelength was 785 nm. The spectral range of the system was from 150–1890 cm^{-1} with a resolution of 5 cm^{-1}. An exposure time of 15 s with no averaging was used.

An Avaspec-ULS2048L spectrometer (Avantes, Apeldoorn, The Netherlands) was connected by a fiber-optic cable to an immersion probe. The light source was an AvaLight Deuterium-Halogen Lamp. All spectra were acquired in the 200-1100 nm spectral range. The exposure time was 1000 ms and 950 ms for the gel and suspension, respectively, with each spectra the average of 5 scans.

The calibration models of the gel were developed using the mean-centered spectra, after standard normal variate (SNV) correction and first derivative transformation, over the regions of 390.1-404.8, 655.0-666.7, 1340.2-1354.9 and 1570.0-1600.0 $cm^{-1}$for the Raman spectra, while for the UV spectra the region 280.1-296.9 nm was used.

The Raman spectra of the suspension were mean-centered and SNV corrected, followed by taking the first derivative and selecting the region between 1390.0-1430.2 cm^{-1}. For the UV spectra, the same preprocessing was applied, except for the SNV correction and the selection of another spectral region (310.1-325.6 nm).

Spectroscopic data were correlated with the actual concentration of active ingredient, through partial least square (PLS) regression. The PLS models were created using the SIMCA software (Version 14, Umetrics, Umeå, Sweden).

Accuracy profiles, introduced by the SFSTP [1-3], were used as validation approach for the quantitative analytical methods.

RESULTS AND DISCUSSION

Gel

The RMSEP values of the Raman prediction set (0.026, 0.024 and 0.039 %) were lower compared to the ones of the UV data (0.058, 0.071 and 0.059 %). Accuracy profiles confirm the better performance of Raman. The β-expectation tolerance intervals exceed the 5 % acceptance limits only at the 1.75 and 2.37 % (w/w) concentration levels, whereas for UV, the β-expectation tolerance intervals exceed the 5 % acceptance limits over the whole concentration range.

Suspension

It can be noticed from the accuracy profile of the Raman method that the low concentration levels (0.077 and 0.086 % w/w) were predicted higher, the middle concentration level (0.091 % w/w) was predicted correct and the high concentration levels (0.095 and 0.104 % w/w) predictions were lower, suggesting that all the different concentrations were predicted the same value. This observation, the spectral data and the high RMSEP values (0.0097, 0.01 and 0.098 %) confirm that it was not possible to discriminate between the different concentration levels with Raman spectroscopy. For the UV calibration model, the β-expectation tolerance intervals exceed the 5 % acceptance limits over the whole concentration range, but fall within the 10 % acceptance limits at the 0.0865, 0.091 and 0.0955 % (w/w) concentration levels. RMSEP values (0.0050, 0.0015 and 0.0017 %) were lower than the RMSEP values of the Raman based quantification method.

CONCLUSION

The study showed that Raman spectroscopy performed better than UV spectroscopy for the in-line quantification of 2 % (w/w) active ingredient in a gel, but both techniques were feasible. For the low dosed suspension (0.09 % w/w), Raman spectroscopy failed to discriminate between the different concentration levels and thus the quantification of the active ingredient. UV spectroscopy was adequate for the in-line quantification, however between the acceptance limits of 10 %, instead of 5 %.

REFERENCES

1. P. Hubert et al. Harmonization of strategies for the validation of quantitative analytical procedures A SFSTP proposal-part I. Journal of Pharmaceutical and Biomedical Analysis. 2004; 36: 579–586.
2. P. Hubert et al. Harmonization of strategies for the validation of quantitative analytical procedures A SFSTP proposal-part II. Journal of Pharmaceutical and Biomedical Analysis. 2007; 45: 70–81.
3. P. Hubert et al. Harmonization of strategies for the validation of quantitative analytical procedures A SFSTP proposal-part III. Journal of Pharmaceutical and Biomedical Analysis. 2007; 45: 82–96.

458314 Process Development for Continuous Extrusion and Pelletization of a Heat and Shear-Sensitive Formulation

Abstract:

Hot-melt extrusion is an emerging technology for pharmaceutical manufacturing. Not only does it offer the possibility to establish amorphous systems to increase bioavailability, but also carries attractive implications process-wise [1]. It is an inherently continuous processes, performs several processing steps in one machine and offers the possibility of continuous monitoring and direct product shaping, e.g. to pellets, strands or films. At the same time, intense shear forces and dissipative heat generation are challenging when using some traditional carrier polymers [2].

In this case an immediate release formulation was developed containing a BCS class II with poor solubility. The formulation consists of the API, Eudragit E and Methocel E5. Methocel E5 was added in order to achieve immediate release from the very dense pellet. Besides the shear and heat sensitivity of Methocel E5, a high pressure drop across the extrusion die is obtained due to the high viscosity and the yield point. Therefore, the extrusion process line consists of (1) a hot-melt extruder with a melt pump, (2) and a strand pelletizer. The strand pelletizer is used to directly produce pellets of small size (x_{50} = 1000 µm) for subsequent continuous tableting.

The process development focused on the split-feeding unit, the intake zone, the temperature profile, the die plate temperature and the die plate design. After developing a stable process, a screening DoE was performed to investigate content uniformity, dissolution behavior, pellet size and material degradation. Moreover, the residence time distribution was determined.

References:

[1] J. Breitenbach, "Feste Lösungen durch Schmelzextrusion - Ein integriertes Herstellkonzept," *Pharm.*

Unserer Zeit, vol. 29, no. 1, pp. 46–49, 2000.

[2] A. Meena, T. Parikh, S. S. Gupta, and A. T. M. Serajuddin, "Investigation of thermal and viscoelastic properties of polymers relevant to hot melt extrusion - II : Cellulosic polymers .," *J. Excipients Food Chem.*, vol. 5, no. 1, pp. 46–55, 2014.

458474 Wet Granulation-Induced Polymorphic Transitions in Piracetam: Raman Spectroscopy As a Potential in-Situ Process Analytical Tool

Abstract:

Wet granulation induced polymorphic transitions in Piracetam: Raman spectroscopy as a potential in-situ process analytical tool

Gayathri Kollamaram, Catherine B. Kelly, Denise M. Croker, Gavin M. Walker

University of Limerick, Synthesis and Solid State Pharmaceutical Centre, Department of Chemical and Environmental Sciences, Limerick, Ireland.

Gayathri.Kollamaram@ul.ie

Wet granulation is one of the most popular techniques used during the post processing of active pharmaceutical ingredients (APIs) due to its ability to improve powder flow and compaction properties however the addition of water can affect the crystal structure of certain APIs exhibiting polymorphism. Piracetam is a highly aqueous-soluble drug which can exist in many polymorphic forms including hydrates. The potential therefore exists for process-induced transformations during wet granulation. Process-induced polymorphic transitions in APIs are a major challenge in the pharmaceutical industry as these unpredictable phase transitions can affect processability of the drugs owing to altered physical properties such as compressibility and can result in regulatory non-compliance due to altered bio-availability. Prior knowledge and understanding of the mechanism of these process-induced polymorphic transitions is therefore of the utmost importance. XRD and DSC techniques were used along with Raman to examine the polymorphic transitions of Piracetam during wet granulation so as to determine the effectiveness of Raman spectroscopy as a process analytical technology (PAT) tool for this system.

Piracetam undergoes polymorphic transitions under the influence of both solvent and temperature. FIII of Piracetam is the most stable form at room temperature. FIII transforms to FI when heated to 140°C which transforms to metastable FII upon cooling to room temperature following the Ostwald's step rule of successive transformation. At ambient conditions FII will eventually transform to the more stable FIII. It is known that polymers can have a stabilizing and/or inhibitory effect on the polymorphic transitions in APIs so different excipients with a range of hydrophilicities, aqueous solubilities and molecular weights were examined. Excipients used were Lactose monohydrate (LMH), microcrystalline cellulose (MCC), Hydroxy propyl methyl cellulose (HPMC), Hydroxy propyl cellulose (HPC) and Polyethylene glycol (PEG). LMH was chosen as the excipient representing low molecular weight; MCC, HPMC and HPC represent the excipients with varying degree of hydrophilicity while PEG is a water soluble polymer.

Piracetam was wet granulated with excipients with drug to excipient ratio of 3:2 as Piracetam is a high dosage drug. The granules produced were dried overnight at room temperature and the dried granules were then analysed at pre-determined times points using XRD, Raman and DSC. Raman was also employed to investigate the use of this technique as an in-situ analysis method. Principle component analysis (PCA) and direct cluster analysis (DCA) were used to process the Raman spectra as shown in (figure 1). Prior to processing all spectra were normalised and the first derivative was taken. XRD indicated that PIR granulated without any excipients transformed from FIII to Piracetam monohydrate (PIR-MH) in the presence of water and then recrystallized to FII upon drying. FII so obtained gradually recrystallised to the stable FIII at room temperature within two weeks. The Raman spectra gained from the granulation and drying of PIR were analysed using PCA and DCA and showed clustering into three distinct spectra which were confirmed to be those of FII, FIII and PIR-MH using XRD. Raman was therefore used in conjunction with XRD to inspect the progress of the transformation both from FIII to PIR-MH and from FII to FIII (Figure 2).

Granules of PIR containing LMH, MCC, HPMC and HPC exhibited similar transformations as pure PIR during granulation, namely that FIII transformed to monohydrate in the presence of water which recrystallized to FII upon drying (Figure 3). The kinetics of solid state transformation of FII to FIII at room temperature was retarded in the presence of excipients in comparison to that of pure PIR and by different degrees depending on the polymer. This was attributed to the interaction of excipients with the active functional groups in the drug and by the hydrophilicity and molecular weight of the excipients. However PEG had a different effect on these polymorphic transitions. Piracetam remained in its stable FIII throughout the processes of wetting and drying. Unlike the other excipients used in the study PEG is freely soluble in water and therefore it is possible for PEG to competitively inhibit the interaction between PIR and water. It is suggested that PEG inhibited the solution-mediated polymorphic transition of FIII to FII through the intermediate monohydrate.

Through the use of the complimentary techniques of XRD, DSC and Raman spectroscopy the polymorphic transformations exhibited by Piracetam during wet-granulation have been investigated. The importance of detecting the monohydrate form of the API as a precursor to the production of the metastable FII, the effectiveness of Raman spectroscopy in this detection, and the importance of understanding the excipients typically used during wet granulation have been demonstrated.

Description: Description: C:\Users\Gayathri.Kollamaram\Desktop\Exp\Piracetam\Images\PCA.tif

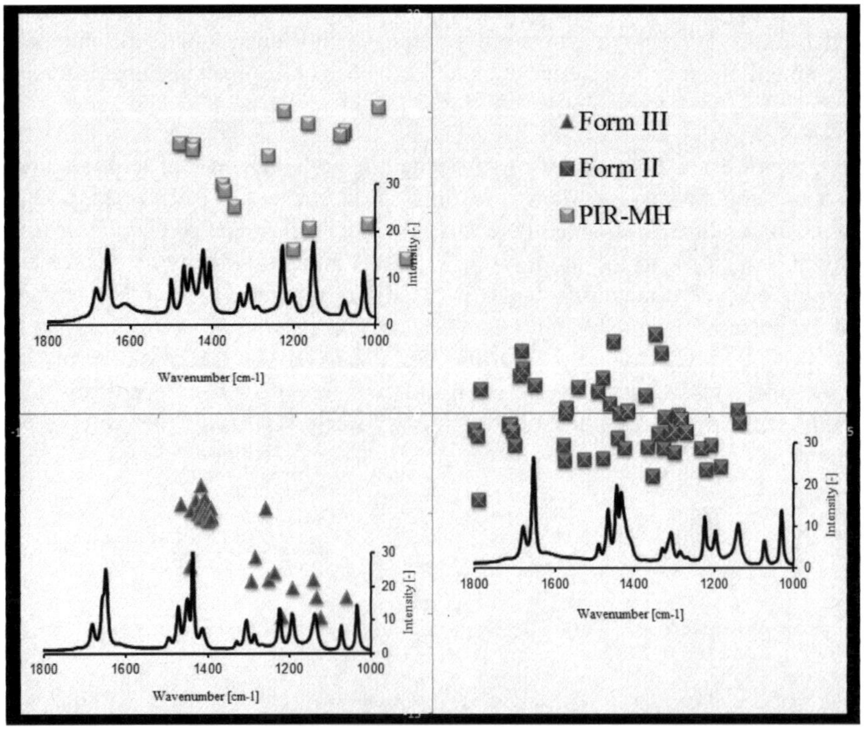

Fig 1: Prinicpal component analysis of the Raman spectra gained from the studies of Piracetam in water indicated clustering of the spectra into three distinct forms.

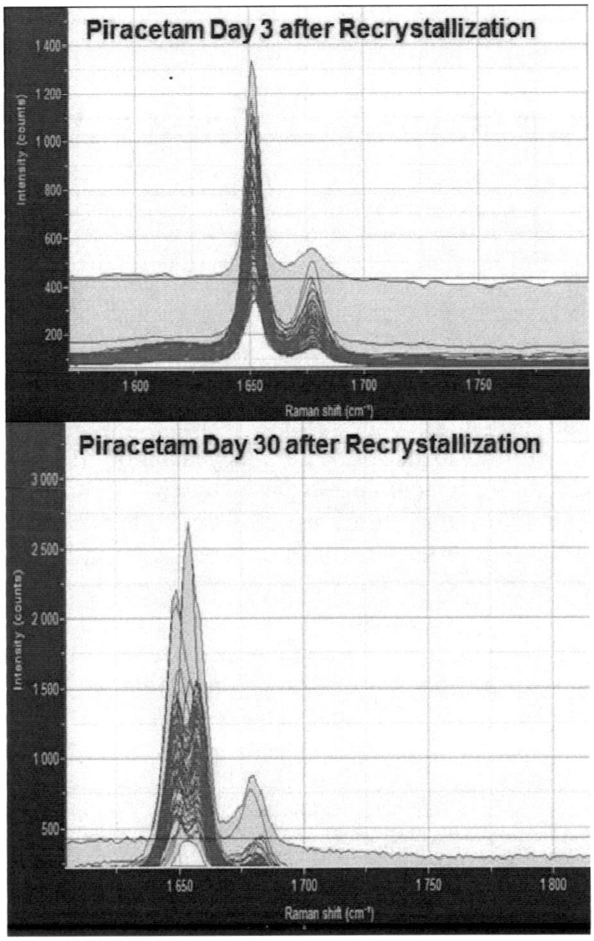

Fig 2: Raman spectra collected from an area on the surface of the API on Day 3 and 30 after recrystallisation from water illustrating the transformation of Piracetam FII to FIII during thirty days at ambient temperature.

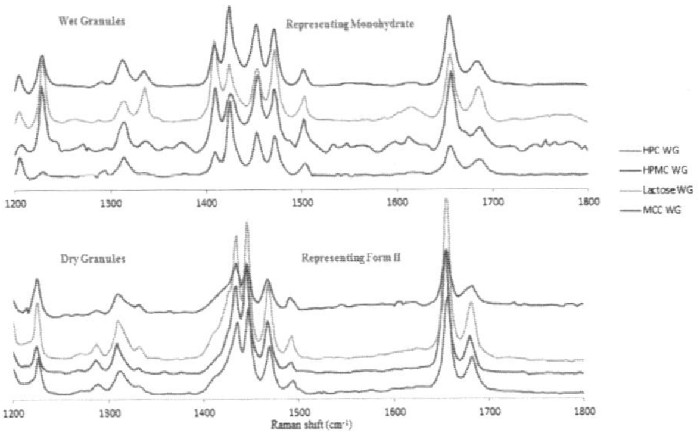

Fig 3: Raman spectra showing transformation of Piracetam from monohydrate to FII upon drying.

459850 Dissolvable Microcarriers for Efficient Cell Production and Recovery

Abstract:

Cell scale-up on microcarriers is an increasingly important technique for bioprocess applications. However, the harvest, separation, and concentration of final product (e.g. cell, virus, or antibody) can be time and resource consuming. To address these issues, Corning has developed a new generation of microcarriers that can be quickly dissolved, allowing easy collection of grown cells or secreted products without the need for microcarrier separation. Here we present unique properties and potential applications for dissolvable microcarriers. Human mesenchymal stem cells (hMSCs) and Vero were grown in spinner flask cultures or bioreactors on dissolvable microcarriers. For all cell types evaluated, growth performance on dissolvable microcarriers was equivalent to other commercially available microcarriers, and high quality cells were recovered without the need for microcarrier separation.

461030 The Effect of Particle Engineering on the Processing and the in Vitro Performance of Inhalation Blends in Dry Powder Inhalers

Abstract:

Introduction

In order to use API (active pharmaceutical ingredient) particles intended to target the tiny airways of the deep lung in dry powder inhalers (DPIs), the range of an aerodynamic diameter of 1 µm to 5 µm is highly preferred. Particles in this size regimen are typically cohesive and possess poor flow properties [1], leading to difficulties concerning volumetrically dosing. To overcome this flowability problem, it is a general practice to formulate carrier-based formulations wherein the API particles are attached to the surface of a larger carrier particle (50 µm – 200 µm).

Materials and Methods

Spray dried and micronized salbutamol sulphate (SS, Selectchemie, Zurich, Switzerland) were chosen as model API. Spray dried active was prepared on a Nano Spray Dryer B-90 (Buechi Labortechnik AG, Flawil, Switzerland). The spray drying conditions were chosen according to our previous work [2]. Micronization was done with an air jet mill (Spiral Jet Mill 50 AS, Hosokawa Alpine AG, Augsburg, Germany) at an injection pressure of 6 bar and a pressure inside the micronizer chamber of 3 bar. Modification of the carrier was done through wet decantation, which was supposed to decrease the fine particles on the carrier material and to modify the carrier surface [3]. Spray-dried (SDSS) and micronized API (MSS) was blended with α-lactose monohydrate (LAC_BD) and α-lactose monohydrate decanted (LAC_AD), so four adhesive mixtures with 2% API were prepared via sandwich method in a tumble blender TC2 (Willy A. Bachofen Maschinenfabrik, Muttenz, Switzerland). The mixing time was 60 min at 60 rpm. Both carriers, the APIs as well as the adhesives mixtures were extensively characterized (e.g. particle size, particle shape, flow properties, surface topography) to compare the different particulate properties.

Subsequent capsule filling was performed with different process setting (Setting 1: 3.4mm dosator, 2.5mm dosing chamber, 5mm powder layer; Setting 2: 3.4mm dosator, 2.5mm dosing chamber, 10mm powder layer) at a filling rate of 2500 capsules per hour (cph) on a dosator nozzle capsule filling machine (Labby, MG2, Bologna, Italy) with a target fill weight of 20 to 25mg. To evaluate the performance of the different mixtures, in vitro lung deposition experiments were carried out with a next generation impactor (NGI, Copley Scientific, Nottingham, United Kingdom), the emitted dose (ED) and fine particle fraction (FPF) were calculated based on the specification of the European pharmacopoeia. The inhalation device used for these experiments was the Aerolizer®/Cyclohaler®, a capsule inhaler.

Results and Discussion

Figure 1 shows SEM images of the spray dried (Fig. 1a) and micronized salbutamol (Fig. 1b) particles. SEM images show that both techniques were able to produce particles that have suitable size for inhalation but with very different shape and morphology. Spray dried particles are spherical, whereas micronized particles are needle shaped. Moreover, spray dried particles are amorphous while micronized particles were largely crystalline. The mean particle size (x50) determined via laser diffraction for spray dried and micronized particles as well as the blends can be found in table 1.

Figure 1: SEM images of spray dried (SDSS) and micronized (MSS) salbutamol sulphate particles

Table 1: Particle size and distribution of engineered actives and adhesive mixtures

	x_{50} [µm]	Span [x_{90}-x_{10}/x_{50}]
MSS	1.99	2.21
SDSS	2.91	1.68
MSS+LAC_BD	183.26	1.05
MSS+LAC_AD	173.45	1.06
SDSS+LAC_BD	193.11	1.09
SDSS+LAC_AD	185.48	1.04

Table 2: Fill weight and fill weight variability (RSD)

	Setting 1 [mg]	**RSD**	**Setting 2 [mg]**	**RSD**
MSS+LAC_BD	19.74	6.57	27.22	2.00
MSS+LAC_AD	24.07	2.37	25.08	1.97
SDSS+LAC_BD	20.85	4.19	27.19	2.33
SDSS+LAC_AD	24.74	2.21	27.33	2.00

Table 3: In Vitro performance: Fine particle fraction(FPF), Emitted dose (ED) and stage reached in the impactor

	Setting 1			**Setting 2**		
	FPF [%]	**ED [µg]**	**max. Stage**	**FPF [%]**	**ED [µg]**	**max. Stage**
MSS+LAC_BD	19.20	829.88	6	22.41	1131.95	7
MSS+LAC_AD	32.53	939.91	7	32.53	939.91	6
SDSS+LAC_BD	6.49	1035.02	5	5.07	1114.91	4
SDSS+LAC_AD	2.39	1261.05	4	3.35	1162.72	5

The FPF for micronized powder blends increased significantly, whilst spray-dried blends showed a decrease for decanted mixtures in comparison to the undecanted. Also the attainable stage in the NGI got improved through micronization of the API, but showed no correlation to the wet decantation of the powder blends. The different powder bed height settings in capsule filling (table 2) led to distinctions in attainable stage and FPF but do not seem to be influencing factors on the emitted dose. In summary, micronization of the API has much more influence on the ability for reaching the deep lung (API detachment from carrier) than wet decantation of the carrier material has. The obtained data are highly useful, to improve the understanding of the relationship between carrier morphology and carrier type, drug detachment and capsule filling efficiency and will help to generate DPI formulations with the desirable performance.

[1] Pilcer, G., Wauthoz, N. and Amighi, K. (2012) *"Lactose characteristics and the generation of the aerosol"* Adv. Drug Deliv. Rev., vol. 64, no. 3, pp. 233–256.

[2] Littringer, E., Zellnitz, S., Hammernik, K., Adamer, V., Friedl, H. and Urbanetz, N.A. (2013) *"Spray Drying of Aqueous Salbutamol Sulfate Solutions Using the Nano Spray Dryer B-90—The Impact of Process Parameters on Particle Size"* Dry. Technol., vol. 31, no. 12, pp. 1346–1353.

[3] Faulhammer, E., Zellnitz, S., Wahl, V., Khinast, J.G., Paudel, A. (2015) "*Carrier-based dry powder inhalation: Impact of carrier modification on capsule filling processability and in vitro aerodynamic performance" Int J Pharm; 491: pp231–242.*

461690 Feasibility Analysis of Flowsheet Models in Continuous Pharmaceutical Manufacturing Processes Considering the Effects of Noise

Abstract:

Continuous manufacturing has gained increasing attention in the pharmaceutical industry due to its potential in reducing cost and producing consistently qualified products [1]. As it is becoming more time-consuming to discover and develop a new drug, and together with the intensified competition from generic companies, continuous manufacturing stands out as a promising alternative for pharmaceutical companies to improve production efficiency and maximize profits within the productsâ€™ patent life [2, 3]. The development of continuous pharmaceutical manufacturing has been facilitated by the improved continuous equipment (e.g. feeding, mixing, tableting, etc. [1]) and advanced process analytical technology (PAT) tools for process understanding and process control [4]. However, in order to systematically generalize the process knowledge and optimize the process design, we need to use the computer-aided process systems engineering (PSE) approach [3].

Recently, extensive modeling work has been done to simulate the dynamic behaviors of unit operations, based on which integrated flowsheet models can be built to predict the process output under a certain set of input values [3, 5]. Such models can not only correlate the materials properties (e.g. flow rate, powder bulk density, mixture concentration, etc.) and operating conditions (e.g. blade speed, turret speed, etc.) to the steady-state process response (mass hold-up, product properties), but also dynamically trace the variation in any input factors and its propagation through the subsequent unit operations. Such information can be useful for risk assessment and sensitivity analysis of the process. Following the requirement from Quality by Design (QbD) [4], it is naturally to consider how to apply the predictability of flowsheet models to the quantification and optimization of the design space of the manufacturing process, which is known as â€œfeasibility analysis problemsâ€ in the PSE community.

Feasibility analysis is to identify the feasible region where the process can meet all operating, quality, and product

constraints under uncertainty [6, 7]. This is important to help understand how much variation the process can withstand while maintaining operability. However, the feasibility analysis problem can be very difficult to solve because of the complexity of the flowsheet model, which may include hundreds of variables and constraints. Moreover, the computational cost for the simulation can be high, which requires a highly efficient sampling strategy for the analysis. In this presentation, we focus on using surrogate-based approaches (e.g. Kriging, Radial Basis Function, etc.) to efficiently approximate the computational expensive flowsheet model and predict the feasible region under uncertainty. More specifically, we apply this approach not only to deterministic models, but also to models that contain noise in the output prediction. The objective is to provide a framework of feasibility analysis for flowsheet models that contain various sources of uncertainty.

References

[1] Schaber S D, Gerogiorgis D I, Ramachandran R, et al. Economic analysis of integrated continuous and batch pharmaceutical manufacturing: a case study[J]. Industrial & Engineering Chemistry Research, 2011, 50(17): 10083-10092.

[2] Plumb K. Continuous processing in the pharmaceutical industry: changing the mind set[J]. Chemical Engineering Research and Design, 2005, 83(6): 730-738.

[3] Boukouvala F, Ierapetritou M G. Surrogate-based optimization of expensive flowsheet modeling for continuous pharmaceutical manufacturing[J]. Journal of Pharmaceutical Innovation, 2013, 8(2): 131-145.

[4] Lawrence X Y. Pharmaceutical quality by design: product and process development, understanding, and control[J]. Pharmaceutical research, 2008, 25(4): 781-791.

[5] Rogers A J, Inamdar C, Ierapetritou M G. An integrated approach to simulation of pharmaceutical processes for solid drug manufacture[J]. Industrial & Engineering Chemistry Research, 2013, 53(13): 5128-5147.

[6] Rogers A, Ierapetritou M. Feasibility and flexibility analysis of black-box processes Part 1: Surrogate-based feasibility analysis[J]. Chemical Engineering Science, 2015, 137: 986-1004.

[7] Grossmann I E, Calfa B A, Garcia-Herreros P. Evolution of concepts and models for quantifying resiliency and flexibility of chemical processes[J]. Computers & Chemical Engineering, 2014, 70: 22-34.

464170 Characterization of Compaction Behaviour of Powders in Solid-Dosage Manufacturing Using Finite Element Analysis

Abstract:

Powder compaction is an essential step in pharmaceutical solid-dosage manufacturing, irrespective of the chosen path for powder pre-processing. The compaction changes the structure of material from a loose arrangement of particles with various shapes and sizes to a condensed structure that behaves like a continuum. An optimised compaction process is necessary to ensure production of high-quality tablets. Experimentally determining the effects of involved parameters and subsequent optimization of the compaction process is very difficult and expensive for this process, as tablets contain the expensive drug substance. Therfore, numerical simulation of the compaction process based on Finite Element Analysis (FEA) is suggested. The FEA is used for calculating forces, deformations, stresses and strains throughout a bonded structure, which is later used to ensure the formation of compact material with uniform density, free from cracks or chips. Hence, these simulations provide a detailed and cost-effective means of understanding and predicting the compaction profiles of the formulation material based on the processing parameters, such as imposed pressure, relative density of the powder and die shape.

In this study, the mechanical properties of a range of pharmaceutical formulations were approximated using the modified Drucker-Prager Cap (DPC) plasticity model and applied for FEA. The DPC model parameters were varied depending on the local relative density using an external subroutine. Additionally, the effect of wall friction on the mechanical behaviour of different formulation compacts during compaction is explored. This numerically predicted density distribution is compared with experimental measurements. The verification results are important for establishing a predictive capability of this model. The results from this study contribute to a better understanding of the impact of powder properties and process settings on the tabletting process and final properties of the produced tablet.

464286 Gender-Dependent PBPK Modeling

Abstract:

Personalized medicine has the potential to improve disease prevention and treatment by tailoring therapies to the unique physiological and environmental factors that describe a subpopulation or even an individual patient. Personalized medicine aims at providing patients with "the right drug at the right dose at the right time" (1) to maximize therapeutic efficiency and minimize adverse events. Anatomical differences (height, body weight, body composition), ethnicity, genotype, disease state, age, and gender, amongst other factors can lead to significant variations in the bioavailability of a drug due to differences in the rates and/or extent of absorption, distribution, metabolism and elimination (ADME) following administration. The success of personized medicine hinges on the industry's capacity to understand the role of these physiological and genetic factors that drive differences in ADME and ultimately the action of a drug. Identifying factors with the potential to impart the greatest effect on bioavailability is a critical step to explaining clinical observations and understanding the usefulness of a particular treatment. One such tool to enable this mechanistic understanding *in silico* is physiologically-based pharmacokinetic (PBPK) modeling.

Unlike traditional pharmacokinetics which calculates key parameters from existing experimental data, PBPK models have the ability to extrapolate bioavailability predictions to new populations or disease states, and for new administration routes or formulation, in addition to explaining physiological differences such as age or gender. PBPK models consist of a system of mathematical equations to describe the ADME processes across the various organs and tissues of the human body (2). These mechanistic models, in comparison to traditional compartmental modeling, expand the usefulness of pharmacokinetics to drug discovery, development and regulation for knowledge-based decision making. These models leverage drug physical chemical properties and body-specific physiological parameters to predict the amount of drug that is absorbed and reaches systemic circulation (3).

Physiological differences between genders are of particular interest given that women respond differently to therapies than males, both in regards to the effectiveness of a particular treatment as well as in the extent of observed adverse effects. These observations are well cited in the literature, indicating that physiological differences between genders can lead to vast differences in pharmacokinetic parameters (AUC, Cmax, clearance, volume of distribution) and potentially variations in pharmacodynamic responses (4-6). Differences in pharmacokinetics and pharmacodynamics has been observed clinically across many systems of the body including the central nervous system, cardiovascular system, energy metabolism, and the immune system (4). The clinical relevance of the differences in pharmacokinetic responses varies with the physicochemical properties of the drug, female physiology, and mechanism of action of the drug. The vast differences in drug behavior across a multitude of bodily systems suggests that sex-dependent pharmacokinetics are highly complex as the potential for differences span across all ADME processes.

Gender dependent responses are not always consistently reported in clinical studies given the strong influence of sex hormones during development, the menstrual cycle and with the use of oral contraceptives to drive higher variability in the pharmacokinetics for women as compared to men. Thus, incorporating gender differences is an important consideration and the next logical step in the expansion of PKPD modeling for more accurate predictions of clinical performance to tease out differences between males and females in well-defined and controlled virtual clinical studies. This expansion begins with an understanding of how the ADME processes vary in males and females in order to accurately predict bioavailability in both genders. In general, males and females are believed to have similar metabolic and signaling pathways with sex-dependent regulation of transcription, translation, and receptor signaling by sex hormones (7). Therefore, the focus of this work is to establish gender-specific PBPK model parameters for male and females that can be used for prediction of absorption and ultimately bioavailability of an oral solid dosage form. The opportunity for gender-dependent pharmacokinetics of an orally administered drug begins with absorption from the gastrointestinal tract. Sex hormones influence absorption from the gastrointestinal tract on several levels including gender-dependent physiology (passive diffusion), expression and activity of intestinal transporters (active transport), and expression and activity of gastric enzymes and cytochrome P450 isozymes (gut metabolism) in addition to gastrointestinal pH and transit time (8).

The properties of the drug (or formulation) for a solid oral dosage form dictate the extent of the observed differences in pharmacokinetics between genders. In a preliminary case study, we chose to study the effect of formulation properties, mainly pH-dependent solubility or *in vivo* dissolution, on the rate and extent of absorption in males and females. The purpose of this study was to establish an understanding of the gender-dependent sensitivity of absorption to changes in *in vivo* release. Using GastroPlus™ software, the Advanced Compartmental Absorption and Transit model parameters were adjusted to reflect male and female physiology. Gender-dependent absorption was predicted given dissolution profiles as input to the software with varying rates and extent of release.

1. Hamburg MA, Collins FS. The path to personalized medicine. The New England Journal of Medicine. 2010:301-4.

2. Yang RS, Mayeno AN, Liao KH, Reardon KF, Reisfeild B. Integration of PBPK and reaction network modelling: Predictive xenobiotic metabolomics. ALTEX 22. 2005(2).

3. Rowland M, Peck C, Tucker G. Physiologically-based pharmacokinetics in drug development and regulatory science. Annu Rev Pharmacol Toxicol. 2011;51:45-73.

4. Soldin OP, Chung SH, Mattison DR. Sex differences in drug disposition. J Biomed Biotechnol. 2011;2011:187103.

5. Ibarra M, Vazquez M, Fagiolino P, Derendorf H. Sex related differences on valproic acid pharmacokinetics after oral single dose. J Pharmacokinet Pharmacodyn. 2013;40(4):479-86.

6. Chen ML, Lee SC, Ng MJ, Schuirmann DJ, Lesko LJ, Williams RL. Pharmacokinetic analysis of bioequivalence trials: implications for sex-related issues in clinical pharmacology and biopharmaceutics. Clin Pharmacol Ther. 2000;68(5):510-21.

7. Regitz-Zagrosek V. Sex and Gender Differences in Pharmacology. Heidelberg: Springer; 2012. 599 p.

8. Freire AC, Basit AW, Choudhary R, Piong CW, Merchant HA. Does sex matter? The influence of gender on gastrointestinal physiology and drug delivery. Int J Pharm. 2011;415(1-2):15-28.

464607 Model-Based Process Analysis in Continuous Manufacturing of Pharmaceuticals: Calibration of a Fluid Bed Dryer Model

Abstract:

The goal of this work is to investigate the dominating phenomena in the fluid bed dryer in a continuous pharmaceutical wet granulation line, which are acting on the Particle Size Distribution (PSD) and moisture content of the product. Knowledge about these phenomena is built with model-based analysis, using a semi-mechanistic model. A thorough and novel experimental data collection campaign is done to create sufficient data for the calibration and validation of this model. Along with that, this work provides clarity into the application of a model for researching the particle property dynamics in a fluid bed dryer, and to produce concrete criteria for the success of every step in the application of the model. In this way, model-based process analysis for the transition to continuous manufacturing in the pharmaceutical industry, becomes more concrete.

The interaction and evolution of PSD and moisture content is investigated in a ConsiGma 25™ semi-continuous fluid bed dryer, which is fed with wet granules from a twin-screw granulator through pneumatic transport in a tube. In the dryer unit, the continuous flow of granules is divided over its six separate drying segments, in which the granules are batch dried. This partitioning into parallel and sequential batch drying processes approaches a continuous mode over the entire dryer.

In total, 49 different drying experiments were performed with wet granules that were produced under constant granulation settings. For the dryer, the following process settings were varied: the dryer inlet air flow rate was varied between 360 and 440 m^3/h and the inlet air temperature between 40 and 60°C. The filling time of a dryer cell, governing the cell load, was varied between 60 and 180 seconds. For each of the combinations of the previous settings and one center point of the settings, experiments were conducted for, at least, five different drying times, ranging from the filling time plus 20 seconds to maximally six times the filling time. As such, the drying times in the experiments span from 80 to 1080 seconds. The high-resolution PSD of the dried granules was collected using a particle size analyser, QICPIC (Sympatec), with a dispensing system, Gradis (Sympatec). The sample is collected after drying and directly sieved into five fractions. The moisture content of each sieve fraction, as well as the total sample, is determined by means of the Loss-On-Drying method (Mettler LP16 moisture analyzer).

An obvious observation can be made that the moisture content of the granules of the total sample decreases when drying time increases. Two phases can be distinguished during drying: (1) a fast phase, where evaporation of the surface water occurs, and (2) a slower drying phase in which water from the core needs to migrate to the surface to get evaporated (Mortier et al., 2012). It can be noticed that smaller granules have a lower moisture content compared to the larger granules, caused by the larger specific surface of the small granules. This phenomenon can only be noticed when shorter drying times are applied. When the drying time is increased, the moisture content of the different size fractions does not deviate significantly, indicating that drying is finished. The experimental data also indicates that most of the breakage occurs during the transport of the granules to the dryer and the loading into the dryer segments. The extent of this breakage was significantly different between the two formulations used in the experiments.

Through the application of a 2D Population Balance Model (PBM), the evolution of PSD and moisture content is simulated. The PBM describes the granules in the system according to a two-dimensional grid of discrete classes of particle size and moisture content, where the last one is given as the wet radius of the particle. Breakage and loss of moisture content of the granules are solved sequentially within each time step, approaching a simultaneous simulation of these phenomena (Mortier, 2014). Breakage is solved using the fixed-pivot discretization scheme. The change in wet radius is modelled according to a drying curve with two regimes, separated by the point where the wet radius equals the particle radius. When the wet radius is smaller, it is assumed that the moisture in the particle resides in its pores, resulting in a lower drying rate. The evolution of the moisture content is solved using the high resolution finite volume scheme (Mortier et al., 2013).

The three steps in the application of the model are model selection, calibration and validation. Model selection denotes selecting the breakage function, which describes the breakage regime in the system. Once a suitable breakage function is found, its empirical components, as well as the drying rates, are calibrated to the system for some of the different process conditions, using the experimental data. The other part of the process conditions are reserved for validation of the model. The methodology of these three steps, including the success criteria, will be presented. The study is done based on criteria from the statistics of comparison of distributions, applying the moments of a PSD to assess calibration and validation success. Each moment contains different information on the PSD. Therefore, incorporating various moments into a success criterion results in an all-round assessment of the PBM prediction. This way, the concrete and statistically sound comparison of PSD is brought into the calibration and validation steps of a PBM, important steps that are often overlooked in practice (Mortier, 2014).

References:

Mortier, S. T. F. C., De Beer, T., Gernaey, K. V., Vercruysse, J., Fonteyne, M., Remon, J. P., Vervaet, C., Nopens, I. (2012). Mechanistic modelling of the drying behaviour of single pharmaceutical granules. *European Journal of Pharmaceutics and Biopharmaceutics*, Vol. 80(3), 682–9.

Mortier, S. T. F. C., Gernaey, K. V., De Beer, T., Nopens, I. (2013). Development of a Population Balance Model of a pharmaceutical drying process and testing of solution methods. *Computers & Chemical Engineering*, Vol. 50, 39–53.

Mortier, S.T.F.C. (2014). *Modelling the drying behaviour of wet granules in the context of fully continuous pharmaceutical tablet manufacturing*. Ghent University. Faculty of Bioscience Engineering, Ghent, Belgium.

466116 Development of a Model Predictive Tool for Continuous Manufacturing of Solid Oral Dosage Forms Via Roller Compaction

Abstract:

Continuous Manufacturing (CM) is currently receiving significant attention in the pharmaceutical industry. In the past, common manufacturing processes for oral solid dosage (OSD) forms relied on batch processing followed by final product testing to ensure the desired quality has been achieved. This approach is considered inefficient due to increased material consumption and prolonged processing times. Therefore, it is crucial to take a step towards model predictive CM to enable prediction of the final tablet critical quality attributes (CQAs). To develop a quality by design (QbD) framework, a compaction and compression integrated work stream has been developed to enable prediction of compaction and compression behaviour of various powders and their blends. The blends consist of varying amounts of active pharmaceutical ingredient (API), 1 wt% lubricant and as remainder microcrystalline cellulose (MCC). These are either directly compressed or compressed after dry granulation via roller compaction. This work develops a real time control tool to predict final tablet CQAs based on powders with varying properties. This tool will ultimately provide control of roller compaction and tablet press settings to produce desired tablet properties. For the development of this tool a detailed understanding of the impact of powder properties on the compression and compaction behaviour was established. Three different model APIs were used in this study to establish the production limits and to define the optimised process settings of the roller compactor and tablet press. Each API displays a different morphological structure (spherical, monocline and needle shaped) that produces varying flow, compaction and compression characteristics. Experimental work was carried out using various (10 – 40 wt%) API loadings to establish the optimal compaction and compression conditions. The resulting experimental data was analysed to identify the key process parameters, which were linked to the model predictive tool. A top fed roller compactor was used under ambient conditions with varying screw speed (12 – 24 rpm), roller speed (3 - 5 rpm) and roller pressure (10 - 20 bar) for

ribbon production. The ribbon density, which is a key factor of roller compaction, was analysed with a GeoPyc 1360 envelope density analyser. The ribbons were granulated in a conical mill (Comil) and analysed for particle size distribution using sieving and a Microtrac Particle Size Analyser. The granule produced was compacted into tablets using various die sizes (6, 8, 10 and 12 mm) and a single punch Manesty F3 tablet press. Displacement measurements were assessed using commercial and custom built compaction testing equipment to determine compaction behaviour of powders and their blends. The tablets were tested for properties, including weight consistency, hardness, thickness, diameter, friability, dissolution and disintegration time.

466270 Application of Model-Predictive Control to a Feeding Blending Unit

Abstract:

Continuous pharmaceutical manufacturing is pursued by the FDA [1][2]. By continuously monitoring the process and by actively taking actions in case of deviating process parameters, the product quality will be improved while the production costs decrease. The proper operation of the continuous production line requires the development and implementation of sophisticated control strategies. On the one hand, the control strategy has to ensure that intermediates of bad quality are discharged. On the other hand, this missing material has to be compensated by means of adjusting the production speed of the upstream and/or downstream unit operation. This operation mode poses the requirement of adjusting the production speed of individual unit operations.

The paper at hand focuses on the design of a model predictive control (MPC) strategy [3][4] for a feeding blending unit (FBU), see Figure 1. The considered setup utilizes two feeders and a blender in order to create a blend of a given composition. The controller aims at tracking a prescribed blender outlet mass flow. Simultaneously, the mass hold-up in the blender, as well as the blender speed should remain within given bounds. A mathematical model of the plant, which is based on physical relations, is given. Model parameters have to be identified from measurement data. Based on the presented plant model, a model predictive control strategy is designed.

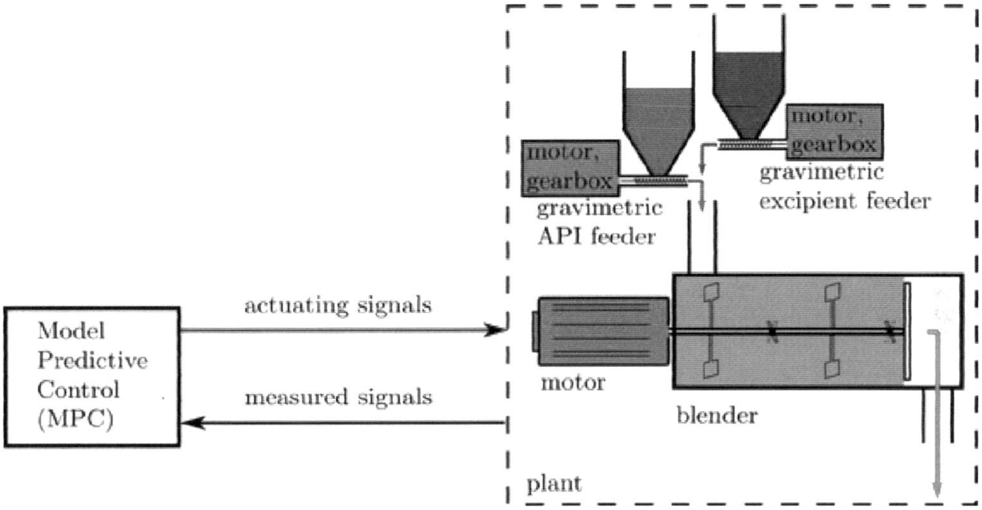

Figure 1: Considered setup

The presented plant model is based upon the idea of dividing the blender into several compartments. For each compartment, the mass balance equation is solved. The inlet mass flow of the compartments is known: for the first compartment, it is equal to the sum of the feeder mass flows, and for the following compartments, the inlet mass flow is equal to the outlet mass flow of the preceding one. The outlet mass flow of one compartment is computed from the hold-up in the compartment and the mean residence time of the material in this compartment. The mean residence time depends on the hold-up and on the blender speed. A simple relation, which relates the mean residence time to the blender speed is fitted from measured data. By considering the mass hold-ups of the individual compartments as state variables, a state space realization [5] of the plant is derived. The number of state variables equals the number of compartments.

Furthermore, the input/output behavior of the blender regarding concentration is modeled by means of a third order low pass filter with additional dead-time. The time constants and the dead time have to be identified from measurements.

A model predictive control strategy, which is based on the presented plant model, is proposed. The formulation of the objective function is discussed. Constraints on the state and the actuating signal are taken into account by the suggested control strategy. Several test scenarios, which investigate the fluctuations of feeder mass flows on the outlet of the blender are set up.

Simulation studies demonstrate the advantages of the proposed concept. The tuning of the controller is very intuitive, multi-input multi-output systems with constraints can be handled naturally.

[1] U.S. Department of Health and Human Services, "Guidance for Industry - PAT - A Framework for Innovative Pharmaceutical Development, Manufacturing, and Quality Assurance," 2004.

[2] S. Chatterjee, "FDA Perspective on Continuous Manufacturing," *IFPAC Annual Meeting*, 2012. [Online]. Available: http://www.fda.gov/downloads/AboutFDA/CentersOffices/OfficeofMedicalProductsandTobacco/CDER/UCM341197.pdf. [Accessed: 21-Mar-2016].

[3] J. B. Rawlings and D. Q. Mayne, *Model Predictive Control: Theory and Design*. Nob Hill Publishing, 2015.

[4] J. Maciejowski, *Predictive Control with Constraints*. Prentice Hall, 2001.

[5] B. Friedland, *Control System Design: An Introduction to State-Space Methods*. New York: McGraw-Hill, 1986.

466304 Systematic Technoeconomic Comparison of Separation Solvents for Green Continuous Pharmaceutical Manufacturing of Diphenhydramine

Abstract:

Continuous pharmaceutical manufacturing (CPM) has been shown to provide significant cost savings and enhanced material efficiencies compared to the currently used, yet inefficient, batch methods [1]. With the highest research and development costs of all industrial sectors, and historically increasing commercialisation costs [2], technological innovation is essential for pharmaceutical firms to remain profitable and sustainable [3]. For the benefits of CPM to be realised by industry, demonstration of its feasibility and viability via systematic process evaluation is paramount [4].

Continuous flow synthesis is the key to all candidate CPM processes, which has been demonstrated for several active pharmaceutical ingredients (APIs) to date. Several promising API candidates have already been identified as suitable for CPM implementation, including ibuprofen and artemisinin, which have already been examined in recent publications [5][6]. Systematic flowsheet synthesis, process modelling and simulation are an essential method for rapid assessment of potential processes and have already been widely applied for CPM case studies [7].

Diphenhydramine, a first-generation antihistamine API with hypnotic and anti-depressant uses [8] has been shown to be a promising candidate API for CPM implementation. Our previous work developed a process model on the continuous upstream processing of diphenhydramine [9] on the basis of a published continuous flow synthesis route which demonstrates that this economically and societally important API can be synthesised in plug flow reactors with molten ammonium salts [10]. This method achieves solvent and waste minimisation, achieving low total costs, enhanced material efficiency, and reduced hazards of chemical waste disposal.

The present paper illustrates a comprehensive technoeconomic analysis of various separation solvents for the continuous purification of diphenhydramine in this novel CPM process. This work provides a systematic methodological framework toward assessing liquid-liquid extraction (LLE) performance and material efficiency of the CPM process using different separation solvents. The inefficiencies incurred by each solvent at varying feed ratios on plant capacity, unit operation sizes and the corresponding capital, operating and total costs are also calculated here and compared to the CPM process with a batch separation train.

It is shown that whilst there is significant variation in process equipment sizes, costs and material efficiencies for CPM with different separation solvents, there is a clear trade-off between greatest cost savings, improvements in material efficiency, solvent toxicity and green, sustainable CPM operation. Original results indicate small computed reactor volumes and considerable heating requirements for the plant capacity. Chloroform emerges as a potent extraction agent in comparison to other candidate solvents, allowing the highest material efficiency (environmental factor of 3.43, total cost savings of 49.5%), but its high toxicity prohibits CPM use. Methyl-cyclohexane is the next strongest performer (environmental factor of 31.06, total cost savings of 37.3%) whose chemical properties render it significantly more acceptable for CPM implementation.

The present study exhibits the potential for technological innovation available *via* CPM in an effort to facilitate the transition from batch production methods in the pharmaceutical industry; by demonstrating the technoeconomic feasibility and economic viability for CPM of diphenhydramine via a rigorous comparison of a wide range of continuous separation options, we highlight the importance of systematic process systems engineering methods toward facilitating promising transitions from batch to CPM processes.

LITERATURE REFERENCES

1. Plumb, K., 2005. Continuous processing in the pharmaceutical industry - Changing the mind set, *Chem. Eng. Res. Des.* **83**(A6): 730–738.

2. Morgan, S., Grootendorst, P., Lexchin, J., Cunningham, C., Greyson, D., 2011. The cost of drug development: A systematic review, *Health Policy* 100(1): 4–17.

3. Behr, A., Brehme, V.A., Ewers, C.L.J., Gron, H., Kimmel, T., Kuppers, S., Symietz, I., 2004. New developments in chemical engineering for the production of drug substances, *Eng. Life Sci.*, **4**(1): 15–24.

4. Teoh, S.K., Rathi, C., Sharratt, P., 2016. Practical assessment methodology for converting fine chemicals processes from batch to continuous, *Org. Process Res. Dev.*, **20**(2): 414-431.

5. H. G. Jolliffe and D. I. Gerogiorgis, D.I., 2015. Process modelling and simulation for continuous pharmaceutical manufacturing of ibuprofen, *Chem. Eng. Res. Des.* **97**: 175–191.

6. H. G. Jolliffe and D. I. Gerogiorgis, D.I., 2016. Process modelling and simulation for continuous pharmaceutical manufacturing of artemisinin, *Chem. Eng. Res. Des.* (in press, available online).

7. K. V Gernaey, K.V., Cervera-Padrell, A.E., Woodley, J.M., 2012. Development of continuous pharmaceutical production processes supported by process systems engineering methods and tools, *Future Med. Chem.* **4**(11): 1371–1374.

8. Richardson, G.S., Roehrs, T.A., Rosenthal, L., Koshorek, G., Roth, T., 2002. Tolerance to daytime sedative effects of H1 antihistamines, *J. Clin. Psychopharmacol.* **22**(5): 511–515.

9. Diab, S., Gerogiorgis, D.I., 2015. Process modeling and simulation for Continuous Pharmaceutical Manufacturing (CPM) of diphenhydramine, *Proceedings of the AIChE Annual Meeting,* Salt Lake City, UT, USA.

10. Snead, D.R., Jamison, T.F., 2013. End-to-end continuous flow synthesis and purification of diphenhydramine hydrochloride featuring atom economy, in-line separation, and flow of molten ammonium salts, *Chem. Sci.* **4**(7): 2822–2827.

466733 Continuous Wet Granulation in a Twin Screw Extruder: Effect of Process Variables on Residence Time Distribution and Granule Characteristics

Abstract:

Wet granulation using a twin screw extruder is a promising continuous processing technique for granulation of pharmaceutical formulations. Although granulating with more consistency than a traditional high shear batch mixer,[1] the understanding of the granulation mechanisms in twin screw granulation is still not complete.[2] Furthermore, the residence time is very important, especially when multiple unit operations are connected in continuous processes.

Here we analyse the dependence of the residence time and granule size distribution (GSD) in a twin screw granulator on process parameters (screw speed, powder feed rate, liquid-to-solid ratio) and screw configuration (conveying, kneading, and cutting elements). The formulation was granulated in a 12 mm twin screw extruder (Three-Tec, Switzerland) with a length-to-diameter ratio of 40, using water as the granulating liquid. GSD was measured using a laser diffraction technique, while the residence time distribution was measured by adding a small dose of dye into the powder feeding port during steady state operation, followed by the sampling cup method described elsewhere.[3] Parameter values were chosen using a d-optimal factorial design to identify the most important parameters, and the results were rationalised to elucidate the granulation mechanisms in the different screw regions/screw elements.

[1] E.I. Keleb, A. Vermeire, C. Vervaet, J.P. Remon, Extrusion granulation and high shear granulation of different grades of lactose and highly dosed drugs: A comparative study, Drug Development and Industrial Pharmacy. 30 (2004) 679â€"691.

[2] M.R. Thompson, Twin screw granulation - review of current progress, Drug Development and Industrial Pharmacy.

(2014) 1â€"9. doi:10.3109/03639045.2014.983931.

[3] A.S. El Hagrasy, J.R. Hennenkamp, M.D. Burke, J.J. Cartwright, J.D. Litster, Twin screw wet granulation: Influence of formulation parameters on granule properties and growth behavior, Powder Technology. 238 (2013) 108â €"115.

467254 Virtual High-Throughput Screening Pipeline: Limitation Characterization and Application to Different Targets

Abstract:

Drug leads comprise a small portion of all known compounds. In an effort to identify promising leads, thousands to hundred thousands of compounds are experimentally tested for biological activity in high-throughput screening (HTS) experiments. However, typical hit-rates are only a few percent at best (Dobson, 2004). To increase efficiency, researchers are using previously obtained data and computational techniques to identify probable leads to focus resources on.

Previously, we developed an iterative virtual HTS (vHTS) pipeline. Using previously experimental data as input, the pipeline will develop quantitative structure-activity relationships (QSAR) models capable of classifying a compoundâ €™s class (active or inactive) and predicting its quantitative activity (Chen and Visco, 2016). Applied to a compound database, we can screen for probable leads. We tested our pipeline using NCBIâ€™s PubChem Bioassay 825 dataset, targeting a receptor implicated in viral diseases. We experimentally confirmed predictions from 2 iterations, with a hit rate of 19% in the 1st iteration and 75% in the 2nd.

To expand on our work, we will examine three key parameters to determine the robustness and applicability of our pipeline. Two parameters are related to the experimental datasets used to train our models: size and distribution. Size refers to the number of entries in the dataset while distribution refers to the classification of each entry and the number of members in each class. These two parameters together dictate the amount and quality of information available for training our models. Determining limitations for these two parameters will further direct experimental efforts towards fruitful pursuits as well as increase prediction confidence. The third examined parameter is pipeline iteration. By iterating, training set data can be updated with the addition of data from the validation experiments. However, there are diminishing returns with additional iterations. Identifying when it happens will increase efficiency.

In this poster, we will present work towards determining the aforementioned limitations. From PubChemâ€™s Bioassay database, we have identified datasets that are useful to determining the effect of each parameter. While controlling for two parameters, we examine the effect of the third by apply our pipeline to these datasets and experimentally validate predictions. Based on the observed changes in prediction accuracy and efficiency, we have a clearer idea what limitations exist and what effect a particular parameter has. We present the results of our work and aim to show our success is repeatable and our pipeline is robust enough to handle different datasets targeting different receptors while also investigating its limitations.

Chen J.J.F., Visco D.P. Jr., Developing an in silico pipeline for faster drug candidate discovery: Virtual high throughput screening with the Signature molecular descriptor using support vector machine models, *Chemical Engineering Science*, 2 March 2016. http://dx.doi.org/10.1016/j.ces.2016.02.037.

Dobson, C.M., 2004. Chemical space and biology. Nature 432, 824-828.

466960 Real-Time Monitoring and Ejection of out-of-Spec Tablets Via NIR Chemical Imaging

Abstract:

Near-infrared chemical imaging (NIR-CI) using high-speed cameras based on the push-broom acquisition principle is a strongly evolving technology in food, waste or mineral sorting. Yet, in the pharmaceutical industry, it is mainly used in lab settings.

Here we present a NIR-CI system, which is capable of real-time monitoring of approx. 500.000 tablets per hour and forwarding aggregated data (mean API content, standard deviation (SD) and number of analyzed tablets) to a process control system (PCS). Alternatively, the obtained chemical information can be used to identify and eject single out-of-spec tablets. The system consists of a push-broom spectrometer (EVK Helios G2 CLASS), a software for image analysis and pneumatic valves for ejection of tablets.

The camera (312 pixels spatial-, 248 pixels spectral resolution, 500 Hz measurement rate) generates 49 MB/s of spectral raw data. To reduce the data, chemometric models developed in Simca 13 are transferred to the spectrometer. They are executed in real-time on the onboard FPGA (Field Programmable Gate Array). The resulting video stream (0.29 MB/s) is further analyzed via openCV, an image processing library. For each detected tablet the mean and SD of API content, predicted via PLS, are calculated and averaged for one second. The results (3×4 byte/s) are written to an OPC-DA server, a standard interface to PCS, enabling a reaction to measured deviations. For ejection of out-of-specs tablets the video stream is analyzed via a DSP/TI System. Objects, which have to be ejected are forwarded to an IO subsystem, which controls the pneumatic valves.

Thereby the overall quality of tablet manufacturing can be raised by removal of potential out-of-spec tablets from the product stream at full manufacturing speed.

467823 Multivariate Statistical Process Control for the Continuous Manufacturing of Solid Oral Dosage Forms

Abstract:

Nowadays a commodity in the industrial software industry, the application of principal components analysis (PCA) to monitor the state of a process with Multivariate Statistical Process Control (MSPC) charts is far from a venture. Its implementation in a commercial plant with a contemporary control and/or automation system requires a modest effort, given a stable process and the availability of a historian data base with abundant data representative of the normal operating conditions (NOC) for the plant. This work presents the development of an MSPC monitoring scheme for a continuous drug product manufacturing process, using information available at the R&D stage. At the time when the process is still being developed, the data representative of the expected NOC may not be abundant. Additionally, the product development efforts may have other experimental priorities, for example running experimental trials at different mass rates and/or configuration of the feeders, mixer, and/or tablet press equipment.

The approach presented is based on the assumption that, regardless of the target set point for the main operating conditions (e.g. mass rate), the common cause variability in the process will remain the same. Hence the covariance structure for the variables in the system can be identified by a PCA model on a matrix resulting from the concatenation of data from multiple â€œtrial runsâ€ that were pre-centered around their mean values. An analysis of the loadings vectors reveals variable correlations that matches the expectations from a process engineering standpoint; and the corresponding movement in the scores space reveals directions of expected normal (and tolerated) variability, as well as patterns that can be matched with the startup and the shutdown of the system.

Our case study presents the analysis of a model built with the data collected from the system of screw feeders that dispense material to the unit, as well as the analysis of the model obtained with data collected from the tablet press. Furthermore, a method is presented to determine the lag time between a measurement vector from the feeders, and a measurement vector from the tablet press; such that a combined lagged model can be built. The combined model eases the holistic identification of contributions in the case of a disturbance, and the use of the three models (one per system and the combined) in tandem provides a complete monitoring approach for the system as a whole.

468412 Modelling Approaches to Spray Drying Process Development, Scale-up and Troubleshooting

Abstract:

Modelling approaches to spray drying process development, scale-up and troubleshooting

P.C. Valente, Í. Duarte, F.Neves & M.N. Temtem

[1] R&D Drug Product Development, Hovione Farmaciência S.A, Sete Casas, 2674-506 Loures, Portugal;

Initiatives such as Quality by Design in the pharmaceutical industry have greatly increased the emphasis on physical and mathematical modelling of the various processes and the focus on science-based development [1]. The shift from the more traditional approach of the Quality by Testing is driven by the rational that better understanding of the underlying physical mechanisms leading up to the final dosage form leads to improved control of the quality of the final product and to accelerated product and process development. Spray drying of drug product intermediates is one of the pharmaceutical processes being shifted from Quality by Testing to Quality by Design and it is also illustrative of the various models that need to be developed in order to reach that goal. For example, when a spray drying process is employed to obtain an amorphous solid dispersion, an adequate understanding of the miscibility/phase separation between polymer and API can accelerate the screening of polymer candidates and optimized process conditions to ensure a homogeneous molecular dispersion between polymer and API. However, due to the complexity of the underlying physical mechanisms together with the fact that multiple physical phenomena are simultaneously at play, the task of creating an adequate model can be quite daunting. To overcome these difficulties it is crucial to have a pragmatic approach to modelling and to adequately formulate the problem statement so that no more complexity than absolutely necessary is fed into the model.

In this work we discuss five case studies where different levels of modelling are used, from statistical to first principles, to address product and process development needs as well as troubleshooting. In the first and second case-studies two models based on first principles are applied to the formulation of amorphous solid dispersions and to forecast the impact of particle design on the dissolution behavior. In the third and fourth case-studies two different levels of modelling, statistical and mechanistic, respectively, are applied to predict particle size of a spray dried powder based on the nozzle geometry, atomization and process parameters, showing that there is an unavoidable compromise between model accuracy and model generality. This case-study is also used to emphasize that the choice of model closely depends on the problem statement. A fifth case-study of troubleshooting a pharmaceutical process using computer fluid dynamics is also presented (Fig. 1).

Finally, the adequateness of the model complexity for tackling a given problem is discussed and two illustrative examples of overkilling or oversimplifying a model are given.

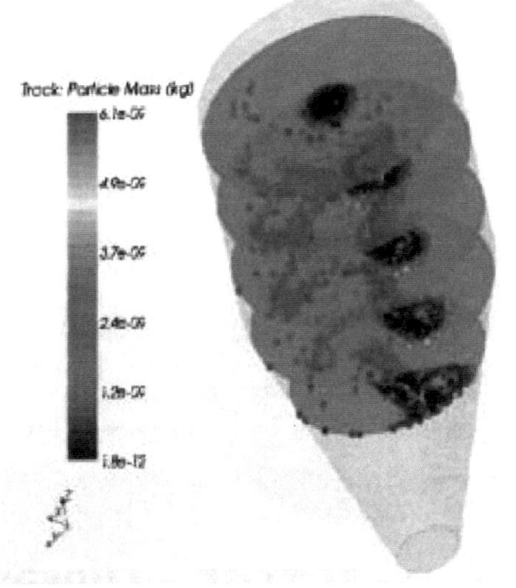

Figure 1: Troubleshooting example of a spray drying process using computer fluid dynamics. (Left) Gas velocity contours, (center) droplet spatial distribution within the drying chamber and (right) gas streamlines in the gas disperser and drying chamber of a spray drier.

References

[1] Lawrence X. Y. "Pharmaceutical Quality by Design: Product and Process Development, Understanding and Control", Pharm. Res., 25 (4) 2008.

468611 Solid-State Reactivity of Drugs with Potential Reactive Excipient Impurities: In silico and Experimental Studies

Abstract:

Demonstrating the chemical (long term) stability of drugs and drug formulations is a regulatory requirement, and guidelines for the evaluation of chemical stability are available. However, the most common formulation of small molecule drugs is the solid oral dosage form, and experimental procedures for the measurement of chemical degradation in the solid state are not well established, the predictive power of existing procedures is limited, and the required experimental effort is considerable.
One issue that is commonly encountered in the development of solid state drug formulations is the degradation of the active pharmaceutical ingredient (API) through a reaction with small molecule impurities present in excipients [1,2]. Here, we present first results of our attempts to use a combination of computational chemistry and experiments to investigate the impact of the solid state, its structure and dynamics, on chemical degradation in such systems.
We studied this type of reaction using a set of small molecular weight drugs that can potentially react with formaldehyde, a commonly observed impurity in polymeric excipients. Comparing the reactivities of these compounds with formaldehyde in solution to the corresponding reactivities of solid state (crystalline powder) samples, we found that the correlation of relative reactivities of different compounds in solution versus the solid state is limited. We performed an analysis of the solid state samples based on the structures of crystal bulk and surfaces using molecular modelling and Molecular Dynamics (MD) simulations, and defined a number of descriptors that can be calculated from first principles. These descriptors were used to explain the different reactivities observed for the studied drug compounds concentrating on the effects of limited accessibility of reactive sites in the crystalline state and relative lattice and surface stabilities and the related degree of disorder. Our results provide insights into the impact of solid state on relative reactivities and pave the way for the development of theoretical methods that can be used to optimize formulations towards improved stability.

References

[1] Hotha, KK, Roychowdhury, S, Subramanian, V. (2016) "Drug-Excipient Interactions: Case Studies and Overview of Drug Degradation Pathways." American Journal of Analytical Chemistry, 7, 107-140.
[2] Wu, Y, Levons, J, Narang, AS, Raghavan, K, Rao, VM. (2011) "Reactive impurities in excipients: profiling, identification and mitigation of drug-excipient incompatibility." AAPS PharmSciTech. 12, 4, 1248-1263.

467448 Prediction of Formulation Effects of Continuously Formulated Solid Oral Dosage Forms on Bioavailability Using PBPK Modeling

Abstract:

The *Quality by Design* (QbD) paradigm posits that in depth understanding of the relationship of a pharmaceutical product's *Critical Quality Attributes* (CQAs), ingredient (including API) *Critical Material Attributes (CMAs)*, and *Critical manufacturing Process Parameters* (CPPs), to the *in vivo* characteristics of the product as defined by the appropriate *Quality Target Product Profile* (QTPP) and *Clinically Relevant Specifications* (CRS), including absorption/bioavailability, will lead to better products more rapidly and economically for the benefit of the patient [1]. As stated, the QbD paradigm essentially requires establishing reliable quantitative relations between the "input" to the system (CMAs and CPPs) and the surrogate outputs (CQAs), which are presumed to be predictive of the "real outputs", (CRS). Ideally, the relationships between CMAs, ingredient ratios, and CPPs to highly predictive CQAs would be established *a priori* without extensive experimentation, especially *in vivo*, or substantial time/cost investments. For establishing such quantitative relations, we need to a) accurately, effectively and efficiently tune CQAs, ideally using predictive relationships to the ingredient proportions, CMAs and CPPs; and b) accurately, effectively and efficiently predict CRS from the CQAs (and perhaps, also using information from CMAs and CPPs).

Meeting this target has remained elusive. In recent years, two major advances have made a significant impact. First, *Continuous Manufacturing* (CM) has clearly demonstrated itself as a superior alternative to batch processing allowing for tight control of product specifications, thus making it possible to target specific CQA. Second, physiology-based mathematical models of dissolution and absorption are becoming increasingly more reliable and dependable for connecting *in silico* CQA and CRS. However, despite previous studies linking *in silico* CQA and CRS, we still lack a clear demonstration of an integrated framework that considers the overall "supply chain" of drug manufacturing, starting from the API, manufacturing process parameters, ingredient proportions and CMAs, and that eventually leads to *in vivo* predictions of the biorelevant profile. The lack of such integrated formalism a) creates sub-optimal designs and delivery systems; and b) does not take full advantage of the opportunities offered by CM. We propose to develop, implement, and demonstrate a systematic computational framework for connecting CAQ with CRS *in silico* in the context of continuous solid oral dose manufacturing processes.

More specifically, we explore computational methods for linking formulation-dependent dissolution characteristics to *in vivo* bioavailability. We achieve this aim in two main steps: (1) using physicochemical and processing parameters we predict *in vivo* dissolution from formulation and process parameters; and (2) using the predicted dissolution profile along with physicochemical and physiological properties we predict *in vivo* bioavailability. These two steps are accomplished by making use of DDDPlus^{TM} and GastroPlus^{TM} respectively. DDDPlus^{TM} simulates the dissolution curve of a drug product based on parameters such as diffusivity, solubility, particle size, and pKa, along with tablet properties such as composition, compression force, and tablet dimensions. The second step consists of feeding the output dissolution curve from DDDPlus^{TM} into a GastroPlus^{TM} model. GastroPlus^{TM}is based on the ACAT (Advanced Compartmental Absorption and Transit model) which has the ability to simulate the expected systemic bioavailability of a drug in the form of a plasma versus time graph [2, 3].

We hypothesize that since formulation regulates the parameters inputted into the dissolution model, for which the output profile serves as an input into the absorption model, it should have a direct impact on simulated bioavailability. An important part of this work will thus consists of quantifying the extent to which formulation can affect bioavailability. To aid in this endeavor, a parameter sensitivity analysis (PSA) will be performed to establish the parameters within the manufacturing design space that have the highest impact [4]. Furthermore, the significance of process variability on bioavailability will be studied in order to determine optimal operation range and to map out a design space for a given target bioperformance.

The advantage of mathematical modeling is the ease with which useful information can be produced compared to more labor-intensive and costly experiments. We seek to demonstrate the tremendous potential of PBPK models in guiding pharmaceutical processing development for oral tablet manufacturing within a Quality by Design framework. Future work also involves utilizing this mechanistic system to advance personalized medicine. By varying physiological properties, the model can be made to account for age, sex, ethnicity, extent of disease, and circadian rhythms, an aspect which could significantly enhance the customizing of drug formulation for specific populations.

1. Kesisoglou, F. and A. Mitra, *Application of Absorption Modeling in Rational Design of Drug Product Under Quality-by-Design Paradigm.* AAPS J, 2015. **17**(5): p. 1224-36.

2. Huang, W., S.L. Lee, and L.X. Yu, *Mechanistic approaches to predicting oral drug absorption.* AAPS J, 2009. **11**(2): p. 217-24.

3. *GastroPlus User Manual - Simulation Software for Drug Discovery and Development*, Cognigen, Editor. 2015, Simulations Plus.

4. Zhang, X., et al., *Utility of physiologically based absorption modeling in implementing Quality by Design in drug development.* AAPS J, 2011. **13**(1): p. 59-71.

*** Formatting Issue - Best Available Paper/Graphic ***

467752 Predicting the *In Vivo* Performance of Amorphous Solid Dispersions Based on Molecular Descriptors and Statistical Analysis

Abstract:

Over the last decade, amorphous solid dispersions (ASDs) have become an increasingly popular strategy to address the solubility constraints of new chemical entities. The main advantage of ASDs is the potential to create and maintain a supersaturated state, once the drug is dissolved in the the GI fluids. While supersaturated, drug concentration in solution may reach values well above its intrinsic solubility, meaning that more drug would be available to be absorbed.

To retard drug precipitation, the selection of the stabilizing polymer (POL) is crucial. From a screening perspective, it is of interest to early evaluate different polymers in terms of their supersaturation potential and precipitation inhibition effect, in order to reduce time, risk and costs during formulation development.

Current screening methodologies to predict ASDs performance in solution generally consist in the implementation of miniaturized kinetic solubility experiments based on pH- or solvent-shift methods, using smaller volumes apparatus or in 96-well plates [1]. An alternative is the use of computational tools particularly the existing physiologically-based pharmacokinetic (PBPK) models, such as GastroPlus™ or Simcyp®. However these models need to be combined with accurate *in vitro / in vivo*dissolution experiments as input parameters. These data is typically obtained at advanced stages of formulation development. Thus, from an early screening perspective, there is lack of a theoretical basis for the ranking of the best polymers with precipitation inhibition effect.

Today, there is a significant number of research papers demonstrating the *in vivo* bioavailability of ASDs when compared with the reference products [2]. Taking advantage of the data available in the literature,

*** Formatting Issue - Best Available Paper/Graphic ***

the purpose of this work was to develop a statistical model, based on multivariate data analysis tools that could help on guiding ASD formulation design to obtain the desired *in vivo*performance.A database with 37 observations (ASDs) and 35 XY variables was constructed as schematically shown in Figure 1. The X variables include simple molecular descriptors of the APIs and POLs (*e.g.* MW, Log P, T_m, T_g, number of H-bond donors/acceptors, etc) and interactions thereof, while the Y variables correspond to experimental data (*e.g.* C_{max}, t_{max}, $AUC_{in\ vivo}$) obtained from the literature. Principal component analysis (PCA) and partial least square (PLS) models were developed using SIMCA-P+ 13.0 (Umetrics, Sweden). All variables from the dataset were mean centered and scaled to unit variance before effective analysis, in order to give variables equal weight.

Figure 1. Schematic representation of the dataset.

First, an outlier's identification stage was conducted using PCA to handle variability in data from disparate sources. An amorphous dispersion was identified as an outlier, which was then removed from the dataset.

After outlier's identification, a two-component PLS model with multiple responses was developed - Log ($AUC_{in\ vivo,\ ASD}/AUC_{in\ vivo,\ REF}$) and Log ($C_{max\ in\ vivo,\ ASD}/C_{max\ in\ vivo,\ REF}$). These Y-variables were identified as being correlated from the PCA analysis (data not shown). Figure 2 shows the correlations between the observed Log ($AUC_{in\ vivo,\ ratio}$) and Log ($C_{max\ in\ vivo,\ ratio}$) and the respective predicted values, together with the predicted test set. These correlations were obtained after variable selection to maximize model performance/minimize prediction error.

A

B

Figure 2 – Observed data *versus* predicted data by the PLS model: A – Log AUC_{ratio} response; B - Log $C_{max,\ ratio}$; training set (red circles); prediction set (blue circles).

The R^2 and Q^2 values for each Y-variable were around 0.7 and 0.5, respectively. Given the existing uncontrolled variability in the data - *in vivo*data obtained disparate sources and different animal models - the obtained model is considered adequate for interpretation purposes. Figure 3 shows the loading plot together with the most important variables (VIP) plot of the model. The loading plot shows the relationships between the inputs and output variables simultaneously, while the VIP shows the variables by descending order of influence in the model.

A

B

Figure 3– PLS loading plot (A) and VIP plot (B) of model with two PCs, based on the training set.

As can be seen from Figure 3, the most important variables for the model included general API-related molecular descriptors, POL-related molecular descriptors and API-POL interaction variables mainly associated with the capacity of the POL to establish hydrogen bonding with the APIs. These results confirm the key role of polymers for the kinetic stabilization of drugs in solution.

Figure 4 shows two scatter plots of two of the most important variables for the model. The size of each point/observation corresponds to the $AUC_{in\ vivo}$ and $C_{max,in\ vivo}$gains which were categorized in three different levels (low, medium and high).

A

B

Figure 4 – Scatter plots of two of the most important variables for the model. The size of the points/observations represent the $AUC_{in\ vivo}$ gain (A) and the $C_{max,in\ vivo}$gain (B).

The importance of hydrogen bonding for improving the *in vivo* performance of ASDs was in line with the observation that polymers with higher solubility parameters also tend to contribute for higher AUCs. In general, cellulose-based polymers (*i.e.* HPMCAS, HPMC, HPMCP) seem to provide better precipitation inhibition across different classes of drugs, when compared with other polymer families.

To conclude, chemometric tools based on multivariate data analysis can be applied to assess correlations between molecular descriptors of the formulation ingredients and performance related output variables, namely $AUC_{in\ vivo}$ and $C_{max,\ in\ vivo}$. These models are useful not only for narrowing the scope of formulations variables to be further explored, but also to rapidly identify suitable systems with synergistic interactions for subsequent clinical evaluation.<span">

References:

[1] Bevernage J. *et al.*, Evaluation of gastrointestinal drug supersaturation and precipitation: Strategies and issues, International Journal of Pharmaceutics, 2013, 453 (1), pp. 25-35;

[2] Kawabata Y. et al., Formulation design for poorly water-soluble drugs based on biopharmaceutics classification system: Basic approaches and practical applications, International Journal of Pharmaceutics, 2011, 420 (1), pp. 1-10;<span">

468110 Behaviour of Magnesium Stearate in Continuous Feeding

Abstract:

Behaviour of magnesium stearate in continuous feeding

Raghu V. G. Peddapatla*[1,2], Caroline A. Blackshields[1,2], Michael F. Cronin[1], Abina M. Crean[1,2, 3]

[1]School of Pharmacy, University College Cork, Ireland

[2]Pharmaceutical Manufacturing Technology Centre, Ireland

[3]Synthesis and Solid State Pharmaceutical Centre, Ireland

*School of Pharmacy, University College Cork, Ireland, Tel: + 353 (21) 4901683,

email: r.peddapatla@umail.ucc.ie

INTRODUCTION

In a pharmaceutical continuous process line, feeding powders constantly and accurately into downstream processing units is challenging [1]. Consistent feeding is particularly challenging for cohesive materials such as the powder lubricant, magnesium stearate (MgSt). Even small changes in the feedrate of powders being fed for a short period of time, could result in products falling off the specification limits[2]. Hence, the initial feeding raw material step is important and must be monitored continuously to achieve consistency in feeding. Feeding of pharmaceutical excipients into continuous process lines is most commonly accomplished through Loss-In-Weight (LIW) or Loss-In-Volume (LIV) feeders which operate in gravimetric or volumetric mode [1,3].

In this study the continuous feeding of magnesium stearate alone and as a component of a tablet blend was studied. Magnesium stearate was selected as it is the most commonly used powder lubricant in tablet processing, and variability in its properties can comprise tablet ejection compaction, and performance (hardness and dissolution)[4] .

The objectives of this study were

1. To identify how the physical attributes of four different grades of magnesium stearate affect its continuous feeding performance.

2. To identify the effect of continuous feeding rate on the compaction and properties tablets prepared from blends (with and without lubricant).

MATERIALS

Four different grades of magnesium stearate obtained from two different suppliers, hereafter referred to as samples 1-4.

METHODS

Magnesium Stearate characterisation

Physical properties such as powder flow, bulk density, % compressibility are analysed using Brookfield powder flow tester (PFT) and Laser Light Scattering (Malvern 3000) for particle size distribution

Blends Preparation

Two blends are prepared in double cone Erweka blender blended for 30 minutes @ 30 rpm. Magnesium stearate is added to one of the blend and blended for extra 1minute @ 30rpm Blend 1 composed of 90% microcrystalline cellulose PH200 and 10% metoclopramide HCl. Blend 2 composed of 89.5% microcrystalline cellulose PH200 and 10% metoclopramide HCl and 0.5% magnesium stearate.

Feeder setup and evaluation

Loss in weight feeder are designed to monitor the changes in the bulk density of the powder in the feeder hopper[5]. Powders are fed through a K-Tron MT12 feeder using self-cleaning screws with fine square screen in gravimetric mode. Magnesium stearate samples are fed using Coarse Concave Screws (CCS) with Fine Square Screen (FSqS) setup, with two different set points of 0.15kg/hr. and 0.25kg/hr. Both blends are fed at 3 different set points 0.2238kg/hr., 0.5594kg/hr. and 1.0069kg/hr. with Fine Concave Screw (FCS) & Fine Square Screen (FSqS).

Mass flow performance was measured independently using a K-Sample Test System for Gravimetric Feeders. The relative standard deviation (RSD %) was determined from data collected at 30 sec intervals over a 30 minute duration.

RESULTS AND DISCUSSION

Continuous feeding performance

Feed factor is the gravimetric speed equivalent for 100% screw speed with a specific screw and screen setting. It gives an indication of the max. feed rates that can be achieved for a powder with specific equipment set up. Table 1 shows the measured bulk density and feed factors for the 4 grades of magnesium stearate studied.

Table 1

Sample	Bulk Density (g/cm^3)	Feed Factor (kg/hr)
MgSt Sample-1	0.255	1.9486
MgSt Sample-2	0.186	1.2784
MgSt Sample-3	0.203	1.3676
MgSt Sample-4	0.154	0.7635

Physical properties of magnesium stearate impact on gravimetric powder feeding behaviour. Samples feed factors are directly proportional to their bulk densities.

A measure of feed performance is the variability in quantity feed which is expressed as the relative standard deviation (RSD %). In this study weights where determined at 30 sec intervals for a duration of 30 minutes. The results for magnesium stearate alone and tablet blends are shown in Figure 1.

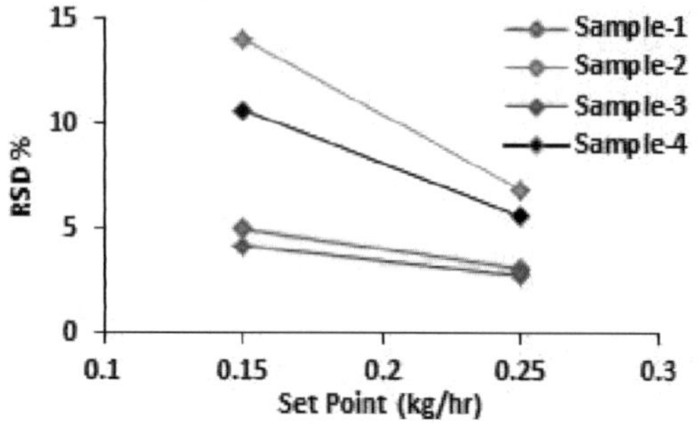

Figure 1: Feeding performance of magnesium stearate samples (left) and blends (right)

Magnesium stearate samples showed less variability at higher set points and greater variability at lower set points (Figure 1 left). Brookfield powder flow testing results showed magnesium stearate samples to exhibit very cohesive to cohesive flow behaviour and no relationship was seen between the feed variability and the measured flow function coefficient.

Blend 2 (which contained magnesium stearate) showed a lower % RSD compared to blend 1 at 3 different set points (Figure 1 right). Two blends fed at the lowest fed rate, 0.2238kg/hr, showed the highest variability.

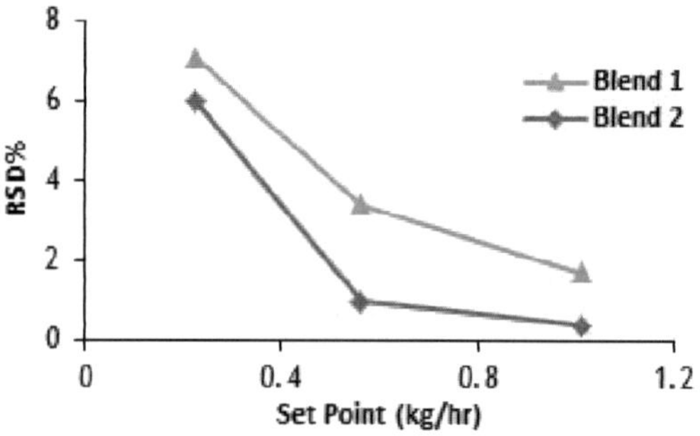

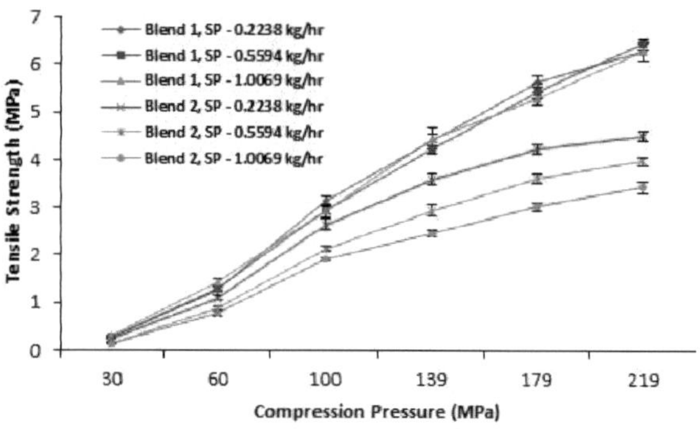

Figure 2: Tensile Strength profiles of both blends fed through the feeder at a range of fed rate set points.

There was no significant difference observed in tensile strength profiles fed at different set points for control blend 1, without magnesium stearate. Whereas the blend 2, with magnesium stearate, showed decrease in tablet tensile strength profiles upon feeding at higher set points (Figure 2). This may be due to increased mixing between the screws, as screws rotate with greater speeds upon increase in set point and would increase blending of the blend.

CONCLUSIONS

The bulk density magnesium stearate effected its feed factor - the higher the bulk density, the higher the feed factor achieved. While all magnesium stearate samples showed higher feeding variability at a lower feed rate, the difference was more pronounced for magnesium stearate samples 1 and 4. The physical properties responsible for differences in magnesium stearate feeding performance are currently being investigated.

The feeding rate of the blend containing magnesium stearate was seen to undermine its compaction properties. The higher the feeding rate the lower the tensile strength of the tablet produced. Future work will look at feeding rate impact on tablet dissolution.

ACKNOWLEDGEMENTS

Research funded by Pharmaceutical Manufacturing Technology Centre (PMTC), an Enterprise Ireland, Irish Development Agency funded centre and Synthesis and Solid State Pharmaceutical Centre (SSPC) and Science Foundation Ireland (SFI) under grant number 12/RC/2275. PH200 was donated by FMC Corporation. K-Tron for support in use of MT12 micro feeder.

BIBLIOGRAPHY

1. Engisch WE, Muzzio FJ. Method for characterization of loss-in-weight feeder equipment. *Powder Technol.* 2012;228:395-403. doi:10.1016/j.powtec.2012.05.058.

2. Pernenkil L, Cooney CL. A review on the continuous blending of powders. *Chem Eng Sci.* 2006;61(2):720-742. doi:10.1016/j.ces.2005.06.016.

3. Simonaho S, Ketolainen J, Ervasti T, Toiviainen M, Korhonen O. European Journal of Pharmaceutical Sciences Continuous manufacturing of tablets with PROMIS-line — Introduction and case studies from continuous feeding , blending and tableting. *PHASCI.* 2016. doi:10.1016/j.ejps.2016.02.006.

4. Gupta A, Hamad ML, Tawakkul M, Sayeed V a, Khan M a. Difference in the lubrication efficiency of bovine and vegetable-derived magnesium stearate during tabletting. *AAPS PharmSciTech.* 2009;10(2):500-504. doi:10.1208/s12249-009-9229-y.

5. Engisch WE, Muzzio FJ. Feedrate deviations caused by hopper re fi ll of loss-in-weight feeders. *Powder Technol.* 2015;283:389-400. doi:10.1016/j.powtec.2015.06.001.

468802 Comparing and Contrasting the Solid-State Esterification of APIs with Excipient: Computational and Experimental Assessment

Abstract:

Chemical stability of pharmaceutical products is one of the key quality attributes. There are limited studies that attempt to construct a molecular level definition of a drug product and predict how solid state factors influence the chemical stability of solid pharmaceutical products. In this study, we aim to combine computational and experimental approaches in an effort to better understand drugâ€"excipient degradation reactions in the solid state.
For our studies, we chose carvedilol (CAR) and codeine phosphate (COP) in formulation with citric acid (CA) as our model systems. Both CAR and COP possess a secondary alcohol and are known to undergo esterification in presence of CA [1,2]. To identify effects that are purely based on differences in the crystal structure, all calculations and experiments were carried out with two different polymorphic forms of, each, CAR and COP. For the morphologically important crystal faces, Molecular Dynamics (MD) simulations were used to calculate a range of descriptors which we expect to account for relevant structural, energetic and kinetic properties of the API crystal surfaces [3].
For the experimental assessment of the solid state stability, we first recrystallized the different crystal forms of CAR and COP and fractionated different sizes by sieving in order to obtain crystals with defined characteristics such as particle size and surface area. We then produced mixtures of the model APIs with citric acid and stored the samples at accelerated conditions. The accelerated storage conditions were chosen in a way that all crystal forms are stable, i.e. no polymorphic transformation and no loss of crystal water would occur. A different set of degradation experiments was performed where disorder was introduced to our APIâ€"excipient systems by ball milling.
By comparing results from our computational and experimental assessments, we gain insights into which material properties crucially determine the stability of pharmaceuticals in formulation with excipients in the solid state.

References

[1] Larsen, J, Cornett, C, Jaroszewski, JW, Hansen, SH. (2009) Reaction between drug substances and pharmaceutical excipients: formation of citric acid esters and amides of carvedilol in the solid state J. Pharm. Biomed. Anal., vol. 15, pp. 11-17.
[2] Silver, B and Sundholm, EG. (1987) Solid-State Esterification of Codeine Phosphate by the Acid Constituent of Effervescent Tablets J. Pharm Sci., vol. 76, pp. 53â€"55.
[3] Brunsteiner, M, Gressl, Khinast, JG Paudel, A 1 C, Davis, AL, Landis, M, Pencheva, K, Scrivens, G, Wood, GPF (2015) Stability by Design (SbD): Utilizing material science principles to predict physical and chemical stability of pharmaceutical solids AIChE 2015.

469699 The Effects of Formulation Factors on the Aerosolization Performance of Metered Dose Inhalers

Abstract:

Introduction: Metered dose inhalers (MDIs) are complex drug-device combination products widely used as portable delivery systems to treat pulmonary disorders, such as asthma and chronic obstructive pulmonary disease (COPD). A typical MDI consists of a canister containing the formulation, a metering valve, and an actuator-mouthpiece.[1] The formulation can be either a solution [active pharmaceutical ingredient (API) dissolved in the liquid propellant] or a suspension (API particles dispersed in the liquid propellant) along with inactive ingredients (e.g., co-solvents and surfactants).[2] The product performance of MDIs depends on a multitude of factors including, but not limited to, the physicochemical properties of API, device geometry (e.g., valve metering chamber volume, actuator nozzle orifice diameter, actuator sump depth, and actuator orifice jet length), and nature and amount of inactive ingredient(s).[3] Under the Quality by Design (QbD) paradigm, systematic investigations are necessary to understand how changes in critical quality attributes (CQAs) of formulation, device and manufacturing process influence the product performance.[4] Although much is known about the effects of changes in device geometry on the aerosolization performance of MDIs,[5],[6] the effects of changes in formulation factors on MDI product performance are not clearly defined. Therefore, the purpose of this work is to provide a better understanding of the effects of different levels of inactive ingredients and API particle size distribution (PSD) on key aerosolization performance parameters of MDI products. The systematic approach applied in this work can be utilized as a QbD tool to develop mathematical models and design spaces, allowing the manufacture of formulations with desired *in vitro* product performance parameters through a proper combination of formulation factors.

Methods: Three commercial MDI products (albuterol sulfate and mometasone furoate suspension formulations, and beclomethasone dipropionate solution formulation) were chosen for this study. Reverse engineering was utilized to determine the qualitative and quantitative composition of the formulations. The aerosolization performance was characterized by measuring the delivered dose (DD) and aerodynamic particle size distribution (APSD) performance. MDI model systems were then developed and used as anchor point for developing a statistical experimental design to study the effects of formulation factors (inactive ingredients concentrations and API PSD) and simulate design spaces of product performance. The purpose of defining the MDI model systems was not to achieve *in vitro* bioequivalence to the commercial MDI products, but to develop systems similar to the commercial MDIs with respect to formulation composition and key aerosolization performance parameters. A reduced factorial statistical design of experiments (DoE) approach was then used to develop a MDI batch manufacturing plan to vary the levels of inactive ingredients (ethanol and oleic acid) and API PSD (D_{50}, the value of the particle diameter at 50% in the cumulative distribution, that is, the median diameter of the PSD). The following ranges were studied:

API	D_{50} (μm)	Ethanol (% w/w)	Oleic Acid (% w/w)
Albuterol Sulfate	1.4 - 2.5	7 - 20	0.005 - 0.1
Mometasone Furoate	1.1 - 2.0	0.45 - 3.6	0.001 - 0.025
Beclomethasone Dipropionate	N/A	7 - 9	0 - 2

API D_{50} was varied for the suspension formulations only; micronized API was prepared from the same mother batch by sizing down to the desired API D_{50} using jet mill process. While the commercial solution MDI chosen for this work does not contain oleic acid, such an inactive ingredient was included in the DoE plan to study its effects in the solution formulation in comparison to the suspension formulations. The MDI batches were manufactured under non-GMP conditions, characterized and the data was statistically analyzed – a formulation factor (ethanol and oleic acid concentrations, or API D_{50}) was considered to have a statistically significant effect on the aerosolization performance parameter if p value < 0.05. Multivariate mathematical models and design spaces were then developed to predict the MDI aerosolization performance with respect to DD and APSD according to the different levels of formulation factors. The potential effects of different propellants [hydrofluoroalkane (HFA)-134a in the albuterol sulfate and beclomethasone dipropionate MDIs, and HFA-227 in the mometasone furoate MDIs] were not studied in this work.

Results: For the albuterol sulfate suspension MDI formulations studied, the effect of ethanol on the DD was statistically significant – as the ethanol concentration increased from 7 to 20% w/w, the DD decreased by 17%. Both ethanol and API D_{50} showed statistically significant effects on the APSD – as the ethanol concentration and API D_{50} increased from, respectively, 7 to 20% w/w and 1.4 to 2.5 μm, the fine particle dose less than 5 μm (FPD<5) decreased by 40%. The effect of oleic acid on the DD and APSD was not statistically significant. For the beclomethasone dipropionate solution MDI formulations studied, the effect of ethanol on both DD and APSD was not statistically significant within the range studied. However, the effect of oleic acid on the DD and APSD was statistically significant – as the oleic acid concentration increased from 0 to 2% w/w, the DD decreased by 11% and the FPD<5 decreased by 65%. For the mometasone furoate suspension MDI formulations studied, the DD was significantly affected only by ethanol – as the ethanol concentration increased from 0.45 to 3.6% w/w, the DD increased by 7%. All three formulation factors (ethanol and oleic acid concentrations, and API D_{50}) showed statistically significant effects on the APSD, but the strongest effects were provided by the ethanol and API D_{50} – for these, a change of approximately 40% in FPD<5 was obtained from the lowest to the highest factor level.

Conclusions: The changes in API PSD had statistically significant effects on the APSD performance of suspension MDI formulations studied, but not on DD. The changes in concentrations of inactive ingredients (ethanol and oleic acid) showed, in some cases, statistically significant effects on DD and APSD performance of suspension and solution MDI formulations studied. However, several cases without effects were also found, despite some large changes in concentrations of inactive ingredients studied. The possible effects of varying these must hence be studied on a case-by-case basis. The outcomes of this study allowed defining design spaces for DD and APSD performance according to the different levels of formulation factors (ethanol and oleic acid concentrations, and API D_{50}). The systematic approach utilized in this work can contribute as a QbD tool to evaluate the extent to which the formulation factors govern the aerosolization performance of MDI products, helping to design MDI formulations with desired *in vitro* product performance parameters.

Keywords: Metered Dose Inhalers (MDIs); Quality by Design (QbD); Factorial Design of Experiments (DoE).

References:

[1] Stein, S.W., *et al.*, Advances in metered dose inhaler technology: Hardware development. *AAPS PharmSciTech*, 2013. **15**(2): p. 326-338.

[2] Bell, J. and S. Newman, The rejuvenated pressurised metered dose inhaler. *Expert Opinion on Drug Delivery*, 2007. **4**(3): p. 215-234.

[3] Smyth, H.D.C., The influence of formulation variables on the performance of alternative propellant-driven metered dose inhalers. *Advanced Drug Delivery Reviews*, 2003. **55**(7): p. 807-828.

[4] Lionberger, R.A., *et al.*, Quality by design: Concepts for ANDAs. *The AAPS Journal*, 2008. **10**(2): p. 268-276.

[5] Smyth, H., *et al.*, Spray pattern analysis for metered dose inhalers I: Orifice size, particle size, and droplet motion correlations. *Drug Development and Industrial Pharmacy*, 2006. **32**(9): p. 1033-1041.

[6] Smyth, H., *et al.*, Spray pattern analysis for metered dose inhalers: Effect of actuator design. *Pharmaceutical Research*, 2006. **23**(7): p. 1591-1596.

469725 Exploiting Solvent and Impurity Effects to Modify Crystal Shape and Bulk Powder Properties of API and Intermediate Molecules

Abstract:

Crystallization design is mainly focused on designing API manufacturing process to deliver drug molecules with desired particles size and shape, as well as bulk powder properties to allow downstream formulation process to be more robust and less expensive. However, similar focus on upstream steps of the API process could benefit from improved physical properties of the API or intermediates that lead to better filtration, impurity rejection, cycle time, and throughput on scale. This work will present examples showcasing merits of exploiting solvent and impurity effect to improve process robustness and productivity.

469947 Continuous Manufacturing of Co-Crystals in Twin Screw Extruder: A Case Study

Abstract:

Co-crystals are crystalline materials, comprised of two or more components that are solids at room temperature and held together by noncovalent forces. A pharmaceutical co-crystal consists of an active pharmaceutical ingredient and a pharmaceutically acceptable co-former. Nowadays, pharmaceutical co-crystals are becoming an integral part in pharmaceutical strategies as they can improve the physicochemical properties of the API.Wet granulation is a widely used process in the pharmaceutical industry and its main goal is to produce granules from fine particles of active pharmaceutical ingredients (API) and excipients using a granulation liquid. In a twin screw extruder, the process is undergone under the high shear forces exerted by the screws on the particles.

The main aim of this study is to investigate and develop novel processes for continuous pharmaceutical processing of co-crystals. Wet granulation was chosen as a manufacturing method since it combined synthesis and granulation in a single continuous operation step. A 12 mm twin screw extruder (Three-Tec, Switzerland) with a length-to-diameter ratio of 40 was used for the manufacturing of the granules.A novel Theophylline-Aspirin (THP-ASP) model system was selected for the study. The effects of different process parameters (feed rate, screw speed, liquid-to-solid ratio) and formulation parameters (API and co-former ratio) on the final granules properties and the extent of conversion were investigated. Particle size distribution was analysed using Microtrac particle size analyser. The solid state form of the final granules was characterised with various techniques, including DSC, and PXRD.

470423 Fault-Tolerant Control Design and Risk MAP Based Resiliency Analysis of Continuous Solid Dose Manufacturing

Abstract:

The recent development and adoption of continuous manufacturing technologies in pharmaceutical industries[1-3] has also created the demand for more efficient process monitoring and control techniques. This is particularly the case with solid dose manufacturing processes, in which unit operations, such as blending, milling, *etc*., that in a classic batch processing are performed sequentially, are now integrated with a higher degree of automation, and operated at steady state with less manual intervention.[4-8] It has also become apparent that fast and reliable control strategy design is a requirement for successful continuous manufacturing technology. Hence, in this study, a systematic framework for fault-tolerant control design and resiliency analysis is proposed for continuous solid dose manufacturing processes based on risk mapping, accessing, and planning (Risk MAP).[9]

First, a risk mapping was designed to characterize the frequency and severity of the potential faults during the design and operation stages, *e.g.*, model-plant mismatch, calibration errors, failing/missing sensors, process disturbances, *etc*. The three levels of hierarchical control strategies design, which ranges from single proportional-integral-derivate (PID) control loop to plant-wide advanced model predictive control (MPC), were then assessed subject to the potential faults by using control key performance indices (KPIs), such as, time to product (T2P), magnitude to product (M2P), and other control measures, such as Morariâ€™s resiliency index (MRI).[10] Mitigation planning for severe faults were also suggested, *e.g.*, to activate or decommission a specific control level, or to divert the off-spec products.

The proposed systematic framework for fault-tolerant control design was demonstrated in a pilot plant for continuous dry granulation process[11] and was shown to be efficient and robust in handling potential faults and achieving a robust processing line for solid dose.

References

1. Ierapetritou M, Muzzio F, Reklaitis G. Perspectives on the continuous manufacturing of powder-based pharmaceutical processes. *AIChE Journal*.2016;62:1846-1862.
2. Lee SL, Oâ€™Connor TF, Yang X, Cruz CN, Chatterjee S, Madurawe RD, Moore CMV, Xu LX, Woodcock J. Modernizing pharmaceutical manufacturing: from batch to continuous production. *Journal of Pharmaceutical Innovation*. 2015; 10:191-199.
3. Buchholz S. Future manufacturing approaches in the chemical and pharmaceutical industry. *Chemical Engineering and Processing*. 2010; 49:993-995.
4. Vervaet C, Remon JP. Continuous granulation in the pharmaceutical industry. Chemical Engineering Science. 2005; 60:3949-3957.
5. Teng Y, Qiu Z, Wen H. Systematical approach of formulation and process development using roller compaction. *European Journal of Pharmaceutics and Biopharmaceutics*. 2009; 73:219-229.
6. Singh R, Ierapetritou M, Ramachandran R. An engineering study on the enhanced control and operation of continuous manufacturing of pharmaceutical tablets via roller compaction. *International Journal of Pharmaceutics*. 2012; 438:307-326.
7. Singh R, Sen M, Ierapetritou M, Ramachandran R. Integrated moving horizon-based dynamic real-time optimization and hybrid MPC-PID control of a direct compaction continuous tablet manufacturing process. *Journal of Pharmaceutical Innovation*. 2015; 10:233-253.
8. Hsu SH, Reklaitis GV, Venkatasubramanian V. Modeling and control of roller compaction for pharmaceutical manufacturing. Part II: control system design. *Journal of Pharmaceutical Innovation*. 2010; 5:24-36.
9. http://www.fema.gov/media-library/assets/documents/18274.
10. Morari M. Design of resilient processing plants III a general framework for the assessment of dynamic resilience. *Chemical Engineering Science*. 1983;38(11): 1881-1891.
11. Hsu SH, Reklaitis GV, Venkatasubramanian V. Modeling and control of roller compaction for pharmaceutical manufacturing. Part I: process dynamics and control framework. *Journal of Pharmaceutical Innovation*. 2010; 5:14-23.

470558 A Comprehensive Statistics-Based Approach to Proactive Process Analysis Including Multivariate Analysis Methods

Abstract:

Following the initial design and validation of a pharmaceutical drug product manufacturing process, the process will undergo further monitoring. For example, those process parameters that are most likely to impact the critical quality attributes, may be monitored over a series of batches. The resulting data can be used to 1) satisfy continued process verification, as part of stage 3 PPQ, or other regulatory needs and 2) ensure the process is in control and increase robustness by proactively improving and optimizing the process (i.e. proactive process analysis).

The approach to process monitoring for continuous improvement purposes can be varied. An example of a comprehensive systematic, statistics-based approach to proactive process analysis is presented here. The approach includes using statistical process control to visualize and identify trends before an impact to quality occurs. Furthermore, the use of multivariate analysis methods, such as principal component analysis (PCA) and partial least squares (PLS), to identify interactions between process parameters and point of cause, in support of root cause investigations, is illustrated.

470746 The Utilization of Amino Acid Consumption Rate Assessment As a Tool for CHO Clonal Stability Evaluation.

Abstract:

<span">CHO cells are commonly utilized for the production of therapeutics.To ensure product reproducibility and quality, assessment of clonal stability is important for both regulatory and economic reasons. We are continuously striving to improve methods for assessing CHO cell bank stability. One new tool we are currently evaluating is specific amino acid (AA) consumption rate and how it relates to CHO clonal stability. By calculating AA consumption data generated at various PDL ages we are able to observe trends between the AA consumption rates per cell per day of multiple clones throughout the process. Clones which exhibit increased AA consumption as they age can potentially result in total depletion of one or more amino acids within the culture. As a result false conclusions may be drawn that a clone is unstable when in reality the feed strategy may need to be further assessed in order to maintain a high level of productivity.

<span"> For this experiment we ran selected clones in an AMBR250 using a platform process. Time point samples were analyzed for AA content using a Hitachi L-8900 amino acid analyzer and titers were analyzed using RP-HPLC. These data were used to calculate specific productivity (Q_P) and specific amino acid consumption (Q_{AA}). The data presented shows that of three high producing clones, C10 exhibits increased AA consumption and decreased titer as the PDL increases. C81 AA consumption and titer production remain constant, while C88 AA consumption levels decrease but titer remains constant across PDL ages. Based on this information, C81 and C88 are better candidates for establishment of stable master and working cell bank. It also points to the potential to prevent the loss of $Q_P$in C10 by exploring additional feeds for late PDL cells.

470932 Powder Tester Comparison to Predict Discharge Rate from Hoppers and Filling Tablet Press Dies

Abstract:

To enhance the productivity in the manufacturing of solid dosage forms it is important to ensure that powders flow from hoppers and into tablet press dies at appropriate rates to enable high throughput and reduce variation to minimize product defects. This paper discusses the results of different flow testers to describe flowability of materials and how they relate to solid dosage form production.

The manufacturing of tablets requires fine powders to flow through a number of hoppers, feeders and finally into dies of a tablet press. The authors evaluated the flowability of fine powders with commonly used powder characterization testers in the pharmaceutical industry namely shear testers, compressibility cells, permeability cells, and funnel, orifice plate and avalanche testers.

For this study the authors selected 10 cellulose ether excipients which are investigated for flowability assessment. The flowablity measured by the different flow testers is compared with the limiting flow rate based on Johanson's work [1]. It was found that the discharge rate or flowability is best described by the permeability and the bulk density. The flow function or the flow factor could not describe the discharge rate appropriately. The Carr index describes the compressibility of a powder but cannot predict the discharge rate. However, the Carr index can be used to evaluate flowability trends if the variation of Carr index measurement is considered.

The impact of flowability of a material on tableting was evaluated experimentally with a Piccola bilayer press. It was found that the standard deviation of tablet weight, compaction force and radial tensile strength can be described as a function of flowability. With increasing flowability, the variability in all three metrics was reduced. Thus, powder flowability can be applied to understand variation in the tableting process. Consequently in this study the flowablity of all cellulose ether excipients was measured and it was shown that Methocel™ DC2 has the best balance between flow properties and the above mentioned tableting metrics.

[1] Johanson, J. R., "Two-Phase Flow Effects in Solids Processing and Handling," ***Chemical Engineering***, pp. 77–86 (Jan. 1, 1979)

470797 PAT Application in Expedited Development of Multi-Step Chemical Syntheses of Active Compounds

Abstract: "PAT application in Expedited Development of Multi-Step Chemical Syntheses of Active Compounds"

Rapid kinetic methods that allow to acquire knowledge of reaction mechanism and rate early-on in the research and development process are becoming widely utilized. They are used to determine the range of operability space in order to ensure a safe, reliable and successful scale-up of complex molecules. One of the limitations to taking such a modern approach to development and scale-up in general is the need for a large number of experiments and an extensive amount of analytical data. In addition, tools are needed for non-experts to interpret, evaluate, and convert the data into chemical reaction knowledge and understanding. PAT, and in particular in situ ATR-FTIR is being increasingly recognized as a key technology to automatically collect a large number of analytical data from chemical reactions in situ and in real time. Since more information is retrieved from each experiment, a smaller number is required to provide concentration dependence of reaction components as well rate and equilibrium constants. This high-throughput approach to data collection can then be complemented with simplified means of representing and analyzing the data, among others, multi-dimensional visuals and Reaction Progress Kinetic Analysis. Literature examples of combining real time ATR-FTIR, simplified kinetic analysis now abound.
The presentation will describe several examples, including the concise scalable synthesis of a chiral dipeptide acid, substructure of the HCV protease inhibitor Faldaprevir. An extensive utilization of PAT early-on provided sufficient process knowledge to complete three steps in one operation in a much shorter time, and afforded a greener and robust process.
scheme 1, 2).

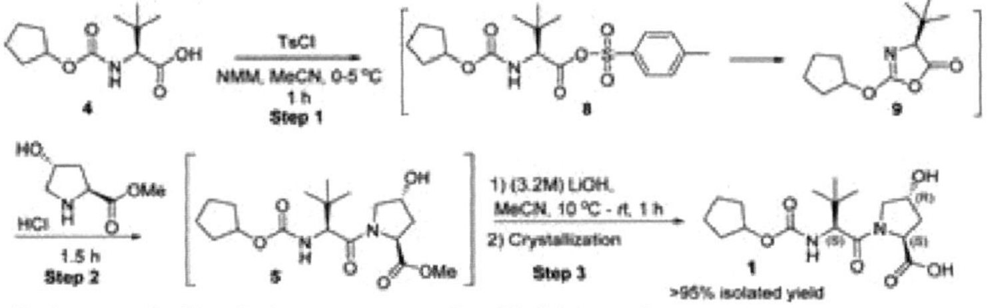

Scheme 1: Scale-up process for Faldaprevir

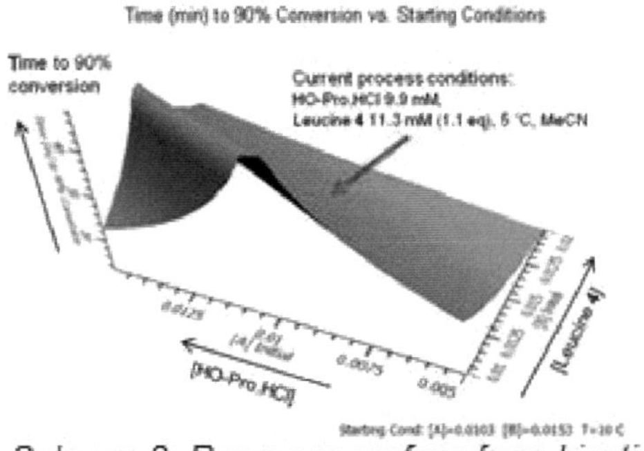

Scheme 2: Response surface from kinetic empirical model

470864 Process Innovation and Optimization of a Commercial Cell Culture Process

Abstract:

Reduction of cadence and optimization of process yield are both required to maximize commercial biopharma manufacturing plant output. Often, these are mutually exclusive and concessions must be made to one or both areas to improve overall plant efficiency or maintain product quality. This abstract describes innovative optimization approaches on two different mammalian cell culture unit operations to improve both cadence and yield.

In the first example, the mammalian cell culture process described requires control of inoculum expansion at a low final viable cell density target, above which viability drop and growth lag are observed in the subsequent stage. The inoculum expansion and seed bioreactor stages require a precise control of transfer conditions to maintain the culture within certain viable cell density range. A flexible and scalable inoculum expansion process with robust control has been implemented. The process is resilient and supports a wide variety of plant cadence targets. In the second example the mammalian cell culture process described showed no increase in final output when production bioreactor titer improved by 20%. Data driven analysis of the production process showed a negative correlation between production bioreactor titer and primary recovery step yield. Multi Variate Analysis (MVA) was performed to identify drivers for lower step yields and process optimization.

471020 Design of Nonlinear Observer for State Estimation of a Bioreactor

Abstract:

Measurement or estimation of process states is critical for process monitoring, advanced process control, and process optimization[1,2]. For chemical process whose state information cannot be measured directly, techniques such as state estimation need to be developed. Model-based state estimation is one of the most widely applied methods for estimation of unmeasured states basing on a high-fidelity process model. However, for some complex process, it requires extensive knowledge and effort to build a high-fidelity model to describe the whole system. Certain unmodeled disturbance or unknown inputs will generate model-plant mismatch, and in some case the mismatch is significant. For a case of bioreactor, there are several situations that can generate model-plant mismatch which can be treated as unknown input or disturbance, such as the effect of nutrient limitation, death of cells due to bust of air bubbles, oxygen delivery at high cell density, carbon dioxide stripping, and accumulation of inhibitive compound[3]. There are several different attempts to address the model-plant mismatch. One of the solutions is to apply parameter estimation for each run of the experiments. However, the adaptive parameter estimation requires extensive measurements of states and cannot be used for state estimation for a new experiment until a new set of parameters is updated. Filter and observer are mathematical methods to extract state information from corrupted measurements, which offer an alternative method for state estimation in the presence of model-plant mismatch.

Unknown input observer was developed to estimate the state in presence of certain types of faults and disturbance[3]. Comparing to other observers, unknown input observer can eliminate the effect of certain disturbance or fault despite of their size when their possible influences are estimated in the process model. To estimate the process state in presence of process disturbance or unknown inputs, a new design of nonlinear unknown input observer is proposed and it is applied for state estimation of a bioreactor. Design of such observer is provided and sufficient and necessary conditions of the observer are discussed.

Experimental study of batch and fed-batch operation of a bioreactor was performed, which used Saccharomyces cerevisiae strain mutant SM14 to produce Î²-carotene. Fed-batch experiment was based on the feeding of glucose and ethanol. Based on the mathematical modeling of the process which is demonstrated in our previous study[5], unknown input observer for the bioreactor was developed. Model-plant mismatch was observed between the original process model and measurements when changing the initial conditions or operating mode of the experiments. State estimation of the process from the designed observer was demonstrated and compared to the experiment measurements. Results indicated that the state estimation from unknown input observer has smaller estimation error compared to the original process model, especially for the fed-batch experiment.

1. Heidarinejad M, Liu J, Christofides PD. State-estimation-based economic model predictive control of nonlinear systems. *Systems & Control Letters.* 2012;61(9):926-935.

2. Ricker NL. Model predictive control with state estimation. *Industrial & engineering chemistry research.* 1990;29(3):374-382.

3. Ozturk S, Hu W-S. Cell culture technology for pharmaceutical and cell-based therapies: CRC Press; 2005.

4.Chen W, Saif M. Unknown input observer design for a class of nonlinear systems: an LMI approach. Paper presented at: American Control Conference, 20062006.

5. M. Carolina Ordoñez, Jonathan P.Raftery, Tejasvi Jaladi , M. Nazmul Karim. Modelling of batch kinetics of aerobic carotenoid production using Saccharomyces cerevisiae. *submitted.*

471193 Computational Fluid Dynamics (CFD) Guided Scale up of Cell Culture Bioreactors from Bench to Pilot to Commercial Scale

Abstract:

Computational fluid dynamics is a valuable tool for better understanding bioreactors and medium preparation tanks used for mammalian cell cultivation. Knowledge generated by CFD, such as on flow characteristics or on impeller mixing, aid in better transfer and scale up of cell culture processes.

This study used single phase simulations to predict power input and mixing time of production bioreactors at different scales. Impeller power numbers generated by CFD were compared across sites and assessed against historical vendor values to align cross-site scale up parameters. Approaches to applying CFD to large scale cell culture bioreactors will be discussed.

472590 Chemical Engineering Principles in Oral Solid Dosage Process Development-Where Are We?

Abstract:

While solid processing is a common engineerâ€™s domain across industries, there are still gaps in the pharmaceutical industry in applying basic chemical engineering principles, from API development to drug product development, especially equipment design and process control.Â General gaps and misconceptions in these areas gathered from field experience will be presented along with case studies. While the barrier to apply equipment modification or to upgrade process control could be complicated and cumbersome in the heavily regulated industry, the benefit of such change cannot be ignored due to the fact that there were (and are) significant amount of development work caused by the lack of basic principles in designing the current system (as simple of sufficient heat transfer). The other element of barrier is rather cultural, i.e., no desire to change as the current system is workable. In summary, opportunities along with challenges in the pharmaceutical process development are waiting for more chemical engineers to explore.

472435 Characterization of Mixing and Heat Transfer in Continuous Lab Scale Reactors

Abstract:

With proper design and optimization, continuous reactors can provide advantages of heat transfer, mixing, and productivity in a smaller footprint. However, batch reactors have the advantage of a flexibility, including slow or variable feed-times. Continuous plug flow reactors with a single feed point provide stoichiometric feed ratio's upfront. This places higher demand on quality of mixing and heat transfer at the point of mixing for continuous plug flow reactors especially if there are possible side reactions. Chemical systems e.g. Bourne reactions were used to assess the quality of mixing in glass Advanced Flow Reactors (AFR manufactured by Corning) to determine the quality of mixing, based on selectivity from parallel competitive reactions as a function of process flow rates. The quality of mixing is compared to lab-scale batch reactors (EasyMax manufactured by Mettler). Heat Transfer through AFR plates will independently be measured. The residence time distribution through the AFR will also be discussed. The methods described herein are generally applicable for characterizing lab-scale continuous reactors.

472802 Screening a Library of Commercially Available Compounds for Drugs with Novel Anti-Biofilm Activity

Abstract:

It is now well established that bacterial infections are often associated with biofilm phenotypes that demonstrate increased resistance to common antimicrobials. Further, due to the collective attrition of new-antibiotic development programs by the pharmaceutical industries, drug re-purposing has become an attractive alternative. In this work, we have screened 1,280 existing commercially available drugs in the Prestwick Chemical Library, some with previously unknown antimicrobial activity, against an organism that has been implicated as a causative pathogen in bacterial infections, such as burn and wound infections: *Staphylococcus aureus* and *Pseudomonas aeruginosa*. From a preliminary screen of the entire Prestwick Chemical Library at a fixed concentration of 10mM, we found 104 drugs that showed activity against planktonic *S. aureus*, and not surprisingly these were mostly antimicrobials and antiseptics. The activity of 18 selected non-antimicrobials observed in the preliminary screen was confirmed in dose response experiments. Finally, a selection of nine drugs with little or no reported antimicrobial activity was tested against pre-formed biofilms of *S. aureus*. Three of these drugs retained their antimicrobial activity against pre-formed biofilms making them attractive candidates for repurposing as novel anti-biofilm therapies. On the other hand *P. aeruginosa* biofilms were resistant to most of the drugs in the library. However, combinations of some of the drugs in the library with sub-inhibitory concentrations of antibiotics and surfactants were effective against the biofilms. Together, our data show that the anti-biofilm activity makes these drugs attractive candidates for re-purposing as novel anti-biofilm therapies, especially in combination with traditional antimicrobials, and warrant further attention to determine their modes of action.

475116 A Small Volume Continuous Facility for GMP API Manufacture

Abstract:

Design concepts for a 10kg/day, Small Volume Continuous facility for the manufacture of active pharmaceutical ingredients using bespoke fume cupboards.

475290 Controlling amorphization versus crystallization of carbamazepine nanoparticles using a supercritical-assisted spray drying process

Abstract:

Spray drying is now regarded as one of the most commonly used technologies in the pharmaceutical industry for API (Active Pharmaceutical Ingredient) particle production in the micron-size range[1]. It is a method of producing dry powders from a solution or a suspension by rapidly evaporating solvent with a hot air stream. This is one of the preferred technologies for material processing and scale-up in the pharmaceutical and food industries because it is a fast, continuous, and a one-step process. However, the solid-state nature of the spray-dried materials is often amorphous due to rapid solidification and is restricted to the micron size scale range. Although amorphous materials provide huge improvements in the APIs apparent solubility, a major limitation of these materials is that the drug is thermodynamically unstable and tends to recrystallize during storage (depending on the stability provided by the polymer matrix in the solid dispersion) where the outcome might be a crystalline form with a much lower solubility.

Pharmaceutical crystalline materials provide a higher potential for enhanced solid state stability of poorly soluble APIs compared to amorphous materials, although providing a poorer solubilities and dissolution rates. Interestingly, producing poorly soluble APIs as nanocrystals provides considerable improvements in their solubility, dissolution rate as well as solid-state stability. One of the most successful top-down methods to obtain APIs nanocrystals or solid nanocrystalline dispersions with the desired size distribution consists of wet milling APIs down to a nano size range, adding excipients to these nanosuspensions to inhibit crystal growth due to Ostwald ripening and spray drying them later on[2]. However, milling limits the control of the API crystal habit and other surface properties, can take long times to achieve the required size, can introduce impurities from abrasion of the milling media, and can lead to thermal/mechanical degradation of temperature-sensitive materials.

In this work a supercritical CO_2-assisted spray drying technology has been used to produce and control the solid-state form carbamazepine (e.g. amorphous, crystalline) in a nano-sized range. This process involves contacting the API solution with supercritical CO_2 in the nozzle and immediately spraying it through the nozzle exit for solvent extraction at atmospheric pressure[3]. By restricting the high pressure exclusively to where it is indispensable, it is possible to control the precipitation of the API before the nozzle (anti-solvent nucleation/crystallization) or after the nozzle (atomization/spray drying) while providing processing flexibility. This also facilitates the adaptation of this technology onto existing spray drying equipments.

Solutions of CBZ (carbamazepine) dissolved in methanol ([CBZ] = 50mg/ml) with or without selected additives (e.g. ethyl cellulose, sodium stearate, maltitol, Kollidon VA64) were spray dried using supercritical CO_2. In this process the carbamazepine solutions are fed to the nozzle exit, while immediately and continuously feeding the solutions or suspensions through the nozzle to obtain the final powders. Depending on the type of spray drying mechanism being used (atomization or antisolvent-assisted atomization) by changing process variables (e.g. pressure, temperature, supercritical antisolvent/solvent ratio) has provided control on the size and solid-state outcome (e.g. amorphous, crystalline) of the carbamazepine nanoparticles. The inclusion of small amounts of selected additives (0.1-10% w/w) in the carbamazepine-methanol solutions has provided an additional step to controlling the formation of nanoparticles in the amorphous form or favoring the crystallization of a particular polymorph.

1. Vehring R. *Pharm. Res.* 25 **(2008)** 999-1022.

1. Paredes *et al. Drug Dev. Ind. Pharm.* (**2016**), DOI: 10.3109/03639045.2016.1151036
2. Padrela *et al. J. Supercrit. Fluids* 86 (**2014**) 129-136.

This work has been supported by Science Foundation Ireland, Grant number: 12/RC/2275.

475659 Evaluation of the Therapeutic Potential of FOXC1 Knockdown for Treatment of Basal-Like Breast Cancers

Abstract:

Basal-like breast cancer (BLBC) accounts for 15% of breast cancer diagnoses, is characterized as one of the most aggressive forms of breast cancer, and exhibits the worst prognosis. The cancer tends to metastasize to the brain and lungs during early stages [1]. Due to the BLBC triple negative phenotype (estrogen receptor-, progesterone receptor-, human epidermal growth factor 2-), chemotherapy remains the only treatment despite its poor success rate and propensity for relapse. The aggressive growth and mobility of BLBC contribute to its difficulty of treatment and result in low survivability [2-3]. This mortality indicates the urgency for new therapeutic approaches to replace or supplement the limited chemotherapeutic treatment options. Since the forkhead-box transcription factor C1 (FOXC1) was isolated as a unique biomarker for BLBC [4], alternative therapeutic approaches focusing on inhibition of FOXC1 have emerged. FOXC1 overexpression in BLBC was linked to excessive cell proliferation and aggressive mobility. Knockdown of FOXC1 using RNA interference, in contrast, was shown to decrease proliferation and mobility of BLBC cell lines *in vitro*. This suggests that the elimination or knockdown of FOXC1 is a potential therapeutic approach to the treatment of BLBC. This project evaluates this therapeutic potential both *in vitro* and *in vivo* by various gene therapy methods.

Knockdown of protein products from aberrant or overexpressed genes can be achieved by delivering repair machinery to act at the DNA level, or RNA interference machinery to elicit degradation of mRNA products of targeted genes. The CRISPR/Cas9 system is a cutting-edge method that has the ability to produce double stranded cuts in DNA at targeted locations in the eukaryotic genome in order to produce controlled deletions or insertions of genetic code. RNAi was used to knockdown FOXC1 in 4T1 cells (mouse BLBC). Knockdown (>95%) was verified via western blot and proliferation studies showed a statistically significant ($p<0.013$) reduction in cell growth up to 96 hours post transfection. The CRISPR/Cas9 system was used to produce clone 4T1 cells exhibiting knockout of the FOXC1 gene

(verified via western blot). Clone proliferation studies *in vitro* showed similar results to knockdown studies. These clones were used to produce tumors in mice models that were compared to 4T1 tumors in control specimens. Knockout tumors exhibited no reduction in metastatic activity compared to controls and showed no statistical reduction in growth rate. The study indicates that therapies targeting FOXC1 are possible and even successful *in vitro*. However, the inherent complexity of physiological systems convolutes the effect of any positive therapeutic response *in vivo*.

1. Smid M, et al. *Cancer Res*. 2008; 68:3108-3114

2. Seewaldt VL, Scott V. *N Engl J Med* 2007; 356: e12

3. Sarrio D, et al. *Cancer Res* 2008; 68:989-997.

4. Ray P, et al. *Cancer Res* 2010; 70:3870-3876

456713 Mass Transfer Characterization of a Biphasic Continuous Reaction System for Improved Process Understanding and Control

Abstract:

Continuous processing, while commonplace in other chemical industries, remains a novel and emerging technology in the pharmaceutical industry and presents potential advantages in manufacturing footprint, quality, and enhanced control of mass and heat transfer limitations associated with scale-up. Recent developments in the regulatory environment and pressure to control costs have encouraged growth of continuous manufacture in the pharmaceutical industry. Despite numerous advantages, continuous processing contains challenges in areas such as understanding laboratory and pilot plant equipment capabilities for improved process control. One fundamental element to process development is the characterization of continuous equipment to enable process understanding in terms of engineering first principles. With process understanding, effective control strategies can be implemented to deliver quality product at the lowest possible cost.

A key asset in the continuous portfolio at GlaxoSmithKline involves process development of a partially miscible chemical reaction mixture, in which the reaction is limited by mass transfer. To support process understanding, control, and scale-up from the laboratory to pilot plant facilities, a protocol was developed to characterize the mass transfer of the system in terms of fundamental engineering first principles. This presentation will detail the characterization of a partially miscible mass transfer limited reaction for continuous processing including observations of equipment specific scale-up parameters.

461095 The Development of Integrated Continuous Drug Substance Processes

Abstract:

As pressure grows on pharmaceutical companies to deliver increasing levels of process performance, reduced capital and manufacturing costs the industry is looking more and more to continuous processing to help tackle some or all of these challenges.

Within this presentation case studies will be presented where unique continuous processing hybrid platforms have been utilised for the development and operation of continuous processes. These processes included heterogeneous and homogenous chemistries, work-up/distillations, crystallization and isolation. Specifically within the case studies presented details will be outlined, via PAT integration, engineering kinetic and statistical models, experimental and lab rig design, required to enable the design, optimisation and operation of these platforms over multiple week long periods. During the presentation the design considerations over CSTRs, PFRs, hybrid or novel reactor design configurations will be presented and how such design selections can impact or tailor the design properties of the intermediates and importantly the final API solid state characteristics.

464864 Process Intensification through Continuous Spherical Crystallization Using an Oscillatory Baffled Crystallizer

Abstract:

Of current interest in the pharmaceutical industry is the development of continuous processes either within or for the manufacture of final drug products. These processes include, but are not limited to, drug synthesis, crystallization, formulation and tableting/encapsulating. Crystallization is the key link in going from drug synthesis to final formulation, and can often be the bottleneck in downstream process efficiency. For these reasons, there has been increased importance in the development and control of continuous crystallization systems of active pharmaceutical ingredients to produce crystals of good physical, processing, and biopharmaceutical properties. Continuous spherical crystallization (CSC) is a process intensification technique that can address many of the present flaws of traditional crystallization systems.

Peña & Nagy[1] studied and showed the benefits of CSC as a process intensification technique whereby both internal (primary crystals) and external (agglomerates) properties can be controlled experimentally through a decoupled CSC approach in a two-stage mixed suspension mixed product removal (MSMPR) crystallizer; providing the means by which both biopharmaceutical (bioavailability, dissolution) and manufacturing (flowability, filtration, drying) properties can be simultaneously adapted to meet desired quality specifications. Tahara et al.[2] also presented the value of the CSC technique using a one-stage MSMPR to make spherical API particles with a recycle stream; allowing for reduced anti-solvent consumption given many crystallization systems require low solution to anti-solvent ratios. Aside from these recent studies, previous CSC of a model drug compound was first carried out by Kawashima et al.[3]

Herein a novel concept and method in the area of process intensification though continuous spherical crystallization is proposed, which is based on performing crystallization/spherical agglomeration in oscillatory baffled crystallizer (OBC) system. OBCs are similar to plug flow crystallizers (PFCs) in that they are both tubular crystallizers, however, the OBC has periodically spaced orifice baffles with oscillatory motion overlapped on net flow.[4] This allows for operations similar to PFCs without the high flowrates, high residence times and long crystallizer length while maintaining good mixing. The system allows for both coupled and decoupled nucleation, growth and agglomeration mechanisms. Decoupling through spatially distributed solution, solvent, anti-solvent and bridging liquid addition will offer more degrees of freedom for the control of each mechanism. This allows for products tailored to biopharmaceutical benefit and efficacy (e.g. bioavailability, dissolution, morphology) and processing efficiency (e.g. filtering, drying, & friability), while continuously producing particles with desired properties in an efficient manner and productivity at the manufacturing level.

In this work, a model oscillatory baffled crystallizer (OBC) consisting of seven segments was evaluated for both coupled and decoupled spherical crystallization systems. For coupled systems nucleation, growth and agglomeration occur from the onset as the solution contains bridging liquid. For spatially distributed (decoupled) systems nucleation and growth occur in the first four segments due to solution and anti-solvent addition, while bridging liquid is added at the end of the fourth segment making the final three segments agglomeration zones. The OBC will be evaluated for feasibility, control of crystallization mechanisms and final product properties.

1. R. Peña; Z. K. Nagy. Process Intensification through Continuous Spherical Crystallization Using a Two-Stage Mixed Suspension Mixed Product Removal (MSMPR) System. *Crystal Growth & Design*, 2015, 15(9), 4225-4236.
2. K. Tahara; M. O'Mahony; A. S. Myerson. Continuous Spherical Crystallization of Albuterol Sulfate with Solvent Recycle System. *Crystal Growth Design*, 2015, 15, 5149–5156.
3. Kawashima, Y.; Kurachi, Y.; Takenaka, H. Preparation of spherical wax matrices of sulfamethoxazole by wet spherical agglomeration technique using a CMSMPR agglomerator. *Powder Technology*, 1982, 32, 155–161.
4. S. Lawton; G. Steele; P. Shering; L. Zhao; I. Laird; X.-W. Ni. Continuous crystallization of pharmaceuticals using a continuous oscillatory baffled crystallizer. *Organic Process Research & Development*, 2009, 13(12), 1357– 1363.

466160 Designing for Continuous: A Step Change in Performance

Abstract:

Traditionally the majority of active pharmaceutical compounds have been manufactured at large plant scales using batch methodologies. Continuous manufacturing is becoming an increasing attractive alternative however due to apparent benefits such as smaller manufacturing footprints, increased process safety, lower manufacturing costs and increased supply-chain flexibility. Realising all of the above benefits using conventional continuous equipment is however quite challenging and sometimes not achievable. In this project, a new approach to the design of continuous manufacturing platforms is being presented which is not restricted to using such conventional equipment but which is based around the bespoke design and manufacture of all unit operations, optimised to fit the specific process being examined. Processes are quantified in detail using methods such as computational fluid dynamics (CFD) leading to tailored equipment designs which can be manufactured with high precision using metallic 3D printing.

The first case study presented in this paper examines a biphasic reaction between an amine and an acid chloride in the presence of an aqueous sodium hydroxide solution. In this type of biphasic reaction, the main design challenge is centred around achieving a sufficiently large interfacial surface area between the two phases. Conventionally, an agitated vessel operating in either batch or continuous mode would be used in order to maximise this area. However when tested in this case study, adequate dispersion between the immiscible phases required very high energy input levels of at least $1kW/m^3$, making such an approach difficult to scale up. Plug flow reactors (PFRs) are the most common continuous alternative to agitated vessels. These typically operate under Taylor flow conditions (where alternating aqueous and organic slugs move sequentially through the reactor), giving only limited contact between the phases. In order to overcome the limitations of these conventional approaches, a bespoke unit operation based upon the specific needs of the reaction is designed in this paper which is capable of achieving extremely high dispersion rates between the two phases under the very low Reynolds numbers typically encountered in the pharmaceutical industry ($Re < 5$). Due to its optimised design, this unit is able to significantly outperform both conventional methods tested, achieving yields of 98% with residence times of less than 0.067 minutes (the batch agitated reactor achieved a 91% yield with a 6 minute residence time while the flow reactor achieved a 49% yield with a residence time of 7 minutes).

In order to demonstrate the suitability of applying this design approach to the entire manufacturing process, a case study is also presented in which two reactions are coupled together, followed by an extraction and wash step and a liquid-liquid separation. All unit operations within this platform are designed using fundamental engineering principles combined with numerical analyses (based upon the requirements of the specific process), prior to being 3D printed and combined to create a bespoke, sealed, hard-piped manufacturing system. The optimised design of the units enables extremely high throughput rates to be achieved corresponding to 580,000 kg/L/day (compared to an approximate batch productivity of 525kg/L/day).

466468 Continuous Drying of Powders and Slurries in an Extruder without Changing Product Particle Size Distribution

Abstract:

Academia, industry and regulators have increasingly demonstrated and addressed the advantages of continuous manufacturing during recent years. Slowly, this is leading to a change in mindset for pharmaceutical manufacturing, moving towards advanced continuous processing based on real-time quality assurance and away from batch processing.

Greater interest is being shown in the pharmaceutical industry for linking primary and secondary manufacturing into a single continuous manufacturing line. This demand is driven by the need of a cost-efficient process development and upscaling phase, which occurs during phase two and three of drug product development. Within the framework of continuous manufacturing, several techniques are commonly used for producing drugs. These include filtration, drying, milling/blending, granulation, and tableting, among others.

Drying is one of the most challenging process steps, if linking primary and secondary production. Not only does drying consume a vast amount of energy, but moreover, particle properties can significantly change during drying. Since the particle structure is carefully considered during continuous crystallization, it is highly desirable to avoid any changes in the particle properties while drying. Two counteracting forces, attrition and agglomeration, constantly occur during the drying of wet powder by shear movement.

When considering continuous manufacturing, not only must the product quality be considered, but also the processability. Therefore, the residence time distribution must be taken into account. Ideally, the residence time distribution should be narrow and no material should be stuck to walls and in dead spaces. Using an extruder helps to overcome many of the challenges normally faced during drying. Agglomeration can be avoided and the residence time distribution can be very narrow if using a twin-screw extruder with forced-feed and self-cleaning screws.

This paper will highlight successful efforts for drying wet powders or slurries, with high solid content, in a twin-screw extruder (Leistritz 27mm twin-screw). This work is focused on the processability of a test substance with a particle size below 100 µm. The screw configuration was optimized to best suit the needs of the drying process. Different process settings where tested and some showed very promising results. Products with different starting moisture contents were also tested. The percent reduction in moisture observed following the drying procedure was 99% with a residual moisture below 0.1% and there was no agglomeration observed during drying.

466532 Continuous Co-Crystallization of Novel Theophylline and 4-Aminobenzoic Acid Co-Crystals from Twin Screw Extrusion

Abstract:

Recently, increased demand for greener and more environmentally sustainable processes has focused the attention on mechano-chemical process. Mechano-chemical synthesis is one in which solid reactants are vigorously ground (milling or grinding) together with minimal or even no solvent [1]. Mechano-chemical synthesis is environmentally friendly, easy, cheap, fast reaction rates and high yields; these advantageous properties made mechano-chemical synthesis a method of choice for co-crystal formation [2 & 3]. Pharmaceutical co-crystals are solids that are crystalline materials comprise of two or more components held together by non-covalent forces [4]. In recent years co-crystals are being studied intensively due to the potential for improved pharmaceutical properties such as increased solubility, bioavailability, chemical stability and hygroscopicity of active pharmaceutical ingredients [1].

High-shear granulation and twin screw extrusion, have recently been used for co-crystal preparation. These methods are environmentally friendly as they avoid, or minimise, the use of organic solvents [2 & 5]. Twin screw extruder mediated co-crystallisation is a promising processing technique for continuous co-crystallisation. Although twin screw extruder mediated co-crystallisation has been reported with various primary experimental parameters (screw geometry, temperature, feed rate and the RPM of the screws) that can affect co-crystal conversion, information on the impact of processing parameters processing temperature (*eg.* at eutectic temperature and room temperature), shear dependent crystallization, residence time and addition of organic solvents is less available. Therefore to enhance understanding of continuous co-crystallisation process in the present study the effect of key extrusion processing parameters that influence extent of conversion to the co-crystals has been studied. A Theophylline-4-Aminobenzoic acid novel co-crystal forming system was selected for the study. Appropriate stoichiometric ratios of

Theophylline-4-Aminobenzoic acid were prepared. Co-crystallization of novel co-crystal system was carried out in a co-rotating 12 mm Hybrid Mini-Extruder (Three-Tec GmbH, Switzerland) with a length-to-diameter ratio of 40:1. The extrudates were mainly characterised by powder X-ray diffraction (PXRPD), differential scanning calorimetry (DSC), and *in vitro* drug release studies. As shown parameters had a significant influence on co-crystallization process, consequently they are important to consider in co-crystallization process. Finally twin screw extruder mediated co-crystallisation can be considered easy, cheap, scalable, and environmentally friendly for continues co-crystallization.

Key words: continuous co-crystallisation, twin screw extrusion, Theophylline-4-Aminobenzoic acid.

References

1. S.F. Chow, M. Chen, L. Shi, A.H.L. Chow, C.C. Sun, Simultaneously improving the mechanical properties, dissolution performance, and hygroscopicity of ibuprofen and flurbiprofen by cocrystallization with nicotinamide, Pharm. Res. 29 (2012) 1854–1865.

2. D. Daurio, C. Medina, R. Saw, K. Nagapudi, F. Alvarez-Nunez, Application of twin screw extrusion in the manufacture of cocrystals, Part I: four case studies, Pharmaceutics 3 (2011) 582–600.

3. James, S.L., Adams, C.J., Bolm, C., Braga, D., Collier, P., Friˇsˇci ´ c, T., Grepioni, F., Harris, K.D.M., Hyett, G., Jones, W., Krebs, A., Mack, J., Maini, L., Orpen, A.G., Parkin, I.P., Shearouse, W.C., Steed, J.W., Waddell, D.C., 2012. Mechanochemistry: opportunities for new and cleaner synthesis. Chem. Soc. Rev. 41, 413–447.

4. FDA Guidance (draft), December 2011: Regulatory Classification of Pharmaceutical Co-Crystals.

5. R.S. Dhumal, A.L. Kelly, P. York, P.D. Coates, A. Paradkar, Cocrystallization and simultaneous agglomeration using hot melt extrusion, Pharm. Res. 27 (2010) 2725–2733.

472415 Streamlined Design for Continuous Production of a Pharmaceutical Drug Product

Abstract:

The pharmaceutical industry is beginning to overcome historical reliance on batch processing and embrace the use of continuous processing - a mainstay of other process industries [1]. Continuous unit operations offer many production, efficiency and quality control advantages, and if designed to address the specific risks of a particular process, can also be achieved at lower cost, smaller footprint and more rapid development. One key to this outcome is the participation of a cross-functional team of development and process engineers; operations and supply chain staff; analytical, PAT, and statistical/modeling specialists; equipment vendors; along with inclusion of representatives from Quality and Regulatory functions early in the design process.

We illustrate how functional perspectives have been used to reach balance in the design of a new facility for continuous manufacture of a direct compression and film coated drug product. Design decisions explored include the definitions of a batch in this context, the use of loss in weight feeders, process-model based controls, continuous blending, semi-batch coating pans and automated on-line tablet testing for real time release. The needs of compliant and robust commercial manufacture as well as new product development flexibility are explored. A risk based analysis of drug product characteristics forms the basis for selection among feasible design, control, and operational choices. We present a proof-of-operations case for an initial implementation of continuous manufacturing for the efficient manufacture of drug product batches, and present considerations related to constraints around global regulatory agency acceptability, future flexibility, the competing needs of commercial manufacturing and process development, complexity of process controls, and adaptability to changing market demand.

The presentation is intended to illustrate our belief that for a well-characterized drug product, a streamlined, robust design can be achieved in the absence of multiple complex layers of redundancy (in instrumentation, controls,

qualification expense, etc) while still achieving the goals of robust assurance of product quality and batch size flexibility.

[1] Sau L. Lee, Thomas F. Oâ€™Connor, Xiaochuan Yang, Celia N. Cruz, Sharmista Chatterjee, Rapti D. Madurawe, Christine M. V. Moore, Lawrence X. Yu , Janet Woodcock â€œModernizing Pharmaceutical Manufacturing: from Batch to Continuous Productionâ€, *J Pharm Innov, 10*(3), 191-199.

468879 Flowsheet Modeling of a Continuous Direct Compression Process at Production Scale

Abstract:

A flowsheet model for a continuous direct compression and tablet coating and drying process is developed and applied to process data generated at production scale. The flowsheet model contains feeding, blending, tablet compaction and tablet coating and drying units. The model is constructed using gSOLIDS process modelling software (PSE Ltd). The feeding units are assumed to have a constant flowrate, the blending units are based on a tanks-in-series (TIS) methodology and the tablet press model produces tablet hardness as a critical quality attribute. A thermodynamic model for the coating and drying process considers mass and energy conversation to describe five segregated sub-processes (loading, preheating, coating, drying and discharging). Process data is collected by Merck & Co using a GEA Group (AG) continuous direct compression and coating and drying process designed for a six ingredient formulation. Parameter estimation is used determine the mean residence time and lag time for the tanks in series representing the blending process. A number of tablet porosity models are added to the tablet press model increasing model accuracy and flexibility. The coating and drying model describes the process exhaust temperature and relative humidity within a relative error of 2.5%. The flowsheet model performs well using the existing data set and will be used for further process optimization as the process approaches commercial production. In addition, the model is used to investigate process variability in the feeding flowrate and compaction force of the tablet press. A major application of the model is to be used in conjunction with process analytical technology (PAT) for process control.

474389 Characterizing a Drug Product Continuous Manufacturing Platform's Residence Time Distribution Using Major and Minor Unit Formula Excipient Step Changes and Soft Sensors

Abstract:

A continuous manufacturing (CM) line's residence time distribution (RTD) must be appropriately characterized to provide product traceability in the event of a process upset or when switching raw material batches. Lilly's current process to determine the CM line's RTD is to characterize the final blend concentration versus time as a result of intentional step changes to ingredient concentration in the formula. Currently, the ingredient concentration is determined using a Near-infrared (NIR) application located in the tablet press feed frame. This presentation explores 1) using step changes in either the major and/or minor excipient ingredients in the formula to produce similar concentration response curves in the feed frame using the NIR application and 2) the option to use soft sensors in lieu of the NIR application to characterize the CM line's response to the ingredient step changes and ultimately determine its RTD.

466624 Establishing a Procedure for Transferring HME Processes Using Physical Similarities

Abstract:

Introduction

Hot Melt Extrusion (HME) attracted increasing interest in the pharmaceutical industry during recent years. The majority of newly developed APIs (active pharmaceutical ingredients) is hardly soluble in aqueous systems, thus, requires adequate formulations and manufacturing technologies to achieve sufficient levels of bioavailability. HME is a potent pathway to overcome this challenge by establishing solid solutions. Since the process parameters (e.g., screw design, temperature, screw speed) are affecting the morphology of the API a transfer of a HME process from one line to another needs proper considerations. In polymer processing such scale up procedures are common practice especially for single-screw extruders by using physical similarities [1]. Although several guidelines for scaling up of twin-screw extruders exist, it is necessary to know the limitation factors (volume scale-up, energy scale-up, temperature scale-up) [2]–[5].

The objective of this study is therefore to establish a suitable transfer of a HME process from a model extruder (*Pharma Extruder ZSK 18* from *Coperion GmbH*) to a target extruder (*PHARMALAB 16* twin-screw extruder from *Thermo Fisher Scientific Inc.*), where the properties of the intermediate products of both processes should be as identical as possible.

Theoretical Basis

As appropriate transfer methods, the existing method from Menges and Feistkorn [3] as well as the one from Rauwendaal [2] were chosen. The first one is based on the model theory and allows a fully worked out transfer of twin-screw extruders (TSEs). The second method does not allow a complete transfer of the extrusion process since it states to hold critical process parameters, (e.g., the specific mechanical energy consumption (SMEC)) constant. By using the suggested equations, the throughput, the screw speed and the SMEC of the target extruder can be calculated. The two methods are compared in this study.

In pharmaceutical HME, product temperature and shear rates can be critical process parameter since pharma polymers and API can be strongly sensitive to both. Due to the high shear rates in the extrusion die and the impact on the processed material, it is essential to consider the extrusion die properly in the transfer procedure, to avoid different pressure or temperature development. This can be achieved by calculating the die conductance of the die of each extruder [6].

Experimental Methods

Materials

The intermediate products consist of a binary material system. As API Fenofibrate *F6020* from *Sigma-Aldrich* and as polymer matrix *Soluplus®*, a graft copolymer from *BASF*, was chosen. A premix (1:9) was blended before the extrusion process.

Experiments

The set-up of the trials on the model extruder is shown in Table 1. The screw and die configuration were not changed during the investigations.

Run	Temperature profile [°C]	Screw speed [1/min]	Feed rate [kg/h]
1	60/80/100/100/100/110/110/110/110/120	120	2
2	60/80/100/100/100/110/110/110/110/120	200	2
3	60/80/115/115/115/125/125/125/125/135	120	2
4	60/80/115/115/115/125/125/125/125/135	200	2

The set-up of the trials on the target extruder is shown in Table 2. The process input parameters have been calculated according to the two existing transfer methods. As scale-factor for the assembly of the screw configuration of the target extruder, the unrolled screw length Z was used.

Run	Temperature profile [°C]	Screw speed [1/min]	Feed rate [kg/h]
1_M	60/80/100/100/110/110/110/120	145	1.6
1_R	60/80/100/100/110/110/110/120	140	1.7
2_M	60/80/100/100/110/110/110/120	240	1.6
2_R	60/80/100/100/110/110/110/120	240	1.8
3_M	60/80/115/115/125/125/125/135	145	1.6
3_R	60/80/115/115/125/125/125/135	140	2
4_M	60/80/115/115/125/125/125/135	240	1.6
4_R	60/80/115/115/125/125/125/135	240	2

The residence time distribution (model and target extruder) was investigated by means of a coloured tracer and video analysis.

Characterization methods

The intermediate products, obtained from the HME process on the model and target extruder, were characterized by differential scanning calorimetry (DSC), dissolution test according to Ph.Eur. – 2.9.3 with apparatus 2 and content uniformity test according to Ph.Eur. – 2.9.40.

Table 3 shows the different glass transition temperatures (T_g) for the runs 1-4 and 1_M. As can be seen especially the shear rate (represented through the screw speed) causes a shift in glass transition to higher temperature, which indicates that the formulation is sensitive to shear. The T_g of Run 1_M is very close to Run 1, which indicates similar thermal properties.

Table 3: Glass transition temperature of different runs

Run	Glass transition temperature [°C]
1	41.2
2	49.3
3	41.6
4	48.8
1_M	40.5

Figure 1 shows the DSC curves of run 1 in comparison to run 1_M, which have been extruded according to the transfer method of Menges and Feistkorn. As can be seen the shape of the two curves are similar with similar T_g (see Table 3) indicating that the transfer process was successful regarding thermal properties.

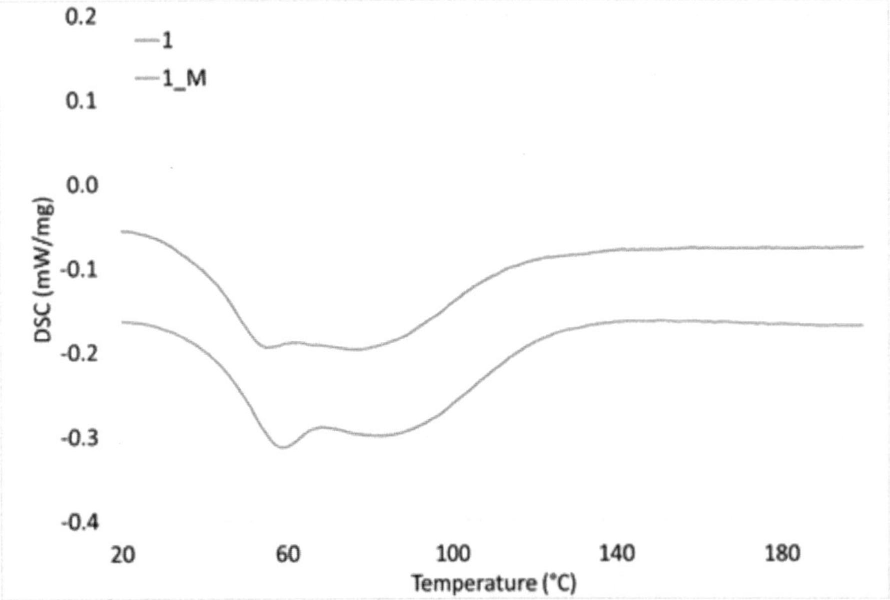

Figure 1: DSC curves (first heating) of Run 1 and Run 1_M

Conclusion and Outlook

By comparing the process output parameters and product characteristics from the model and target extruder, it will be shown if the selected transfer methods are suitable. First results showed that the transfer regarding thermal properties was successful for run 1 using the method of Menges and Feistkorn. Planned analysis will reveal if the transfer was successful regarding dissolution behaviour and content uniformity for all runs. The gained results and the established procedure will help to build up profound knowledge and understanding about scale up of hot melt extrusion processes in pharmaceutical applications.

References

[1] M. Zlokarnik, "Dimensionsanalyse," in *Scale-up: Modelluebertragung in der Verfahrenstechnik*, Second Edition., Weinheim: Wiley-VCH Verlag GmbH & Co. KGaA, 2005, pp. 3–15.

[2] C. Rauwendaal, "10 - Twin Screw Extruders," in *Polymer Extrusion (Fifth Edition)*, Fifth Edition., C. Rauwendaal, Ed. Hanser, 2014, pp. 697 – 761.

[3] G. Menges and W. Feistkorn, "Scale-Up of twin screw extruders application and verification with the example of PVC," *Adv. Polym. Technol.*, vol. 4, no. 2, pp. 123–129, 1984.

[4] A. Dreiblatt, "13 - Technological Considerations Related to Scale-Up of Hot-Melt Extrusion Processes," in *Hot-melt Extrusion: Pharmaceutical Applications*, Chichester: John Wiley & Sons Ltd., 2012, pp. 285–300.

[5] K. Kohlgrueber, *Co-rotating Twin-screw Extruders: Fundamentals, Technology, and Applications*. Munich: Carl Hanser Publishers, 2008.

[6] W. Michaeli, *Extrusionswerkzeuge für Kunststoffe und Kautschuk: Bauarten, Gestaltung und Berechnungsmoeglichkeiten*, 2. Auflage. Munich, Vienna: Carl Hanser, 1991.

468639 Investigation of the Influence of Continuous Blending Process Parameters on the Lubrication of Pharmaceutical Powder Blends

Abstract:

The lubrication of blends of pharmaceutical powders containing 1% magnesium stearate was evaluated in a continuous blender across a broad range of mass flow rates and blender speeds. Significant changes in both the bulk density and tensile strength of powder compacts were observed as a function of blending conditions. The results demonstrate that for a given blender configuration, various final degrees of blend lubrication can be achieved. An optimal combination of mass flow rate and blender speed for maximum lubrication is found to reside at intermediate process values, consistent with results from other studies. The relationship between blending conditions and blend lubrication state are subsequently described by a numerical model, which describes changes in final blend attributes as a function of key blending process parameters. This model, however, collapses at higher blender speeds, where powder conveyance dominates and mixing dynamics are significantly diminished.

473846 "Correlating Raw Material Properties to Critical Processing Parameters to Critical Quality Attributes in a Direct Compaction Continuous Line."

Abstract:

In the last few years, Pharmaceutical manufacturing is experiencing unprecedented scientific and technological innovation as it has been involved in a fundamental transformation from batch to continuous. This revolution has the potential of reducing both the manufacturing time and cost, increase process control and improve overall process quality.

This study demonstrates a systematic methodology to find the critical processing parameters affecting the critical quality attributes of tablets produced in the continuous direct compaction manufacturing line. The methodology is demonstrated in a case study where the API is fixed, the formulation is varied along the DOE but keeping the API at low concentrations (2 to 10%).

The approach follows a screening bi-block two level fractional factorial DOE to elucidate the critical process parameters and corresponding failure modes of the continuous line which depends on the material properties of the formulation. The screening study involves eight processing and two formulation parameters but it only consists on twenty four experiments. A three level factorial DOE on the remaining parameters is then run for characterization and optimization purpose with the ultimate goal to relate material properties to the critical processing parameters and finally to the critical quality attributes of the tablets produced.

The study is run in Rutgerâ€™s direct compaction continuous line and consists of two feeders (API and excipient) connected to a mill and to a blender with a third feeder (lubricant) connected also to the blender, bypassing the mill to prevent blendâ€™s over lubrication. After the blender, the mix flows through a chute to the feed frame of the rotary tablet press where tablets are compacted. The quality attributes measured on tablets are content uniformity, weight variability, dissolution and hardness, but only the first is expected to have a relevance in this case study at low API concentration (2 to 10%).

461124 A Novel Analysis of Residence Time Distributions for Continuous Powder Mixing

Abstract:

Continuous pharmaceutical manufacturing has attracted the attention of industry and academia for a wide variety of reasons. Because this type of manufacturing can utilize smaller manufacturing facilities, it can alleviate bottlenecks related to scale-up and improve control of product quality. With regards to materials, this type of process can eliminate the isolation and storage of intermediate products as well as decrease raw-material and final-product waste. By reducing the number of manufacturing steps, continuous pharmaceutical manufacturing can shorten manufacturing cycle times, provide greater response capacity, increase production yields, and lead to an overall product manufacturing efficiency (Lee et al., 2015).

The residence time distribution (RTD) is a probability distribution function that describes the length of time a material spends in the system. The RTD can be used to address the traceability of raw materials in continuous pharmaceutical manufacturing processes (Engisch & Muzzio, 2016), providing useful data for cases relating to out of specification product investigations, consumer complaints, product recalls or any other situations that may have public health impact. Knowledge of the RTD enables the prediction of when an affected material will reach a point downstream in the process, as well as how much of the adjacent material it may have affected through dispersion. The RTD of convective powder mixers has previously been experimentally studied and correlated to lumped parameters such as the Peclet number and mean residence time in the Taylor dispersion model (Gao, Vanarase, Muzzio, & Ireapetritou, 2011) . However, the RTD measured/fitted for one set of operating conditions may not predict the system RTD when the conditions are altered.

In this work, we conducted an analysis using previous experimental RTD data characterizing continuous powder mixing (Marikh, Berthiaux, Mizonov, Barantseva, & Ponomarev, 2006; Vanarase & Muzzio, 2011). Vanarase and Muzzio (2011) conducted experimental RTDs on a Gericke GCM 250 continuous mixer for eight different conditions (four different impeller rotation rates: 40, 100, 160, and 250 rpm at two throughput levels: 30 and 45 kg/h). Similarly, Marikh et al. (2006) measured eight experimental RTDs on a Gericke GCM 500 continuous mixer (four different throughputs: 40, 60, 80, 100 kg/h at two different impeller rotation rates: 15 and 60 rpm). The experimental RTDs were converted to RTDs in dimensionless time by scaling to each distribution with its respective mean residence time. It is observed that RTD in dimensionless time is invariant with throughput at different impeller rotation rates under certain conditions. In other words, the shape of RTD is the same but the RTD itself is just shrunk or stretched based on the mean residence time. Next, we introduced a novel concept of a time-dependent RTD to contrast with RTDs measured under steady-state conditions. This approach predicts transient changes in the RTD as throughput varies arbitrarily with time. It is anticipated that this effort will enhance process understanding to facilitate extending empirically measured RTDs for solid oral continuous drug product manufacturing processes in a straightforward manner.

References

Engisch, W., & Muzzio, F. (2016). Using Residence Time Distributions (RTDs) to Address the Traceability of Raw Materials in Continuous Pharmaceutical Manufacturing. *J Pharm Innov, 11*(1), 64-81.

Gao, Y., Vanarase, A. U., Muzzio, F., & Ireapetritou, M. (2011). Characterizing continuous powder mixing using residence time distribution. *Chemical Engineering Science, 66*, 417-425.

Lee, S. L., Oâ€™Connor, T. F., Yang, X., Cruz, C., Chatterjee, S., Madurawe, R., . . . Woodcock, J. (2015). Modernizing Pharmaceutical Manufacturing: from Batch to Continuous Production. *J Pharm Innov, 10*(3), 191-199.

Marikh, K., Berthiaux, H., Mizonov, V., Barantseva, E., & Ponomarev, D. (2006). Flow Analysis and Markov Chain Modelling to Quantify the Agitation Effect in a Continuous Powder Mixer. *Chemical Engineering Research and Design, 84*(11), 1059-1074.

Vanarase, A. U., & Muzzio, F. J. (2011). Effect of operating conditions and design parameters in a continuous powder mixer. *Powder Technology, 208*(1), 26-36.

471433 Model Based Loss-in-Weight Feeder Design and Optimization

Abstract:

Over the past year and half the Pharma Industry has seen the first two approvals for the continuous manufacturing of Oral Solid Dosages. This being further evidence that it is no longer a question of "if" continuous manufacturing will come to the pharma, but rather "when/how" the transformation will take place. From a technical point of view it has been acknowledged by most that Loss-in-weight (LIW) feeding is the most critical operation to continuous Oral Solid Dosage products. In recent years a much deeper fundamental understanding of the LIW feeding process has been developed. This understanding includes a material property driven modeling frame work for predicting performance. The goal of this talk is to present a framework for model based design and optimization of feeders. The key deliverable of this approach would be the ability to tailor each feeder for the specific material properties of the raw material it is feeding.

458840 Use of Real-Time Analytics and Apoptosis Assays for Enhanced Understanding of Cell Growth and Viability in a CHO-Based Process for Monoclonal Antibody Production

Abstract:

In a cell culture process for biopharmaceutical manufacturing, monitoring cell growth and viability is crucial for understanding cell behavior, as well as for determining cell culture duration and timing of harvest. Standard industry practice typically includes daily sampling of bioreactors for cell count and metabolite profiles. However, daily measurements are often too infrequent to detect rapid changes in metabolite concentrations and viability, and may not capture early indications of decline in cell health. We have evaluated the use of apoptosis assays as well as in-line monitoring techniques to gain a more complete understanding of cell growth, health and viability throughout the cell culture process.

In this study, we have applied markers of early, middle and late apoptosis to evaluate changes in cell growth, health and viability during a fed-batch process in 5L bioreactors. The markers that were monitored included 7-AAD, Caspases, and TUNEL assay for DNA nicking. Multiple early and mid-stage apoptotic markers were compared for their effectiveness in evaluating cell health and predicting cell growth trends in a bioreactor. Additionally, to provide more frequent cell growth measurements and metabolite readings to align with apoptotic assays, in-line probes were incorporated for real-time data analysis. Through calibration and method development, a capacitance-based Incyte probe was used for real-time monitoring of viable cell density (VCD), and an in-line Raman spectroscopy probe was used for real-time monitoring of glucose and lactate.

The results demonstrated that apoptotic markers can provide additional insight into cell health beyond what is learned from daily cell counts, thereby greatly deepening our understanding of cell culture conditions and process parameters during development. Markers of early apoptosis were observed several days prior to a drop in viability as measured by a commercial cell counter, while late-stage apoptotic markers were observed hours in advance or concurrently with the VCD and viability drop. Further insight into the health and productivity of the culture may be obtained by correlation of in-line metabolite and cell growth profiles with cell apoptotic progress.

In summary, optimization and application of real-time analytics and apoptosis assays in a CHO cell process can provide a rich and complete data set for better visualization of cell growth and health throughout the culture process, thereby enabling process improvement to maximize product titer and quality.

468928 Providing a Molecular Basis to Modeling Protein Adsorption to Multimodal Carbohydrate-Based Ligands

Abstract:

Recognition and binding of proteins to carbohydrate-based multimodal ligands plays a critical, but poorly understood, role in both fundamental cellular (e.g., cell glycocalyx and viral protein binding) and industrial processes (e.g., antibody drug bioseparations). Manufacturing costs of biopharmaceutical drugs can be drastically reduced through development of inexpensive multimodal (or mixed-mode) chromatographic bioprocesses for protein isolation and purification. However, unlike classical affinity or ion exchange based bioseparations that have been well studied, multimodal chromatographic (MMC) systems are still in their developmental stages [1]. Inexpensive and biocompatible MMC matrices/ligands derived from carbohydrates have tremendous potential for industrial bioseparations. However, we currently have a poor understanding of the molecular mechanisms driving protein binding to traditional carbohydrate (e.g., sepharose) and/or glycoprotein (e.g., concanavalin A) derived chromatographic matrices. This has been further stymied by the lack of suitable experimental tools and predictive adsorption models that explain the heterogeneous nature of protein-carbohydrate interactions. One of the challenges with traditional methods (e.g., solution depletion) used for monitoring protein adsorption is that bulk ensemble measurements provide limited insight into the underlying molecular mechanisms driving protein-ligand binding. Studying binding interactions of proteins and carbohydrate-based ligands at the single-molecule level is critical to explaining observed macroscale measurements, which will ultimately lead to the development of mechanistically-relevant protein macroscale adsorption models.

Here we study the adsorption of carbohydrate-binding proteins to two distinct model polysaccharide substrates with identical molecular composition but varying surface hydrophilicity using complementary bulk and single-molecule based adsorption techniques. We have modified a recently developed optical tweezers force spectroscopy technique [2] to characterize the single molecule protein binding-unbinding force dynamics to polysaccharide surfaces and probed

the protein-ligand bond lifetimes at the single molecule level. Analysis of our records was used to identify trends in the force-binding lifetime datasets to find that the unbinding force and total binding lifetimes reflects the underlying heterogeneity of the available binding sites for carbohydrate-derived ligands. This distribution reflects the spectrum of varying affinity sites present on the polysaccharide surface and ultimately provides a molecular basis for using various multi-site Langmuir-type adsorption models to characterize protein adsorption to polysaccharide surfaces. Interestingly, simple regression analysis of bulk adsorption data using multi-site Langmuir-type models reveals that traditional protein bulk adsorption measurements are unable to provide data with sufficiently high resolution to discriminate between the low and high affinity protein binding sites for polysaccharides if the difference in the protein-carbohydrate ligand unbinding force is relatively small. In summary, we now have the necessary tools to systematically characterize protein adsorption to multimodal carbohydrate-derived ligands for optimizing industrial bioseparations using a first-principles approach.

References

[1] Pinto, I. F., Aires-Barros, M. R. & Azevedo, A. M. Multimodal chromatography: debottlenecking the downstream processing of monoclonal antibodies. Pharm. Bioprocess. 3, 263â€"279 (2015).

[2] Brady, S. K., Sreelatha, S., Feng, Y., Chundawat, S. P. S. & Lang, M. J. Cellobiohydrolase 1 from Trichoderma reesei degrades cellulose in single cellobiose steps. Nature Communications. 6, 10149 (2015).

472914 Operational Control Strategy Development Via First Principles Mapping

Abstract:

Product and Operational control strategies are essential elements in efficient and effective manufacturing of quality medicines. Many forms of guidance exist (e.g., PQLI) to help define and structure documented control strategies. Most control strategies are good at identifying the controlled variables and critical parameters that are essential to ensure product quality. What is often missed, however, is identifying the underlying first principle transformations that are the core of every deliberate step that is taken in making medicine. Each process step has an objective to be met. In fact, when we look at each process step, there are multiple objectives to be met. We must ensure the product quality is delivered, but we must also ensure that we do it cost effectively, protect the operator, protect the community, protect the environment, meet demand, etc. To manufacture the Product, we must design a Process Map as a collection and sequence of Process Stages and Process Operations that combine to deliver certain features and requirements of the product. The process operation is simply stated as an action (or transformation) on an object with a qualifier, e.g. *Disperse powder to less than 1% variation.* The mechanisms for transformation can always be traced to first principles of science. Thus, the Control Strategy is based upon key first principles and associated attributes and variables, at the most fundamental level. To satisfy the criteria of the transformation we must choose and implement effective Control Methods. For the example of "disperse powder", we might choose a vertical agitated tank that will deliver an adequate environment for the first principles attributes of shear and fluid velocity. For each process operation, we must consider objectives and control methods for all Features, such as quality, safety, cost, etc. In the case of Safety, for example, we can provide control methods for preventing dust explosions based upon the first principles of ignition source control.

This presentation will define and demonstrate the means to construct and implement a structured product control strategy via process mapping using first principles-based transformations and chosen control methods.

460708 Nanoextrusion: A Platform Enabling Comparative Assessment of Nanocomposites Vs. Amorphous Solid Dispersions for Drug Dissolution Enhancement

Abstract:

A relatively novel nanoextrusion process was used to produce extrudates of griseofulvin (GF), a poorly water-soluble drug, in a polymeric matrix with the goal of enhancing the dissolution rate and thus the bioavailability. Unlike a traditional hot melt extrusion (HME) process, nanoextrusion process (Baumgartner et al., 2014; Khinast et al., 2013) uses a wet-milled suspension of the drug as a feed along with an extrusion polymer and disperses the drug nanoparticles in the polymeric matrix while evaporating water, thus yielding extrudates in the form of nanocomposites. Simply, nanoextrusion process acts as a continuous drier unlike the traditional HME process. Being a continuous process and having the capability to handle viscous drug suspensions, it could be advantageous over various pharmaceutical drying operations in the manufacturing of solid nanoparticle-based formulations. In this study, we propose that as another potential advantage of the nanoextrusion, extrudates with different forms of the drug can be prepared by the same process via use of different extrusion polymers: nanocrystalline drug dispersed in the polymeric matrix (nanocomposites) or amorphous drug molecularly dispersed with the polymer (amorphous solid dispersion, ASD). It is hypothesized that the drug nanocrystals in the nanocomposites dissolve faster than the amorphous drug in the ASD when the drug dose is so low not to cause supersaturation in the dissolution medium (low drug dose). Specifically, hydroxypropyl cellulose (HPC) and Soluplus were used to stabilize wet-milled suspensions and to form the matrix of the extrudates. Feed suspensions of GF were prepared by wet stirred media milling using HPC and Soluplus in the presence of sodium dodecyl sulfate (SDS), an anionic surfactant. The milled suspensions along with additional polymer (HPC/Soluplus) were fed to a co-rotating twin-screw extruder, which dried the suspensions and formed various extrudates. The extrudates were milled into powders via mortar–pestle for further analysis. The effects of drug particle size in the feed suspensions, stability of drug suspensions, and polymeric matrix type–size on the drug dissolution from the milled extrudates were examined. The suspensions were characterized via laser diffraction, while the extrudates and their milled powders were characterized by laser diffraction, SEM, and XRD. Digital microscopy was used to visualize the changes of different polymeric matrices when exposed to water. Statistical analysis with Korsmeyer-Peppas model helped us to identify the drug release mechanisms. By using polymers with different interactions/miscibility with the drug, two different forms of the drug were formed: extrudates with nano/micro-crystalline drug particles dispersed in the HPC matrix as a secondary phase (nanocomposites or microcomposites) and extrudates with amorphous drug molecularly dispersed within the Soluplus matrix (amorphous solid dispersion). In the composites, drug particle size was the dominant factor that affects the dissolution performance, whereas the matrix size appears to play a significant role in the drug dissolution performance in the amorphous solid dispersion. We have demonstrated that for low-dose drugs, e.g. highly potent poorly soluble drugs, drug nanoparticles in the nanocomposites may actually dissolve faster than amorphous form of the drug in an amorphous solid dispersion, depending on the size of the polymeric matrix.

References

Baumgartner, R., Eitzlmayr, A., Matsko, N., Tetyczka, C., Khinast, J., Roblegg, E., 2014. Nano-extrusion: A promising tool for continuous manufacturing of solid nano-formulations. Int. J. Pharm. 477, 1 11.

Khinast, J., Baumgartner, R., Roblegg, E., 2013. Nano-extrusion: a one-step process for manufacturing of solid nanoparticle formulations directly from the liquid phase. AAPS PharmSciTech 14, 601 604.

468257 Evaporation of Binary Mixtures and Shell Formation in Spray Dried Droplets

Abstract:

Evaporation of binary mixtures and shell formation in spray dried droplets

P.C. Valente[1], Í. Duarte[1], T. Porfirio[1], X. Liu[2], F. Zhang[2] & M.N. Temtem[1]

[1] R&D Drug Product Development, Hovione Farmacência S.A, Sete Casas, 2674-506 Loures, Portugal;

[2] College of Pharmacy, University of Texas at Austin, U.S.A;

Amorphous solid dispersions (ASDs) of poorly water soluble active pharmaceutical ingredients (APIs) in hydrophilic polymeric matrices are a commonly used strategy to increase the solubility and dissolution rate of oral-dosage forms [1]. However, since the polymer is hydrophilic and the API is poorly water soluble, it is often difficult to find a common solvent for both and a mixture of co-solvents is used instead. Solvent mixtures also offer the possibility of influencing particle morphology and drying rate [2] and have been reported to offer advantages in the solid state drug-polymer miscibility and on the stability of the ASDs [3,4].

To fully explore the potential of spray dried ASDs from co-solvent drug-polymer solutions, a fundamental understanding on the solvent chemistry, droplet composition history throughout drying and its impact on the drug-polymer interaction, stability and phase behavior are required. These mechanisms are, in turn, strongly influenced by the process conditions, such as thermodynamic layout, atomizing characteristics and spray drier geometry.

The focus of the present work lies on the evaporative dynamics of binary solvent droplets and how they are influenced by process conditions. Unlike the single component counterpart, modelling the evaporation of binary mixtures depends not only on the process parameters, but also on the mass (and momentum) transfer within the droplet [5]. For example, if one considers solely the molecular diffusion for solvent mixtures with starkly different vaporization enthalpies, the evaporation rate of each solvent will be proportional to their initial mass fraction and, counter-intuitively, will be independent of the vaporization enthalpies [6]. However, whenever there is a non-negligible relative velocity between the droplet and the surrounding drying gas, the internal circulation within the droplet can lead to a steep increase in the effective mass diffusion rate. The evaporation rate of each solvent will also strongly depend on their vaporization enthalpy (see Figure 1). This leads to the more intuitive case where the more volatile solvent tends to evaporate first.

We show that these two limits lead to dramatically different drying kinetics and can strongly affect the onset of drug/polymer precipitation and shell formation. After shell formation the drying kinetics tend to be strongly modified, highlighting the importance in predicting the onset of shell formation to bound the applicability of the drying models. These two limiting cases of infinitely slow or infinitely fast droplet component diffusion may be used to formulate scalar (0D) evaporation models [5,6]. In practical spray drying situations, both limits can be attained depending on the binary mass diffusion coefficients, atomization characteristics, temperature profile or drying gas flow velocity, to name a few parameters. These evaporation models are particularly useful when coupled to a numerical solver to predict stability and phase separation of amorphous solid dispersions [7,8] or a numerical solver of the Navier-Stokes equations, commonly denoted as computer fluid dynamics (CFD) [9,10].

We extend the previously developed platform for phase separation prediction of drug-polymer-solvent systems to systems with binary solvent blends in an effort to generalize a screening methodology for ASDs and compare the data against experimental results.

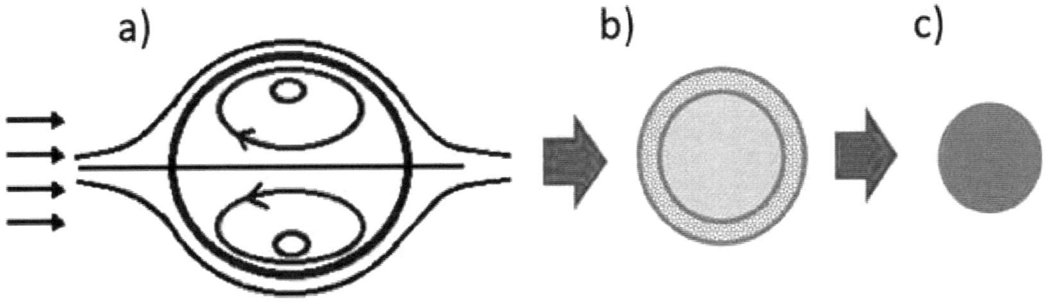

Figure 1: Sketch of various stages throughout the droplet drying history: a) droplet with internal circulation caused by the relative velocity between the droplet and the surrounding gas, b) shell formation and c) the final particle.

References

[1] Perrie Y., Rades T. "Pharmaceutics - Drug Delivery and Targeting", Pharmaceutical Press, 2010.

[2] Wulsten, E., Kiekens, F, van Dycke, F., Voorspoels, J. & Lee, G., "Levitated single-droplet drying: Case study with itraconazole dried in binary organic solvent mixtures", Int. J. Pharmaceutics, 2009, 378(1-2) pp. 116-121

[3] Janssens, S, Anné, M., Rombaut, P. & Van den Mooter, G., "Spray drying from complex solvent systems broadens the applicability of Kollicoat IR as a carrier in the formulation of solid dispersions", Eur. J. Pharm. Sci. 2009, 237, pp. 241-248.

[4] Paudel, A. & Van den Mooter, G., "Influence of Solvent Composition on the Miscibility and Physical Stability of Naproxen/PVP K 25 Solid Dispersions Prepared by Cosolvent Spray-Drying", Pharm. Res. 2012, 29, pp. 251-270.

[5] Sirignano, W. A. "Fluid dynamics and transport of droplets and sprays", Cambridge University Press (2nd Ed.), 2010.

[6] Law, C. K. "Recent advances in droplet vaporization and combustion", Prog. Energy Combust. Sci, 1982, 8 pp. 171 – 201, 1982.

[7] Duarte, Í., Santos, J. L, Pinto, J. G & Temtem, M., "Screening methodologies for the development of spray-dried amorphous solid dispersions", Pharm. Res., 2015, 32 pp. 222 – 237, 2015.

[8] Prudic, A., Ji, Y. & Sadowski, G., "Thermodynamic Phase Behavior of API/Polymer Solid Dispersions", Mol. Pharmaceutics, 2014, 11(7), pp. 2294-2304

[9] Sazhin, S.S et al., "A simplified model for bi-component droplet heating and evaporation", Int. J. Heat Mass Transfer, 2010, 21 pp. 4495-4505

[10] Maqua, C., Castanet, G. & Lemoine, F., "Bicomponent droplets evaporation: Temperature measurements and modelling", Fuel 2008, 87(13-14), pp. 2932-2942.

455382 Impact of Tooling Design during Development and Risk Evaluation of Tri-Layer Tablet Formulations

Abstract:

Multi-layer tablet is one of the formulation strategies utilized to mitigate chemical and/or physical incompatibility between different drug substances when delivering fixed dose combination drugs. Compressing multi-layer tablets has significant challenges because the interface(s) between neighboring layers are subject to defects such as delamination. It is therefore important to understand the factors that contribute to the strength of the interfaces. During early stages of drug development, feasibility studies are often conducted by making multi-layer compacts with round flat-faced tooling as the exact size and shape for the commercial tablets may not be finalized at that point. However, the link between different tooling designs and multi-layer tablet performance is not well established. This study examines the major engineering considerations when using mini-piloting tools to gauge the risk of interfacial defects in a tri-layer tablet. The first layer (layer A) of the tri-layer tablet was a wet granulation formulation of 56% drug load. The middle layer (layer M) was the inert layer consisting of a placebo blend. The third layer (layer B) was a dry granulation formulation of 33% drug load. The impact of tablet weight and dimensions (diameter and thickness) were evaluated to gain understanding during scale up/down as needed in early feasibility assessment and in troubleshooting. The factors in tooling selection, including tablet shape (round vs. oval), cup depth (0 vs. 0.054â€), and embossing (small vs. large), were evaluated to gain insight on the impact of tooling design on the interface strength of the tri-layer tablet. Tablets were compressed under different main compression forces while keeping the first and second layer tamping forces constant. Axial test was performed to quantify the interface strength under tensile stress. Â The thickness, diameter, and aspect ratio of the tablet were found to have a profound impact on the interface strength at a given compression pressure and on the compression pressure under which the interface start to fail, causing tablet delamination. These were the results of combined factors of the compressibility of the individual layers, the force transmission through the layers during compression (effect of tablet thickness), and the ejection force (effect of aspect ratio). The presence of a 0.054â€ cup depth was found to lower the interface strength to less than 50% of that of a flat

tablet. This cup depth effect is consistent with the observations made in the monolithic tablet regime and can be explained by the non-uniform density distribution caused by the curvature of the tablet interface. The oval shaped tablets were found to reduce the interface strength by 30 N or more compared to the round shaped tablets, suggesting that compacts made with round flat faced toolings may provide false positive results when used to assess the interface risks of a multi-layer tablet. Large embossing on the tooling appeared to help enhancing the interface strength, probably by providing more surface areas for the neighboring layers to interlock. This study elucidates the effects of scale-related factors like tablet weight and dimensions on the interface strength of a tri-layer tablet. These effects, not previously well studied or even realized during development, can significantly impact the feasibility evaluation of a multi-layer tablet. It also highlights the importance of tooling design selection on the performance of a tri-layer tablet that must be accounted for when using minipiloting tools for accelerated development of fixed dose combination products.

471306 End-Point Determination and Scale-up/Scale-Down of High Shear Wet Granulation Process Using an in-Line Di-Electric Probe

Abstract:

The main goal of this study is to demonstrate anovel in-line process analytical technology (PAT) for detecting end-point of granulation in a high shear wet granulation (HSWG) process. This technique enabled us to monitor liquid distribution and granule densification across different scales. While many reports in the literature have focused on indirect techniques such as power consumption or torque (which are scale and geometry dependent), in this method a physical property (dielectric permittivity of medium) dependent on the liquid amount and granule porosity (or density) is monitored directly during the course of granulation.

A di-electric probe was modified for recording complex dielectric permittivity of wet mass. The output of the probe was monitored and recorded during liquid addition and wet massing, and samples were taken at corresponding time intervals to correlate the final property of granules to the instrument output. Dependence of the real part of the effective dielectric permittivity on solid fraction of moist powder was suggested by Louge et. al. (1997) for non-spherical particles (snow). We adopted this model for high shear wet granulation process and monitored (in-line)Â density of wet mass during granulation.

Use of this in-line characterization tool enabled us to successfully transfer a 50% drug load API to a 600 L top-driven granulator without the need for any development batch by monitoring granulation end-point based on physical properties of wet mass. Specifically, we could understand the impact of changing process parameters (impeller speed, spray time and wet massing) and API properties (particle size /density) on final granule attributes and tablet CQAâ€™s which was consistent during scale-up/down to geometrically similar or non-similar granulators.

M.Y. Louge, R.Â Steiner, S.C. Keast, R. Decker, J. Dent, and M. Schneebeli, "Application of capacitance

instrumentation to the measurement of density and velocity of flowing snow", *Cold Regions Sci. & Tech.*, 25, 47-63 (1997).

465003 Selection Support Methodology Between Batch and Continuous Technologies for Pharmaceutical Tablet Manufacturing Processes

Abstract:

In solid drug product manufacturing, continuous technology is being developed as an alternative to the conventional batch technology. There are many differences between these two technologies, which may affect critical objectives such as cost-effectiveness, quality, process safety or occupational health. However, general tools are still missing, which can support the choice between batch and continuous technologies, and furthermore, process design of solid drug product manufacturing.

This work presents the selection support methodology between batch and continuous technologies for pharmaceutical tablet manufacturing processes. The methodology consists of four steps and includes iteration: (I) process modeling, (II) evaluation, (III) sensitivity analysis, and (IV) interpretation. The initial input of step I is a design situation, such as a launch of a new product to be manufactured in either technology that needs to be decided. In step I, the processes are modeled with covering the main processes such as weighing, granulation, blending, compression, and coating, and supporting processes such as testing, disposal, cleaning, maintenance of process analytical technology, and heat ventilation and air conditioning. Models are created so that the characteristics of two technologies in the actual manufacturing situations are reflected, such as numbers of operators, product yield or required area for manufacturing. Using these models, evaluation is conducted in step II, which produces results of the objective functions such as net present cost (NPC) after the launch. In step III, sensitivity analysis is conducted in order to quantify the effects of varying input parameters in the evaluation results. Finally in step IV, the results of evaluation and sensitivity analysis are interpreted, and the necessity of iterating the previous steps is investigated. The final outcome of the methodology is the appropriate technology to be chosen, i.e., batch or continuous, for manufacturing, for the tablet manufacturing.

A case study was performed for demonstrating the methodology. The design situation was the launch of a new product for which the estimation is available regarding the API price as well as the maximum demand quantity over the production period of 30 years. It was also assumed that the tablet product can be produced by either technology which needs to be specified based on the economic evaluation. After performing the steps I and II, the quantitative relation between demand quantity/API price and NPC over 30 years was clarified. The NPC did not include the investment cost, since the plant was assumed to have both technologies already. The result showed that continuous manufacturing was more lucrative in the areas of high demand quantity with low API price or low demand quantity with high API price. On contrary, batch manufacturing showed lower NPC in the areas where both API price and demand quantities are both high. Step III identified the number of campaign production in batch manufacturing as one of the most sensitive parameters in the model, for which the step II was iterated. In step IV, an economically appropriate technology was suggested based on the interpretation of the evaluation and sensitivity analysis results.

467693 Development of a Continuous Production Line for Solid Dosage Forms Via Dry Granulation

Abstract:

During the last years, academia, industry and regulators have increasingly demonstrated and addressed the advantages of continuous manufacturing. More and more, this is leading to a shift in paradigm for pharmaceutical manufacturing, towards advanced continuous processing based on real-time quality assurance and away from batch processing. As a result, turn-key solutions of continuous manufacturing systems are provided by various equipment suppliers. The available equipment is ranging from raw material feeding, to blending, granulation as well as tableting and coating. Even though several systems have been studied and discussed in the literature there are still some remaining questions open. Now with the first continuous plants being implemented for continuous production of solid oral dosage forms, interest is shifting towards system robustness, integration of unit operations, in-line measurement and process control. Such â€advancedâ€ turn-key solutions are required to fulfill a wide variety of demands, when processing different raw materials with varying properties.

In this talk we will highlight a control strategy that was validated by system tests of the integrated plant. The objective of the control strategy was to consider all possible process settings and raw material properties, thereby guaranteeing constant product quality. The described manufacturing line is dedicated to the continuous and automated production of solid-oral-dosage forms via a dry granulation route. In order to enable continuous process control and real-time-release, several PAT instruments were implemented. Proper control actions are based on actual process information from the in-line monitoring systems. Process parameters were screened for utilization as soft sensors, to enable even better process understanding and control. Out-of-spec material is detected and removed from the process at different stages. The control strategy enables continuous production even if out-of-spec material has to be removed. By setting the proper process parameters the mass flow can be modified, while still producing in-spec products. A concept will be presented, which allows variation of the mass flow of single unit operations to keep the entire production line within specified limits.

464713 Considerations in Developing a Control Strategy for Continuous Direct Compression

Abstract:

In this work a continuous direct compression (CDC) process was investigated for a low dose drug product for an API with cohesive material properties. The CDC process was developed on the GEA CDC-50 system. A risk based approach was followed during the entire process development of CDC. The process parameters for feeding and blending were selected such that the feeding variability is sufficiently filtered by the available residence time distribution (RTD) of the blending system. The holistic control strategy for CDC involved specifications for input materials, in-process controls for various drug product attributes including blend concentration, tablet weight and hardness, and a product release strategy. The blend concentration was monitored using an in-line NIR method and a RTD model of the CDC line as two redundant methods. In the event when the blend concentration falls outside the given specification limits, non-conforming tablets were diverted. For using the RTD model as a process monitoring method, sensitivity of RTD model to material variability and method of feed rate measurement were identified as the main risks. In this presentation, the development studies conducted to address these risks will be discussed. In addition the pros and cons of using each process monitoring method as well as considerations in designing a diversion system will be discussed.

463948 Applying Supervisory Process Control Practices to Continuous Pharmaceutical Manufacturing

Abstract:

Successful supervisory process control implementations in the pharmaceutical industry have been hampered due to the batchwise nature of most process operations. In fact, batch processes present some specific challenges because the main steps operate discontinuously. Consequently, temperatures, concentrations and other properties vary in time which requires complex non-linear and/or time-varying models and control strategies. In contrast, continuous manufacturing is one step closer to the "holy grail" for process control engineers, i.e. a linear time invariant (LTI) system. The availability of common practices and numerous examples for such a LTI system representation significantly reduce the complexity of process identification and control implementation. Hence, dynamic process operation with feedback and automatic manipulation of quality attributes is proposed as the way forward for continuous manufacturing [1]. A technical framework for this approach is already available in continuous oil refining, and in the chemical and food industries.

In order for the pharmaceutical industry to embrace fully continuous manufacturing with supervisory process control, numerous challenges still exist. This said, the current work tries to face them and apply existing classical and modern continuous control laws to manipulate quality attributes of a commercial continuous wet granulation unit within a continuous from-powder-to-tablet installation, being the ConsiGma^{TM}-25. The idea is to have stable automatic control of species concentration in the granules formed during twin-screw wet granulation. In this case the water content is considered, as it is directly related to both chemical stability and physical characteristics of the granules.

A first encounter in this effort was the inline estimation of moisture concentration from the particulate stream using the possibilities offered by high-speed NIR measurements. This includes process-instrument interfacing, industrial data communication, time-alignment of instrumentation and process data, latent variable modeling and signal filtering. These steps are typically highly customised to the situation, but some general comments will be presented. In addition, control of a process requires some form of model, and here a black-box mathematical model, of the type auto-regressive with moving average and exogenous inputs (ARMAX), was identified to predict the fast dynamic behaviour from system actuator to the actual measurement. This allowed us to tune, simulate and validate empirical (proportional-integral) as well as model-based (extended prediction self-adaptive) controllers for the granule moisture concentration loop under study. Another key point that will be addressed is that real-time calculations, here for analysis and control, require computationally inexpensive methods to keep up with the dynamics of the system under study.

[1] Myerson, A. S., Krumme, M., Nasr, M., Thomas, H. and Braatz, R. D. (2015), Control Systems Engineering in Continuous Pharmaceutical Manufacturing. May 20–21, 2014 Continuous Manufacturing Symposium. J. Pharm. Sci., 104: 832–839. doi: 10.1002/jps.24311

456403 Control of a System of Loss-in-Weight Feeders in a Continuous Manufacturing Drug Product Application

Abstract:

Dispensing materials using loss-in-weight feeders is a critical unit operation in most continuous drug product manufacturing processes. As feeding is typically one of the furthest upstream operations, its output has a large influence on the rest of the process. Through experience continuously processing several pharmaceutical formulations, several common feeding problems have been identified including high variability feeding, drifting from set point, and transient disturbances. This work discusses the control of a system of feeders in addition to individual feeder control, showing examples of how a system level ratio control loop can be used to mitigate these common feeding issues and provide consistent output to the downstream process. In addition, identifying special cause transient disturbances from the feeding operation using different types of feeder data will be discussed.

461837 Dynamic Optimization of Continuous Manufacturing of Pharmaceuticals

Abstract:

Continuous manufacturing (CM) of pharmaceuticals is being explored as an alternative to traditional batch-wise production. This shift holds great promise to increase production efficiency, enable smaller production facilities, minimize waste, energy consumption, and raw material use and to enable drug quality monitoring on a continuous basis. However, implementing CM in the pharmaceuticals industry implies short operational campaigns, with significant transient phases (start-up and shutdown), constituting up to 30% of the entire time horizon. This significantly hampers the plant's economic feasibility. Efforts to optimize the overall production process involve mathematical modeling and solution of hybrid (discrete-continuous) systems embedding differential-algebraic equations (DAEs), along with the associated parametric sensitivity trajectories to be used in rigorous gradient-based optimization algorithms.

In this contribution, these efforts will be described, addressing the numerical issues related to the hybrid and non-smooth nature of the mathematical model. A case-study pilot plant model incorporating several synthesis, separation and purification steps to produce final tablets [1] is used for demonstration. The dynamic optimization approach that is employed maximizes the accumulated on-spec production directly over the entire time horizon (so-called Economic-Optimization), while guaranteeing differentiability with respect to the controls [2]. Optimization results will be presented followed by a discussion on the process design and formulation and the intelligent choice of the open-loop control decision variables.

[1] S. Mascia et al., "End-to-End Continuous Manufacturing of Pharmaceuticals: Integrated Synthesis, Purification, and Final Dosage Formation", *Angew* Chem. Int. Ed. Engl., **2013**, 52(47), 12359-63

[2] A. Sahlodin and P.I. Barton, "Optimal Campaign Continuous Manufacturing", *Ind. Eng. Chem. Res.*, **2015**, 54(45), 11344-59.

468095 Extended Iterative Optimzation Technology for Mixture Composition Estimation in Moving Powders

Abstract:

The calibration and maintenance effort for a near infra-red sensor and calibration based solution can be a time, labor and material-intensive exercise. A calibration-free method (iterative optimization technique or IOT) was recently published and applied onto mixture component prediction.[1]. IOT estimates fractions of individual components via optimization by rigorously imposing Beer-Lambert's law and mixture constraints to the measured spectra, given the known spectrum for the pure materials. Advances in IOT theory propose the use of massive blind-search evolutionary algorithms to select spectral regions that can improve the performance of IOT in a real-time scenario.

The application of IOT onto powder flows is severely affected by scattering and uneven sample presentation (among others) which results in large deviations from Beer-Lambert in the mixture spectra. Other application-specific artifacts (like peak overlap) can also influence the performance of IOT, resulting in predictions with trends and biases that can range from mild, to completely unreasonable and inaccurate. This work presents an extension to IOT that preserves the calibration-free nature of the method while improving its real-time performance in the estimation of fractions of individual components in the powder stream of a continuous drug product manufacturing line.

Although still calibration free, EIOT requires at the very least measured spectra at a minimum of two levels of concentration for the active ingredient – this data set is readily available if a step change study is done in the line to determine mixing behavior of the system. The power of the method resides in the spectral differentiation across materials, as such, a data set consisting of spectral measurements at several levels of the individual components is much preferred [2].

EIOT fundamentally challenges the application of Beer-Lambert in relating the fractions of each material through the mixture spectra in the process and pure spectra collected in a static sample. As such, EIOT starts with the estimation of the dynamic pure component spectra that is mathematically necessary such that, the concentrations and the dynamic mixture spectra in the "training set" is explained by Beer's law. Multiple methods are presented to produce this estimate. It has been observed that the dynamic pure spectra estimated differ only in small features from the measured pure spectra. Once these pure spectra are estimated, the procedure continues as IOT would. These differences, although small play an important role in the performance of the method. Several examples are presented, contrasting the performance of a PLS model, IOT and EIOT in estimating the fractions of ingredients from spectra measured with a NIR instrument installed in the feed frame of a tablet press, at the end of a continuous drug product manufacturing line. The trends produced by EIOT lack undesired trends that are attributed to process effects onto the spectra, and represent a useful metric to monitor the real-time fraction of the materials in the feed-frame. Such an estimate can be used in combination with a deterministic model of the system to confirm mixture composition and confirm the state of control of the process.

[1] Muteki, K., Blackwood, D.O., Maranzano, B., Zhou, Y., Liu, Y.A., Leeman, K.R.and Reid, G.L. Industrial & Engineering Chemistry Research, 52, 12258 (2013).

[2] Eriksson, L.; Johansson, E.; Wikstrom, C. Mixture design--design generation, PLS analysis, and model usage. Chemometrics and Intelligent Laboratory Systems, 43, 1-2, 1-24. (**1998)**

458141 Enhancing Process Understanding and Control with Real-Time Analytics for a Challenging Solid Form System of a Pharmaceutical Compound

Abstract:

In the recent years, process analytical technology (PAT) has been widely used on the pharmaceutical development and manufacturing for better process understanding and control. For example, the information of the in-situ particle size distribution by Particle Track (Lasentec ® FBRM) is invaluable on the crystallization development and scale-up. This paper discussed a case of solid state development of a drug candidate that has high tendency to form solvates in most process solvent systems. The only feasible pathway to produce the desired non-solvated form was through a slurry conversion after adjusting the solvent composition through solvent swap distillation. Multiple PAT tools have been evaluated for process understanding and in-line monitoring of a form conversion process, for example, 1) the use of PVM to monitor the conversion of morphology, 2) the use of the FTIR to monitor the progress of de-solvation 3) the use of the combination of Particle Track and Raman for solid form control.

471843 Simultaneous Fast Measurements of Tablet Porosity/Moisture: Real-Time Release Testing for QC Dissolution Prediction

Abstract:

The motivation for this work comes from the dissolution/disintegration sensitivity of a solid dispersion formulation with an erosion based mechanism of dissolution to compression force (which affects tablet solid fraction and hardness), moisture, and the interaction between the two. Solid dispersions were processed by spray drying and subsequent roller compaction which can generate granules with a wide range of quality attributes that impact compactability and compressibility of granules during compression and can be a challenge for obtaining similarities in dissolution rate for proven acceptable range of tablet hardness. We found that for this case dissolution strongly correlates to disintegration time and can be controlled by tablet moisture and density (solid fraction or porosity), but these properties as well as the disintegration test traditionally require separate off-line long measurements.
A novel at-line process analytical technology (PAT), di-electric probe, was developed for simultaneous measurement of solid fraction and moisture content for the tablets. The basic measurement principle of the probe is to estimate the physical properties of bulk solids (such as solid fraction and moisture) by standard correlation with the measured electrical properties (i.e. dielectric permittivity). It was demonstrated that while both real and imaginary component of di-electric permittivity depend on moisture and solid fraction, the phase lag (ratio of imaginary to real part of di-electric permittivity) and di-electric constant of the active formulation depend only on moisture content or relative humidity (RH%). A statistical model was then developed to show dependence of dissolution rate and disintegration time to solid fraction/moisture measured by di-electric probe.

467706 Real Time Prediction of Tablet Tensile Strength By Ultrasound in Continuous Manufacturing

Abstract:

Tablet tensile strength is a critical property that needs to be well controlled in order to produce an acceptable pharmaceutical product. Methods that allow real time and non-destructive prediction of tablet properties can greatly help the development efforts in a continuous manufacturing line. Real-time measurement of tablet strength by ultrasound increases assurance of product quality and is economically beneficial in continuous manufacturing. A model for predicting tablet strength by ultrasound will enable rapid techniques in continuous assessment of tablets ensuring the desired quality attributes.

The objective of this research project was to develop a statistical model to predict tablet tensile strength in a direct compaction continuous manufacturing line. Tablets consisted of model API, Acetaminophen (APAP), lactose monohydrate as the filler, and magnesium stearate as lubricant. Multivariate effects were analyzed using Quality-by-Design methods. The experiment was based on a fractional factorial design with 30 experiments, including 3 repeated center points, to study the influence of four process variables â€" drug concentration (%), compaction force (KN), blending speed (rpm), and feed frame speed (rpm) â€" on the final product. Ultrasound transmission speed was determined by measuring the sound wave time-of-flight (TOF) using transducers with a frequency of 2.25 MHz. The speed of sound (SOS) was calculated by dividing the measured tablet thickness by the TOF. The same tablets were characterized for diametrical crushing strength using a mechanical hardness tester. The ANOVA results indicated that compaction force was the main parameter affecting tensile strength and SOS of the tablets in this process. However, ultrasound was sensitive to drug concentration and feed frame speed. Therefore, a predictive model was developed to correlate tensile strength with SOS and relative density. The model was verified using an independently manufactured validation set. The predicted values obtained from the model were in agreement with the experimental results with a correlation coefficient of 0.92. It was further concluded that by using the ultrasound measurement technique it is possible to assess the quality characteristics of tablets non-destructively and in real time.

472375 Toward Predicting Tensile Strength and Characterizing the Acoustic Properties of Pharmaceutical Tablets on Continuous Manufacturing Platforms

Abstract:

Purpose. An ultrasound measurement system was employed as a non-destructive method to evaluate its reliability in predicting the tensile strength of tablets and investigate the benefits of incorporating it in a continuous line, manufacturing solid dosage forms.

Methods. Cylindrical tablets were prepared either continuously or in batch. The formulation was kept constant, while the compaction force and level of shear strain varied. Ultrasound testing was used to evaluate the strength of tablets by measuring the time of flight. The tensile strength of the same tablets was then determined using a mechanical hardness tester.

Results. The effect of two processing parameters, compaction force and level of shear strain were examined. Ultrasound speed of sound was found to be sensitive to the relative density and the level of shear strain. Ultrasound testing could detect even small differences between tablets that a hardness tester failed to do so.

A strategy for hardness prediction is proposed that uses the existing models for Youngâ€™s modulus and tensile strength of porous materials. A clear correlation between Youngâ€™s modulus and tensile strength at zero porosity is presented.

Conclusions. Ultrasound testing is a good candidate to be placed on/at-line to measure the mechanical integrity of tablets non-destructively. These results provide information about the behavior of processing parameters on the

performance of tablets and the ability to engineer product properties.

466106 Overcoming the Challenges of Noisy Data during the Calibration of NIR As a PAT Technique to Measure Roller Compacted Ribbon Density

Abstract:

INTRODUCTION

A number of research groups have investigated the use of NIR spectroscopy for monitoring ribbon density [1-4]. These studies have taken advantage of the fact that NIR spectra are affected by the density of the material being analysed and offer the potential of measuring/monitoring roller compacted ribbon density in real time [1-3, 5]. In order to develop a calibration equation relating density to the NIR spectra Gupta [2] fitted a regression line through all the scanned wavelengths and used the fitted slope of the line. However, this approach does not take into account issues which arise when there are multiple scans on multiple samples of the same type of material, and each scan records simultaneous NIR spectra for several probes. The spectral dataset can contain erroneous outliers. Data processing in real time is challenging as sampling error outliers must be either filtered or cleaned from the data set before further analysis. [6]. Visual inspection of the data and removal of sampling error outliers though justifiable is not practical for real time data processing. Existing data cleaning and filtering models are complex and are essentially off-line operations [6]. A simple solution is required.

This study aims to highlight how variability in roller compacted ribbon quality can impact on NIR spectral measurement and proposes a simple method of data analysis to deal with this.

<>**Materials and Methods**

Roller compaction: A Fitzpatrick CCS220 roller compactor was used for this study. The roll speed was kept constant at 2 rpm, roll gap at 0.5 mm and the roll pressure was adjusted for each run to 2, 4, 6, 8, 10, 12, 14 and 16 kN. Blends of microcrystalline cellulose, lactose monohydrate and magnesium stearate were compacted. Each of the three blends differed in MCC % moisture content. MCC was stored at 43% for blend 1, blend 2 at 11% and blend 3 at 75% relative humidity. The variation in moisture was introduced to assess its impact on the ribbon quality and thus challenge the NIR technique. Ribbon envelope density was determined by the GeoPyc™ 1360.

NIR spectral collection: Four MultiEye™ NIR probes scanned samples at-line in the manner in which they were produced by the roller compactor i.e. samples were not cut to a specific size. This set up ensured outliers were detected in a similar manner to which they would be detected in-line when monitoring the process. The Perkin Elmer Spotlight 400 FT-IR was also used and the off-line results were compared to the at-line MultiEye™ data. Samples were cut to a specific size to facilitate off-line Perkin Elmer Spotlight analysis.

Statistical data analysis: Three different calibration methods were developed between spectral slopes and envelope density and compared; (1) using the entire data set unfiltered, (2) a visual discard method and finally (3) a 33% Trimmed Mean method.

RESULTS AND Discussion

Calibration of the NIR method to ribbon density was challenging due to the presence of spectral data from broken, split and curved ribbons in the data set particularly for blend 3 which had the highest moisture content (6.96 % w/w). All initial processing of the full data set indicated that some form of data cleaning was required in order to process the MultiEye™ at-line data set. The visual discard method was applied to the full data set and was found to be particularly successful for blends 1 and 2 (**Figure 1**). After using the visual discard data cleaning method MulitEye™ data the calibration correlation was found to be as good as the off-line higher scan resolution Perkin Elmer spectrometer. Blend 1 off line r=0.92, visual discard r=0.95.

However it is necessary to replace the visual method of spectra cleaning with a non-subjective method which was capable of screening for these erroneous probe readings. The trimmed mean eliminates a specified percentage of the data from both the high and the low ends of the dataset before evaluating the standard mean of the remaining data. For this data set the trimmed mean method sets a limit on how data is cleaned from the data set allowing for the removal of a faulty probe reading (25% of data) or a poor sample (33% of data). The 33% Trim Mean in place of the Full Mean reduced the impact of spectral variation or misreads between samples or probes. The Trim Mean method was found to be at least as successful as the Visual Discard method at cleaning the data set prior to development of the calibration equation (**Table 1 and Figure 2**). The Trim Mean method was preferred as the decision on data to include in the calibration model as it is not subjective and simple to apply. Major variation is apparent in the spectral data for Blend 3 (75%) which renders it unsuitable for prediction and calibration. These processing variations were attributes to the high moisture of MCC in this compacted blend.

CONCLUSIONS

The 33% Trim Mean method offers a simple and practical solution to dealing with the NIR spectral challenges when applying this PAT technique to roller compacted ribbon in-line/at-line. Further development of this method when used for calibration could investigate the optimal number of ribbon samples to % data trimmed to achieve a statically robust balance of data cleaning. Increasing the number of samples used to during calibration may allow for a greater retention of data in the TrimMean e.g. 5 samples and keeping 3 giving a 20% Trim Mean. The Trim Mean method is therefore no less successful then other simple calibration methods but offers the advantage of a simple, easy to follow and non-biased means of removing NIR data collected due to probe sampling error at-line/in-line.

ACKNOWLEDGMENTS

Research is funded by the Synthesis and Solid State Pharmaceutical Centre (SSPC) under grant number 12/RC/2275, the Mathematics Applications Consortium for Science and Industry funded by Science Foundation Ireland Investigator Award 12/IA/1683 and the Pharmaceutical Manufacturing Technology Centre, Ireland. Microcrystalline cellulose PH102 donated by FMC Corporation. Innopharma Laboratories for use of the MultiEye™ NIR Spectrometer.

REFERENCES

1. Acevedo, D., et al., *Evaluation of Three Approaches for Real-Time Monitoring of Roller Compaction with Near-Infrared Spectroscopy.* AAPS PharmSciTech, 2012. **13**(3): p. 1005-1012.

2. Gupta, A., et al., *Nondestructive measurements of the compact strength and the particle-size distribution after milling of roller compacted powders by near-infrared spectroscopy.* J. Pharm. Sci., 2004. **93**(4): p. 1047-1053.

3. Gupta, A., et al., *Influence of ambient moisture on the compaction behavior of microcrystalline cellulose powder undergoing uni-axial compression and roller-compaction: A comparative study using near-infrared spectroscopy.* J. Pharm. Sci., 2005. **94**(10): p. 2301-2313.

4. Lim, H., et al., *Assessment of the critical factors affecting the porosity of roller compacted ribbons and the feasibility of using NIR chemical imaging to evaluate the porosity distribution.* Int. J. Pharm., 2011. **410**(1-2): p. 1-8.

5. McAuliffe, M., et al., *The Use of PAT and Off-line Methods for Monitoring of Roller Compacted Ribbon and Granule Properties with a View to Continuous Processing.* Organic Process Research & Development, 2014. **19**(1): p. 158-166.

6. Liu, H., S. Shah, and W. Jiang, *Online outlier detection and data cleaning.* Comput. Chem. Eng., 2004. **28**(9): p. 1635-1647.

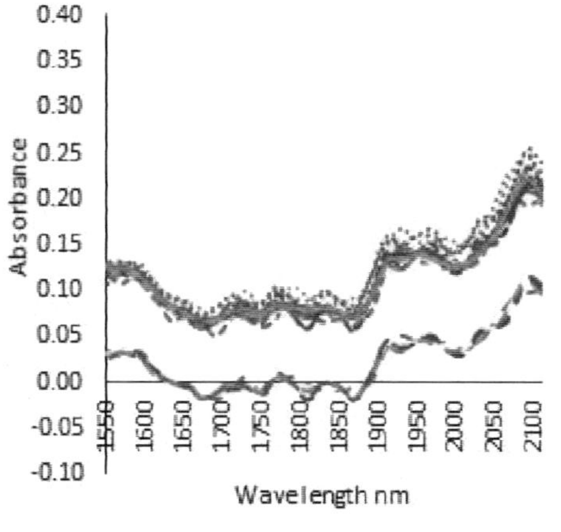

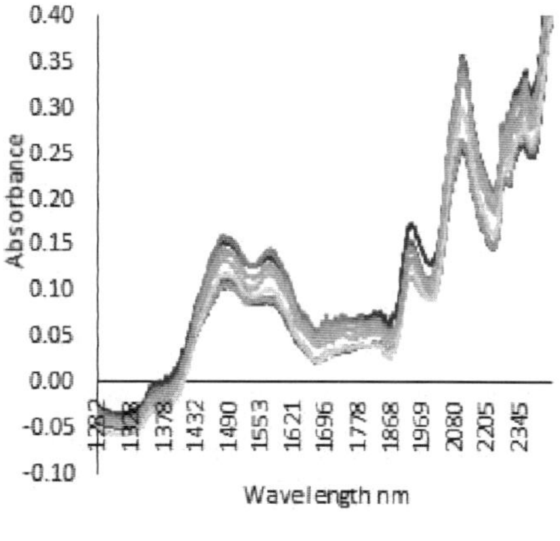

Figure 1 Left panel - Representative spectral data for one piece of ribbon as scanned by the MultiEye™ NIR probes at-line. Probe 1 (continuous line), Probe 2 (dashed line), Probe 3 (dotted line) and Probe 4 (dashed line in lower absorbance region), each of the 5 scans for each probe is seen as a different color i.e. red, yellow, blue, green and purple. When applying the visual discard method to this spectral data set, probe 4 was removed from the data set. Right panel –Representative spectral data for ribbon samples from the same batch as measured by Perkin Elmer Spotlight off-line.

Table 1. Calibration models and the respective correlation coefficients (r) for each of the three blends as calculated by the three different data discard methods. ED indicates envelope density.

Calibration Method	Blend	Calibration model	r
Trim Mean	Blend 1	ED=0.416 + 3584 x Average Spectral Slope	0.96
Visual Discard	Blend 1	ED=0.409 + 3628 x Average Spectral Slope	0.95
Full Mean	Blend 1	ED=0.420 + 3788 x Average Spectral Slope	0.93
Trim Mean	Blend 2	ED=0.319 + 3788 x Average Spectral Slope	0.97
Visual Discard	Blend 2	ED=0.170 + 4257 x Average Spectral Slope	0.88
Full Mean	Blend 2	ED=-0.041 + 5587 x Average Spectral Slope	0.89
Trim Mean	Blend 3	ED=0.465 + 3021 x Average Spectral Slope	0.74
Visual Discard	Blend 3	ED=0.734 + 1515 x Average Spectral Slope	0.92
Full Mean	Blend 3	ED=0.474 + 3115 x Average Spectral Slope	0.74

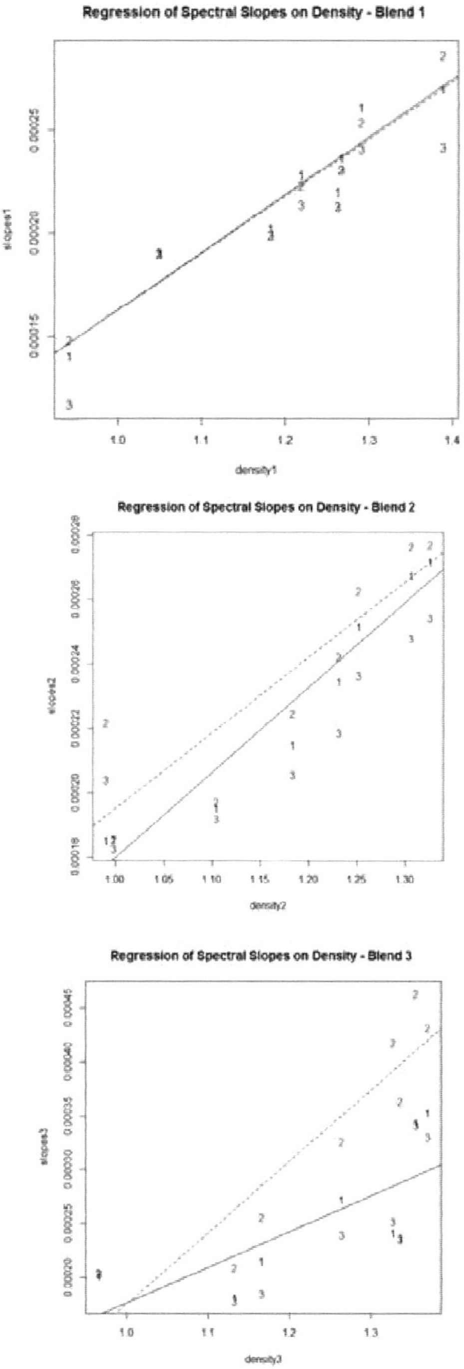

Figure 2. Regression of NIR spectral slopes on Envelope Density. Note: symbols 1=Trim Mean samples, 2= Visual Discard samples and 3 = Full Mean samples. Solid line is fitted line for Trim Mean samples, dotted line is for Visual Discard samples.

457254 PAT Based Batch Statistical Process Control for Pharmaceutical SEMI-SOLID and Liquid Manufacturing Processes

Abstract:

PAT BASED BATCH STATISTICAL PROCESS CONTROL FOR PHARMACEUTICAL SEMI-SOLID AND LIQUID MANUFACTURING PROCESSES

N. Bostijn[1,*], C. Vervaet[2], J.P. Remon[2], T. De Beer[1]

[1]Laboratory of Pharmaceutical Process Analytical Technology, Ghent University, Ottergemsesteenweg 460, 9000 Ghent, Belgium

[2]Laboratory of Pharmaceutical Technology, Ghent University, Ottergemsesteenweg 460, 9000 Ghent, Belgium

Introduction

Pharmaceutical semi-solids and liquids are produced in a batch-wise manner with off-line time-consuming and less efficient quality control (1). The aim of this study is to simulate the batch process on a laboratory scale of five commercial available semi-solid and liquid formulations and to control the critical quality attributes of the process by monitoring the chord length distribution (CLD) of the particles, viscosity, temperature and active pharmaceutical ingredient (API) concentration/homogeneity in-line. For each formulation, reference batches were used to develop a reference batch partial least squares (PLS) regression model. From this model, score control charts were created describing the average batch trajectory. These control charts were used to monitor test batches and to detect any deviations from the expected batch trajectory. This approach allows early fault detection and real-time process adjustments, hence avoiding batch loss.

MATERIALS AND METHODS

Materials

For this study, five commercial available pharmaceutical formulations were selected. Two liquids (shampoo and suspension) and three semi-solids (ointment, cream and gel).

Methods

All experiments were performed using a customized IKA LR2000 modular mini plant reactor system (IKA, Staufen, Germany). The mixing vessel was equipped with a double wall for controlling the temperature of the process. A peristaltic pump (Watson-Marlow, Falmouth, Cornwall, UK) was used for pumping water from a water bath (Type 1032, GFL, Burgwedel, Germany) to the mixing vessel. In-line Raman spectra were recorded using a Raman Rxn1 spectrometer (Kaiser Optical Systems, Ann Arbor, MI, USA), equipped with an air-cooled CCD detector and a fiber-optic immersion probe. The laser wavelength was the 785 nm line from an Invictus NIR diode laser. Every 30 s a spectrum was acquired in the 0–1900 cm^{-1} region with an exposure time of 15 s and a resolution of 4 cm^{-1}.

There were also Raman spectra recorded using a Raman Rxn2 spectrometer (Kaiser Optical Systems, Ann Arbor, MI, USA), equipped with a fiber-optic PhAT probe. The laser wavelength was 785 nm. Spectra were collected every 30 s with an exposure time of 15 s and a resolution of 5 cm^{-1}. The spectral range of the system was from 150–1890 cm^{-1}. Viscosity was measured using an in-line viscometer (Sofraser, Villemandeur, France), equipped with an internal temperature probe. The viscometer had a range from 0.1–1000 mPa.s. Each second the viscosity and temperature were recorded. A focused beam reflectance measurement (FBRM) probe (model C35, Mettler-Toledo, Columbus (Ohio), USA) was used for monitoring the chord length distribution of particles in the system. The probe window was cleaned every 10 s by a pressurized air activated scraper and each 2 s the averaged chord length distribution was saved. The batch model was developed and evaluated using the SIMCA software (Version 14, Umetrics, Umeå, Sweden).

RESULTS AND DISCUSSION

To investigate whether the reference batch models were able to distinguish good from bad batches, process and formulation deviations were induced in the test batches. Only the results of the ointment will be discussed in this abstract.

For test batch 1 the API was added in three different steps. For all the reference batches, API was added in one step after ± 1200 s. In this test batch, 50 % (w/w) of the API was added after 1200 s, 45 % (w/w) after 2520 s and 5 % (w/w) after 3780 s. This induced error resulted in a decrease of the PLS component 1 (PLSC 1) scores in the middle part of the batch evolution scores plot. To monitor and evaluate the overall performance of new batches, a Hotelling's T2 plot can be calculated. When detecting deviations in the Hotelling's T2 plot, it is possible to identify the variables responsible for the deviations via contribution plots. The contribution plots at 1800 s (50 % API) and 3030 s (95 % API), showed that the deviation was determined by a lower viscosity and API concentration. After addition of all the API, the viscosity and API concentration recovered to the average values of the reference model.

In test batch 2 the set point of the water bath was set 10 °C higher compared to the reference batches and corrected after 2580 s. The PLSC 2 scores fell out of the control limits but evolved towards the average batch trajectory at the end of the process when the water bath temperature reached it correct value. At 1800 s the temperature was higher and the viscosity lower compared to the average observations, as seen from the contribution plot.

CONCLUSION

The batch model, developed from four reference batches, was able to detect both the process and formulation induced deviations in the test batches by monitoring the temperature, viscosity, CLD and API concentration in-line.

CHALLENGES AND FUTURE WORK

A next step is to investigate the possibility to produce the same formulations with a continuous production system and to monitor the process using the same and/or new PAT tools.

REFERENCES

1. De Beer T et al. Near infrared and Raman spectroscopy for the in-process monitoring of pharmaceutical production processes. Int. J. of Pharm. 2011; 417: 32–47.

*** Formatting Issue - Best Available Paper/Graphic ***

472690 Real Time PAT Monitoring of Bioreactor Mab IgG3 Cell Culture Process Dynamics Via Fourier Transform Infrared Spectroscopy: Implications and Challenges

Abstract:

Compared to small molecule Process Analytical Technology (PAT) applications, biotechnology product PAT applications have certain unique challenges and opportunities. Understanding the process dynamics of bioreactor cell culture processes is essential to establish an appropriate control strategy for biotechnology product PAT applications. Inline spectroscopic techniques for real time monitoring of bioreactor cell culture processes are highly amendable to PAT approaches to monitor the manufacturing of biotechnology drug products. However, the use of inline Fourier Transform Infrared (FTIR) spectroscopic techniques for bioreactor cell culture process monitoring has not been reported. In this work, real time inline FTIR Spectroscopy is applied to a lab scale bioreactor mAb IgG3 cell culture fluid biomolecular dynamic model. The technical feasibility of using FTIR Spectroscopy for real time tracking and monitoring four key cell culture metabolites (including glucose, glutamine, lactate, and ammonia) and protein yield at increasing levels of complexity (simple binary system, fully formulated media, actual bioreactor cell culture process) is evaluated via a stepwise approach. The FTIR fingerprints of the key metabolites are identified. The multivariate Partial Least Squares (PLS) calibration models have been established to correlate the process FTIR spectra with the concentrations of key metabolites and protein yield of in-process samples, either individually for each metabolite and protein or globally for all four metabolites simultaneously. Applying the 2nd derivative pre-processing algorithm to the FTIR spectra helps to significantly reduce the number of PLS latent variables needed and thus simplify the interpretation of the PLS models. The validated PLS models show promise in predicting the concentration profiles of glucose, glutamine, lactate, and ammonia as well as protein yield over the course of the bioreactor cell culture process. Therefore, the technical feasibility of real time monitoring of the bioreactor cell culture process via FTIR spectroscopy has been demonstrated in this w

475688 Food, Pharmaceutical & Bioengineering Division Award: Control, Sensing, and Chemical Analysis using Microfluidic Devices

Abstract:

The potential uses of microfluidic devices in chemical processing and sensing is almost unlimited. Construction of these devices is currently relatively easy, and there are a large number of published "lab on a chip" systems constructed from a variety of substrates using different actuation, sensing, and control components. In our work, we have focused on components and integrated systems that can be used to measure physical, chemical, and biochemical properties. The applications for these devices have ranged from disease diagnosis to pneumatic computing, and the construction complexity for each device can be very different. For instance, one of our simplest devices is a Newtonian and non-Newtonian viscometer that uses a single microfabricated channel made from a single substrate material. In contrast, our most complex device requires multiple substrates, metal depositions, and ion implants to form a complete system that detects and subtypes the influenza virus. Also, for some of our devices we have needed to redefine the typical low volume and velocities normally seen in microfluidic systems: we have made chips to monitor gallons of fluid per minute or which have fluid velocities approaching a Mach number of 1. In my talk, I will describe some of these devices and discuss the successes and challenges we have encountered. In addition, I will also describe some new areas we are exploring such as reducing water use in food production and continuous flow monitoring of process fluids.

475689 Area 15A, Food - Plenary: Bioelectronic Nose: Integration of Biotechnology and Nanotechnology

Abstract:

Human have five senses. Among the five senses, three of them are physical senses, which are vision, hearing, and touch. The other two senses are chemical senses, which are smell and taste. The science and technology for the physical senses have been advanced enormously, and electronic devices have been developed for these senses, e.g. camera, audio recorder, and tablet PC. However, we do not have any device which can capture smell or taste. In order to mimic the function of human nose and tongue a novel concept for sensor devices functionalized with odor and tastant-recognizing biomolecules was suggested. They are the bioelectronic nose and tongue. A bioelectronic nose consists of primary and secondary transducers. The primary transducer is a biological recognition element such as olfactory receptors. The secondary transducer is a highly sensitive nanodevices that convert biological events into measurable signals. In this presentation, the basic concept and principles of bioelectronic nose will be presented. In addition, applications of bioelectronic nose and the current issues will be discussed.

475690 Area 15B/PD2M, Pharmaceuticals - Plenary: Human Multi-Organ-on-a-Chip Platform for Pharmaceutics

Abstract:

Drug discovery is a lengthy, risky, and expensive process hampered by systemic side effects. Thus, identification of potential liabilities of new molecular entities has become essential in safety assessment throughout drug discovery and development. However, there is no effective in vitro assay currently available for accurately detecting cytotoxicity at early stages of drug development, whereas animal models used at later stages are often ineffective at predicting human responses to candidate drugs. Therefore, there is an increasing demand for improved in vitro models that can better predict the physiological responses of the human body to novel pharmaceutical compounds, particularly with respect to tissue/organ toxicity. With the development of the microfluidic organs-on-a-chip systems, multiple physiological and biochemical parameters of the organ models representing their in vivo counterparts can now be simulated. However, while many drugs can create chronic cellular reactions or delayed cell responses, there is a serious unmet demand of analytical methods for long-term, continual, sensitive analysis of cell-secreted biomarkers from an organ model. In tackling these major challenges facing the pharmaceutical industry, we have innovated a multi-organ-on-a-chip platform integrated with a novel sensing system capable of continual monitoring of physicochemical parameters and soluble biomarkers to enable long-term analysis of drug toxicity in microfluidic organs-on-a-chip. This talk will focus on the development of such an integrated platform, from engineering biomimetic organoids using bioprinting/microtechnologies to sensor integration and automated system operations over extended periods of time.

475691 Area 15C, Bioengineering - Plenary: Status of Biocatalysis: Protein and Process Engineering

Abstract:

Biocatalysts are often highly active and extremely selective but are only slowly gaining general acceptance for catalyzing most applicable transformations. However, there has been significant progress: (*a*) novel protein engineering tools have significantly enhanced biocatalyst development; (*b*) The deactivation of biocatalysts under various stresses can be i) described quantitatively via rational models and ii) remedied via accumulating all stabilizing mutations found in earlier rounds; (c) A host of novel biocatalysts in the past few years has considerably broadened the toolbox for synthetic chemistry.

The presentation discusses both status of and perspectives for biocatalysis. Examples are drawn both from othersâ€™ and the presentersâ€™ own work, for applications ranging from transformations of renewable resources to highly selective synthesis in life science applications.

475692 Area 15D/E, Engineering Fundamentals in Life Science - Plenary: Cellular Bioengineering: Past, Present, and Future

Abstract:

Availability of cultured mammalian cells in the late 70â€™s provided *in vitro* models that enabled pursuit of novel scientific directions, promoted new trends in biomedical research, and initiated the field of â€œcellular bioengineeringâ€.

Recognition that the cellular/molecular events at the microenvironment of the tissue/biomaterial interface dictate the fate of implants *in vivo* resulted in use of cellular *in vitro* models in search of elucidating the underlying mechanisms associated with successful implant-related outcomes. Pertinent findings were subsequently incorporated in the design and formulation of the next generations of biomaterials. Moreover, cellular, *in vitro* models proved valuable alternatives to animal models in examining, for example, the cytocompatibility of biomaterials and in determining the conditions/stimuli (*i. e*., chemical, mechanical and electrical) needed to promote cell functions pertinent to new tissue formation and/or tissue regeneration.

Advances in the medical sciences as well as the need for new therapeutic methodologies in health-care practice expanded the scope of cellular bioengineering applications which now include (but are not limited to) the regenerative capabilities of functional tissues, use of pluripotent stem cells, and changing concepts of biocompatibility. Developments in biology, molecular biochemistry, genetics, and wound healing which are pertinent to new tissue formation/regeneration provide challenges and, at the same time, great opportunities for biomedical engineers. These areas are a substantive part of the present cutting-edge cellular bioengineering milieu and, undoubtedly, will be a major component of future endeavors in biomedical engineering. The scope and creativity of these endeavors are impressive. Researchers working at these frontiers are blazing new trails and are exploring approaches which were undreamed of in the past.

This presentation will briefly review past accomplishments and current advances as well as highlight trends in cellular bioengineering which had an impact but also have the potential for further contributions to the clinical milieu as well as to the biotechnology, tissue engineering, and tissue regeneration fields.

461485 Evaluating Mixing Performance in a Stirred Tank Reactor with Different Impeller Geometries and Reactor Size By CFD

Abstract:

Stirred tank reactors are ubiquitous in the pharmaceutical industry and their mixing efficiency is paramount on the rate of heat and mass transport processes and on the reaction rate as well. Traditional experimental measurement techniques are restricted to the limited number discrete of points were the measurements are taken. Moreover, measurements are difficult in reactors functioning at extreme conditions involving cryogenic temperatures or high pressures.

The numerical approach circumvents this experimental limitation and allows for the flow pattern to be evaluated to its full extent. Furthermore, CFD minimizes the risk associated with scaling up processes involving complex geometries where classical empirical correlations might not ensure the mixture components are well mixed. CFD also allows to assess the influence of several parameters and operational conditions to define an operational design space, very much aligned with the Quality-by-Design (QbD) approach. This enables an enhancement of process knowledge and to easily troubleshoot potential issues occurring in the production trains.

The objective of this work is to evaluate the mixing performance of a stirred tank reactor by assessing the distribution of a solute starting from a stagnant layer at the bottom of the reactor. Different impeller geometries and reactor scales were tested using Computational Fluid Dynamics (CFD). These simulations were implemented to study a single phase incompressible fluid by means of a RANS-type finite-volume (FV) method. Turbulence was modeled by the two-equation standard high-Re k-epsilon turbulence model, as implemented in OpenFOAM, in tandem with both a Moving Reference Frame (MRF) and Arbitrary Mesh Interface (AMI) approaches to account for the effects due to impeller rotation. New solvers were developed to determine the concentration field of the tracer solute: a steady state solver for MRF and a transient state one for AMI. The passive scalar transport approach was applied to simulate the transport of a scalar quantity in an incompressible fluid flow. The primary assumption of this approach is that the solutes are non-reactive, with no influence on the flow field, acting therefore as numerical tracers in the flow and thus are used to evaluate the mixing efficiency of the impeller.

The steady state and transient flow fields were calculated for given flow conditions using new solvers based on simpleFoam (Semi-Implicit Method for Pressure-Linked Equations) and pimpleDyMFoam (merged PISO-SIMPLE) , respectively. These new solvers include the transport equation for a passive scalar that is solved along with the velocity and pressure fields. The new solvers were used to compare the suitability of either approach (MRF and AMI) in the description of mixing of the passive scalar in typical pharmaceutical reactors. In this study a 95 % mixing uniformity was considered as an indication of a well mixed system.

466987 CFD Simulation on Assessment of Biologic Product Mixing Performance in Single Use Technology

Abstract:

The biopharmaceutical industry and has undertaken radical change in the design and operation of many bioprocess. These changes provide more robust process and increased production flexibility. The benefits of Single Use Technology (SUT) for equipment train flexibility and reduced cleaning logistics are now widely acknowledged. This trend has resulted in rapid growth of SUT(equipment) manufactures and product offerings.

In order to select the right SUT from multiple vendors for protein drug product mixing operation, a CFD simulation was applied to assess the impact of the size and shape of tank and impeller,, fill level,Â and agitation rate on the flow and mixing behavior. An approach was developed by introducing two concepts: the highest accumulated shear rate and mixing time safety factor. These parameters, respectively, quantify the acceptable boundaries of shear and mixing time validated within the operational ranges established for a commercial product based on product specific stability profile.

The model simulations were used to assess the SUTs mixing performance, the mixing time, highest accumulated shear rate, and range of acceptable agitation rate for three different SUTâ€™s vendors. Compared with the existing mixing operation for the commercial product, the mixing performance was classified and the decision was made to select the optimalÂ SUT for the product being considered.Â

467302 Characterization of Lab and Plant Reactors Using Fast Competitive Reactions and a Predictive Model

Abstract:

Scale-up of reaction and crystallization conditions often involves assumptions about equivalence of mixing between lab and plant. Scale-down of processes for lab studies require similar assumptions. A question that often arises in pharma applications is how small can the lab equipment be, so that the results translate to scale. With modern lab synthesis workstations providing good control of addition rates and temperature in reactors that resemble their larger scale equivalents, scale-up and scale-down work sometimes takes place in lab systems with volumes as low as 100 mL. It is possible using mixing-sensitive test reactions to quantify the liquid mixing environment in vessels at any scale and to thereby identify lab and plant conditions that are more quantitatively equivalent.

The fast competitive reaction system involving neutralization and ester hydrolysis (a so-called 'Bourne Reaction', ref 1) among other techniques has been used by Pfizer to study mixing characteristics in several common lab reactors used for routine development of chemical processes and also for a series of plant reactors. The final product distribution after running the reaction in fed batch mode reflects the history of relative mixing and reaction rates in the plume or reaction zone that forms near the feed location.

In general, a critical feed time can be identified and longer feed times represent micromixing-controlled conditions. It is possible using a predictive model containing both kinetics and mixing parameters to estimate the product distribution as a function of feed time and mixing energy dissipation rate. This latter parameter, often referred to as epsilon, with units of W/kg, is of interest for a wide range of mixing phenomena. Using the predictive model together with measured product distribution, it is possible to estimate the corresponding epsilon.

We will present and discuss results for different reactors and operating conditions and also how best to use predictive models to design and analyse the experiments, which require a good dynamic range of selectivity to be informative. We will compare mixing performance at operating conditions that are commonly used in the lab and plant. Finally, we will extract some guidelines towards obtaining equivalent mixing between lab and plant reactors.

1. Investigation of micromixing in stirred tank reactors using parallel reactions, John R. Bourne , Shengyao Yu, Ind. Eng. Chem. Res., 1994, 33 (1), pp 41–55

469803 Computational and Experimental Determination of the Hydrodynamics in a Scaled-Down Pharmaceutical Vessel Provided with Angle-Mounted Impellers

Abstract:

In most industrial applications, stirred tanks and reactors are typically provided with baffles to improve their mixing characteristics. However, in a number of pharmaceutical production facilities unbaffled vessels are commonly used. The absence of baffles is preferred in such cases because it reduces the potential for contamination and makes cleaning the vessel between batches easier. However, the lack of baffles also has a negative impact on the system's hydrodynamics since it often results in poor mixing of the batch, especially if the impeller is centrally placed, since the liquid in such a system is subject to a strong tangential flow, but low axial and radial flows. This hydrodynamic regime is highly undesirable if effective top-to-bottom liquid recirculation is required, and especially if a second phase (e.g., solid particles, immiscible liquids) must be dispersed and incorporated into the liquid bulk. For this reason, impellers in unbaffled pharmaceutical vessels are typically placed off-center, and, additionally, they may be mounted on a shaft that is angled with respect to the vessel vertical centerline. This lack of symmetry introduces some degree of "baffling" effects, in that it promotes stronger vertical recirculation of the liquid, and improves the suspension of settling solids and the incorporation of floating solids.

Despite their industrial relevance, especially for the pharmaceutical industry, extremely limited information is available on the hydrodynamics of vessels provided with angle-mounted impellers. Therefore, in this work, experimental and computational tools were used to determine the hydrodynamics of fluids in a Plexiglas, custom-made, scaled-down version (diameter: 304 mm) of an industrial vessel with an elliptical bottom and provided with two angle-mounted (by 5 off the vertical) A-310 Lightnin impellers. The system was operated under different operating conditions in order to replicate the mixing characteristics of the industrial system. Computational Fluid Dynamics (CFD) was used to quantify the hydrodynamics of this system under different geometric configurations (such as liquid level), agitation speeds, and for different fluid rheologies. In all cases a multiple reference frame (MRF) computational approach was used and turbulence was modeled using the k-ε method. In addition, Particle Image Velocimetry (PIV) was used to experimentally determine the velocity flow field in water for some of these configurations in order to validate the CFD predictions, thus providing guidance on the optimal operation of these industrial systems.

The results obtained here indicate that there is substantial agreement between the CFD predictions and the PIV experimental results. The flow in the vessel appears to be very complex. The axial pumping action of the impellers produces a downward flow impinging the bottom of the vessel resulting in flow splitting and in the formation of a less well-mixed zone near the vessel bottom that persists even when the impeller velocity is substantially increased. This zone could be the preferential location for the sedimentation of settling solids in the liquid.

471722 Experimental Study of Wet Milling in a Rotor-Stator Mixer

Abstract:

While wet milling of active pharmaceutical ingredients (API) and other crystalline substances has commanded considerable attention, there are few fundamental studies on which to base mill design, scale up and operating strategies. Often an in-line rotor-stator mixer, fed from and recirculated to a well-mixed holding tank, is used to achieve the ultimate particle size or to precondition a feed to a higher intensity device such as a ball mill. In this study, a systematic framework is developed and employed with definitive experiments performed in a Silverson L4R rotor-stator mixer, to investigate the effect of milling conditions, mixer geometry, and particle physical properties on ultimate particle size and particle breakage kinetics. By varying operating conditions, mixer geometry and crystal mechanical properties, it is possible to discriminate among various breakage mechanisms (fluid shear, particle-particle collisions, particle-wall collisions, etc.) to develop mechanistic relations to aid in process scale up and to serve as inputs to population balance and other models for the evolution of the crystal size distribution.

Based on the literature, it was reasoned that along with size and shape, stress intensity factor (or fracture toughness) and hardness are the important physical properties affecting breakage rate. To vary these properties, several crystalline materials were milled in antisolvents (e.g. sugar in isopropyl alcohol) to allow discrimination between mechanical size reduction and solubility effects. Collision rates were varied by changing particle concentration; and coatings and small geometric changes were used to consider interactions with solid surfaces. Rotor speed and volumetric throughput were adjusted to independently vary energy input and mill head residence time. Crystal hardness and stress intensity factor were measured with a Hysitron TriboIndenter. Particle size distributions were measured by several methods. A Lasentec focused beam reflectance measurement (FBRM) probe and Particle View Microscope (PVM) were placed in the holding tank to continuously monitor particle size and suspension uniformity. Grab samples, acquired in the holding tank at well-defined time intervals, were analysed using a Horiba laser diffraction instrument and by microscopy with an automated image analysis technique. The results from the various measurement techniques will be compared and discussed. Mechanistic models that account for particle properties and milling conditions will be presented for the various fracture mechanisms. The mechanistically correlated experimental results will be discussed in detail, as well as application to process scale up. One intended use of the correlations is as mechanistic inputs (breakage kernels) to population balance models.

472697 Practical Large Eddy Simulation

Abstract:

A novel approach for performing three-dimensional CFD simulations of bioreactors and fermenters is presented. This approach, which is based on the lattice-Boltzmann model (LBM), eliminates the manual grid meshing process associated with CFD and runs orders-of-magnitude faster than conventional modeling tools. Since the LBM approach conserves energy, it provides direct access to the energy dissipation rates and turbulent energy spectrum at arbitration points across within the domain. Moreover, owing to the explicit time-integration of the flow field, the trajectories and suspension times of bubbles and solids inside the reactor can be modeled directly. Used here to model a 200L and 2000 L agitated fermenter, we predict the effects of impeller speed and off-bottom displacement on blend time, maximum energy dissipation rates, and power draw. We then model the evolution of gas bubbles through the tank and the associated bubble residence time distribution.

474214 Scale-up Rules for the Generation of Solid-Stabilized Emulsions

Abstract:

Effects of scale-up on the generation of solid stabilized emulsions (SSEs) were investigated experimentally. Considering unbaffled systems of 0.25, 5, and 35 Liters with geometrical similarities, emulsions were prepared using off-centred pitched blade and Rushton turbines. Standard scale-up criteria allowing for the calculation of impeller speed at a larger scale based on $N_{large} = N_{small}(D_{large}/D_{small})^a$ and commonly used for surfactant-stabilized emulsions (energy dissipation, Weber number, Reynolds number, impeller tip speed and circulation time) were applied for concentrated and diluted formulations of SSEs. Silicone oil (50 cSt) was used as the dispersed phase in water and solid glass microspheres ($d_{32} \sim 3\mu m$) as stabilizers. Emulsions were characterized by means of droplet size measurements using Mastersizer 3000 from Malvern. Sauter mean diameter (d_{32}), dv_{10} and dv_{90} were used to compare the resulting emulsions. A well-established decolorization technique was used to evaluate mixing times. The results from the experiments showed that none of the standard scale-up criteria is suitable for scaling-up Pickering emulsions processes. Weber number, Reynolds number and tip speed as scale-up criteria lead to larger diameters with increasing scale while the opposite tendency is observed with circulation time and energy dissipation. Results tend to indicate that constant drop size distribution can be obtained using an intermediate between the energy dissipation rate and the tip speed suggesting that the suitable scale-up exponent a of $N_{large} = N_{small}(D_{large}/D_{small})^a$ would be between 2/3 (energy dissipation rate) and 1 (impeller tip speed). The presentation will present the results and their interpretation.

453122 Developing Therapeutic Medicines through Chemical Engineering Innovations

Abstract:

Over the past 40 years, Genentech scientists and engineers have pioneered and exploited the use of genetic engineering to develop and manufacture recombinant proteins and monoclonal antibody therapeutics that have revolutionized medicine. Building on this legacy of innovation, chemical engineers in the laboratory, in pilot plants, and in manufacturing continue to play a key role in transforming the latest scientific breakthroughs into innovative products for patients. This presentation highlights various applications of chemical engineering in the development of technologies to enable novel and breakthrough therapeutic medicines and products.

469260 A Regulatory Perspective for Manufacture of Complex Drug Substances

Abstract:

The quality of chemically synthesized, small-molecule drug substances usually can be characterized by routine analytical testing. In contrast, for drug substances that are polymeric molecules (e.g. peptides), heterogeneous mixtures (e.g. heparins, glatiramer acetate, etc.), or molecular complexes (e.g. iron sucrose), the structure and impurity profile are more complex than small-molecule drugs, and generally cannot be characterized easily by analytical means. As a result of this complexity, quality control of these complex drug substances cannot solely rely on analytical testing. The identity, impurity profile, and content of complex drug substances are highly dependent on their manufacturing process as well as raw material and in-process controls. This presentation will discuss a regulatory perspective for manufacture of complex drug substances in support of both new drug and generic drug applications.

473006 CQA Control Strategy for a Synthetic Peptide Therapeutic Candidate Etelcalcetide

Abstract:

Application of innovative continuous improvement strategies to ensure consistency and robustness of pharmaceutical drug substance (DS) manufacturing processes can add significant value over a productâ€™s lifecycle. A case study from etelcalcetide, a synthetic peptide therapeutic candidate, will be presented highlighting Amgenâ€™s modality-independent approach to process design. Application of QbD and QRM principles were fundamental in establishing a solid understanding of the impact of variability on DS critical quality attributes (CQAs) and overall process performance. Holistic frameworks to proactively identify, assess, and mitigate CMC risks was leveraged during the process design phase. Examples of specific DS CQAs will be used to show how establishment of appropriate controls required a combination of material input controls, procedural controls, in-process testing, and release testing.

457444 Development and Implementation of a Mathematical Model As a Key Component of a Palladium Control Strategy for a Semi-Continuous Process to Produce a Small Molecule Drug Substance

Abstract:

The manufacture of a small molecule drug substance employs a coupling reaction catalyzed with soluble palladium in the final synthetic step of the sequence. The control strategy around palladium removal involves selective adsorption of the palladium onto a functionalized silica resin in a continuous fashion followed by a continuous deprotection reaction and the selective crystallization of the drug substance. During the registration stability campaign, at-line and off-line samples were collected. To minimize cost associated with the selective adsorbent, and to ensure a robust control strategy for palladium, a predictive model was developed to aid in the appropriate design, scaling and setup of the manufacturing equipment. Use of the model and the selected control strategy for palladium, alleviated the need for continuous sampling for palladium during processing and enabled a robust operation during validation with significant production cost reduction.

467794 Monitoring the Trace Levels of Free Sulfate Release to Terminate a Depolymerization Reaction Using ATR-FTIR

Abstract:

The synthesis of a pharmaceutical product involves a critical depolymerisation reaction which determines the final product quality attributes. Hence, termination of this reaction at the desired end point is utmost important to achieve a consistent product with required specifications. However, monitoring this reaction is challenging due to low chemical specificity of the reaction species involved. In this work, we developed a novel PAT method based on ATR-FTIR combined with chemometrics to monitor this reaction. To achieve this, we first identified the reaction mechanism and the rate-controlling step and monitored the reaction progress by tracking the free sulfate release which forms as a by-product during the reaction. The PAT method has been validated at lab and kilo scales and subsequently transferred to the plant and successfully implemented at manufacturing scale.

474335 FDA Review Chemist's Participation in Pre-Approval Inspections: Case Study from Drug Substance Facility

Abstract:

In 2014, FDA initiated a new approach to incorporate review chemist's expertise in the inspection of drug substance/product manufacturing facilities. A pilot was launched with the objective to evaluate the feasibility, effectiveness, and efficiency of Office of Pharmaceutical Quality's vision for the integrated team-based quality assessment of A/NDAs. A risk analysis was conducted jointly by the compliance officer, drug substance, drug product, and facility reviewers to determine the need for a pre-approval inspection (PAI) and prioritize the issues to be covered on the inspection. The integrated team-based approach resulted in improved knowledge sharing among the reviewers and compliance officers, and in better identification and evaluation of high-risk areas during the application review. These high-risk areas identified by the review team were covered during the PAI. Collaboration between the review chemist and investigators helped uncover significant issues and resulted in the issuance of deficiencies to the firm.

Based on outcomes of the Agency's pilot integrated PAI program, other inspections were conducted using this paradigm including the recent PAI of a complex drug substance manufacturing facility. Participation on this inspection provided the reviewer with an opportunity to examine the equipment and process closely, to gain a better understanding of the development and optimization of manufacturing process and controls, and in turn allowed the inspection team to identify the CMC-related observations. Furthermore, the firm provided positive feedback regarding the presence of a review chemist on inspection deemed particularly helpful to better understand and improve regulatory submission of future applications. In summary, implementation of this new integrated approach for conducting a PAI has enhanced the efficiency and quality of review process, and improved communication between FDA and industry.

473022 Panel Discussion: Current Issues and Challenges for CQA Monitoring and Control for Drug Substance Manufacturing

Abstract:

In this panel discussion, all of the speakers in this session and invited FDA leaders will join for an interactive Q& A session. It will highlight current issues and challenges in the CQA monitoring and control for drug substance manufacturing, from development, manufacturing, to regulatory review and inspection, and life cycle management.

461930 A Rapid, Materials-Sparing Approach to the Exploration of a Roller Compaction Process Design Space

Abstract:

Roller compaction (RC) is a time and material intensive process commonly used to improve the overall processing behavior of formulations. The typical approach to identifying target processing conditions on a pilot- or production scale system can easily consume kilograms of formulation, and the material consumption associated with system startup and shutdown further preclude effective in-process measurements. These issues can be magnified during efforts to establish a design space according to the principles of QbD, where, for example, a broad range of material property inputs may be evaluated to assess their impact on both downstream process behavior and product performance.

Here we demonstrate the application of a well-established scale down approach to roller compaction simulation in order to evaluate the influence material attributes of a spray dried dispersion (SDD) incur on RC processing and tablet compression behavior. The approach couples a compaction simulation of the ribbon formation process with a ¼-scale replica of an in-line rotary mill to yield granules with densities and size distributions analogous to what is produced on a Gerteis Mini-Pactor roller compactor. Intragranular blends of a constant composition were prepared using SDD lots of varying bulk densities and particle size distributions, and each blend was compressed to form simulated RC ribbons of varying target porosities. All ribbons were subsequently milled under common granulation conditions, and tablets were prepared from these granulations to evaluate both the compression behavior and *in vitro* dissolution performance of each final system. This materials-sparing methodology enables the exploration of the relationship between SDD properties and simulated roller compaction conditions across a broad design space, resulting in the establishment of statistical models which describe the influence of these factors on meaningful attributes of the final drug product.

452650 Advances in Discrete Element Modeling of High-Shear Wet Granulation Process Using Rocky-DEM

Abstract:

High-shear wet granulation process is widely used in the pharmaceutical industry to improve the bulk density and flow characteristics of a formulation. There are different particle-level phenomenon and mechanisms at play during a granulation process, such as nucleation, agglomeration and breakage. The challenge is to ensure that these mechanisms occur similarly at different scales of the granulator to keep consistency in the quality of the granules. The ability to accurately predict the stress and collisional behavior in a granulator, would not only help our fundamental understanding of the process but also help alleviate experimental costs during scale-up studies.

Discrete element method (DEM) is a valuable tool to model such particle-level interactions, but in the past, has suffered from practically restrictive run times. In order to circumvent this, assumptions on particle size and shape (modeled as spheres or glued-spheres) were routinely made. However, more recently, advances in DEM technology, such as that provided by software Rocky-DEM, have addressed some of these limitations. Rocky-DEM provides a step-change in the ability to perform large scale DEM simulations significantly faster by utilizing the large number of cores available in NVIDIA® graphics card (GPU processing). The simulation run times using the GPU were compared with traditional and more expensive CPU solvers. In addition, several capabilities unique to Rocky-DEM in comparison with other commercially available DEM codes were utilized in this study; a few of them include the representation of Avicel using non-spherical particle shapes (polyhedral shape representation) and the ability to include cohesive forces and breakage without loss of mass or volume.

Rocky-DEM was used to predict the movement of Avicel particles in a Diosna® granulator at 50% volume fill (Figure 1). The simulations were validated by comparing results to experimentally-obtained surface velocity profile measurements using video-imaging. The validated DEM model was then used to establish a relationship between various parameters on simulation measurements of bulk properties (such as average velocity and stress distributions in the powder). Some of the variables included the particle stiffness (low and high), particle shape (spherical vs Avicel-shaped), particle size (0.8, 1.8 mm), impeller speed (Froude number vs tip speed), granulation type (wet vs dry particles), and granulator scale (1L, 10L and 150L). Thus, by numerically predicting granular pressures, velocities, and stresses in different regions of the granulator, this study provides guidelines on achieving reproducible particle environments with changes in scale and impeller speeds, and establishes the effect of particle shape and size. The Rocky DEM platform enabled the utilization of actual particle shapes and sizes in these studies and therefore are a closer reflection of a real-case scenario.

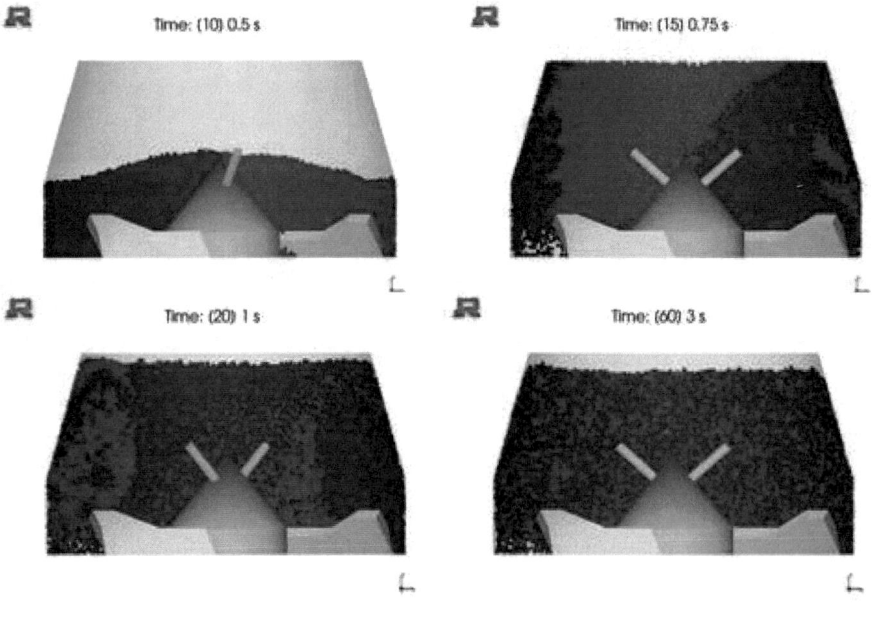

Figure 1. A cross-section of the granulator colored by the particle initial fill position to show mixing profiles

458336 Scale-up of HME Process from a Trilobed 16mm to a Bilobed 27mm Extruder; SPH Simulations and 1D Mechanistic Modelling

Abstract:

Introduction

Hot melt extrusion (HME) is a continuous manufacturing process primarily using co-rotating intermeshing twin-screw extruders (TSE). The continuous manufacturing nature of the process allows for a steady product quality and a decrease in production costs. Furthermore, the process has the potential of increasing the bioavailability of poorly soluble drugs by enabling manufacturing of amorphous solid dispersions, making it interesting to the pharmaceutical industry. Although the process has been used in the polymer and rubber industry for years it is still not well enough understood and fairly new in the pharmaceutical industry. Especially the technical transfer between different extruder scales (scale-up) is still an open question.

Methodology

Simulation of extruders via traditional CFD is prohibitively complex, due to the small gaps, the complex rheology and, especially, due to the free surfaces that occur in the not-fully-filled screw elements. In order to better understand the process conditions and product quality relationships, 1D mechanistic and 3D smoothed particle hydrodynamics (SPH)[1] simulation approaches and the established scale-up theory were used. Standard 3D mesh-based computational fluid dynamics (CFD) models struggle with the complex rotating intermeshing TSE geometry, and lack the ability to investigate partially filed TSE elements[2]. The 1D mechanistic model is able to simulate the HME process as a whole, giving information about the filling degree, pressure and temperature distribution, as well as the specific mechanical energy input (SMEC) across the screw configuration[3]. In order to setup the 1D model, the individual screw elements have to be characterized by determining the inherent conveying capacity and the pressure build up capacity[4]. The characterization is done using the mesh-free SPH simulation approach, giving additional insight into the mixing and sheer rate distribution for every screw element, allowing for a comparison between triple- and double-threaded extruder screw elements.

Results

An HME process on a trilobed 16mm and a bilobed 27mm extruder, with five different materials, was investigated, providing recommendations for the process setup. Additionally, a scale-up scenario was investigated, using the proposed scale-up laws with an in-silico process DoE, with an experimental validation.

Literature

[1] J. J. Monaghan, "Smoothed particle hydrodynamics," *Reports Prog. Phys.*, vol. 68, no. 8, pp. 1703–1759, Aug. 2005.

[2] M. Bierdel, "Computational Fluid Dynamics," in *Co-Rotating Twin-Screw Extruders*, K. Kohlgrüber, Ed. Munich: Carl Hanser Publishers, 2008.

[3] A. Eitzlmayr, G. Koscher, G. Reynolds, Z. Huang, J. Booth, P. Shering, and J. Khinast, "Mechanistic Modeling of Modular Co-Rotating Twin-Screw Extruders," *Int. J. Pharm.*, vol. 474, no. 1–2, pp. 157–176, 2014.

[4] K. Kohlgrüber, *Co-Rotating Twin-Screw Extruders*. Munich: Carl Hanser Publishers, 2008.

455147 Scale-up and Blender Change Model for the Pharmaceutical Lubricated Mixing Process

Abstract:

In the pharmaceutical lubricant mixing process, lubricity of granules after mixing was evaluated by measuring the contact angle with water and tablet hardness. The granules were prepared using three different types of popular tumble mixers; V-blender (effective volumetric capacity: 70 litters), Container blender (36 litters) and Bin blender (20 litters) with different powder loading rates (10-88%), blender rotation speeds (18-37 rpm) and mixing times (5-120 min) for model placebo formulation. Contact angle measurement and tablet hardness are useful as alternative characteristics for evaluation of granule lubricity. Based on the experimental data, Mixing Performance Index (MPI), which is an empirical equation including blender type, powder loading ratio, mixing rotation speed and batch size as functional parameters, were developed. Mixing Ability (MA) was defined by the combination of MPI and mixing time. Calculated MA exhibited good correlation with tablet hardness (correlation coefficient is 0.89) in all datasets. MA was verified by nine different drug product data with different manufacturing scales, to enable the MA model to support the formulation researcher to set mixing process parameters when the batch size or blender type changes.

462399 Scaling Intra-Tablet Coating Variability

Abstract:

Pharmaceutical tablets usually exhibit variability in coating thickness, both across individual tablets as well as over a single tabletâ€™s surface. In this work, the coefficient of variation in coating thickness, referred to as intra-tablet coating variability, is examined as a function of processing scale. The discrete element method (DEM) is used to simulate the dynamics of two different tablet shapes in coating pans of four different sizes.Â The asymptotic intra-tablet coating variability is observed to increase with increasing pan size. Studies designed to explain the observed trend are also presented.

454707 Improving Workflow for Modeling of High-Shear Rotor-Stator Wet Milling for Pharmaceutical Applications

Abstract:

Wet milling is an important unit operation for the size reduction of pharmaceutical powders. The benefits of wet milling to the pharmaceutical industry relies on its ability to control particle size distributions (PSDs) while managing containment concerns and avoiding dry powder handling of potent active pharmaceutical ingredients (APIs). Due to the high-value nature of APIs, a trial-and-error experimentally-driven scale-dependent approach to obtain optimal milling conditions would not be cost-effective.

The use of modeling-aided scale-up of high-shear rotor-stator wet milling processes using a standardized population balance equation (PBE) approach has been reported by this group in the past.[1-2] In this work, recent model improvements as well as the creation of templates for easy data collection and the actions taken to improve the workflow will be discussed. A few case studies from early to late phase small molecules and multiple scales of mills from lab to manufacturing will be used to exemplify the approach.

[1] Luciani et al. PBE-Aided Design of Slurry Milling Process to Reduce Particle Size of Intermediate and Active Pharmaceutical Ingredients, AIChE Annual Meeting 2012 (366b)

[2] Luciani et al. PBE-Aided Design of Slurry Milling Process to Reduce Particle Size of Intermediate and Active Pharmaceutical Ingredients. *Org. Process Res. Dev.*, 2015, *19* (5), 582â€“589

468475 Characterizing Mixing Time Scales in an Exothermic Mixing-Limited Reaction

Abstract:

Exothermic mixing-limited reactions are commonly encountered in the pharmaceutical and fine chemicals industry, and are usually executed in semi-batch reactors. Their scale-up, however, requires a keen understanding of the heat and mass transfer aspects of the reaction and can be influenced by process parameters such as reagent dosing rate and agitation. In this contribution, we will present a case study where scale-up of a highly exothermic reaction resulted in variable yields in different reactor geometries due to heat and mass-transfer limitations. A systematic analysis of local meso-mixing rates with correlations by Bourne et al.[a] and computational fluid dynamics in laboratory and pilot plant reactors revealed presence of local temperature hot-spots, which is attributed as the main cause of variable reaction yields. As a result, a scale-independent meso-mixing rate based parameter was applied to obtain a high reaction yield at different scales.

[a]Bourne, John R. "Mixing and the selectivity of chemical reactions." *Organic Process Research & Development* 7.4 (2003): 471-508.

The work presented here was sponsored by AbbVie. AbbVie contributed to the design, research, and interpretation of data, writing, reviewing, and approving the publication. All authors are employees of AbbVie.

462192 Pharmaceutical Crystallization: A Manufacturing Process Perspective

Abstract:

Typically a number of critical quality attributes (CQAs) such as crystal structure and crystal component identification, purity, crystallinity, crystal size distribution (CSD), polymorphism and impurity, and stability are evaluated to assess the robustness, controllability, repeatability, and consistency of a pharmaceutical crystallization process. There are different ways to achieve in-depth process understanding and to realize rationale process control. This presentation provides several FDA research case studies to demonstrate the technical feasibility of real time process monitoring approaches for better understanding, design and control of pharmaceutical crystallization processes and product characterization. The case studies are: 1) Combined real-time PAT monitoring and process chemometrics for mapping the state of a pharmaceutical crystallization process; 2) Combined real-time PAT process monitoring and first-principle modeling for elucidating the nucleation mechanisms of a dynamic pharmaceutical crystallization process; 3) Combined real-time PAT process monitoring, Design of Experiments (DOE), and General Linear Modeling (GLM) to establish a hybrid approach for process characterization and process design space development; and 4) Integrated PAT approach for nucleation induction time measurement. This presentation will briefly address some of the current opportunities in pharmaceutical crystallization, including switching from batch to continuous crystallization and several promising, novel technologies in the area of crystallinity monitoring, from a process engineering perspective.

456484 Cooling Crystallization within Micromixers: Mathematical Modeling, Theoretical Analysis, and Experimental Validation

Abstract:

A key objective in crystallization is to control crystal size distribution (CSD), to improve product quality (e.g., pharmaceutical and biopharmaceutical) and process efficiency. Micromixers have been an effective technology to enhance the control of nucleation, which is the key step towards final product CSD in continuous [1] and semi-continuous crystallization processes [2]. Common micromixers combine solution and anti-solvent streams such as in a dual-impinging-jet (DIJ) micromixer (e.g., [3]–[5]).

Rather than anti-solvent crystallization as commonly used, a DIJ mixer has been demonstrated to combine hot and cold saturated solutions of the same solute/solvent to generate small uniform seed crystals [2], which appears to be the first reported cooling DIJ nucleation. The product crystals were less than 10 microns in length, placing them in the correct size range for direct application in inhalers [6]. Theoretically, if the mixing was perfect (that is, concentration and temperature were completely mixed at the molecular scale in the mixer), the average supersaturation level would be too low to nucleate crystals in a cooling DIJ mixer. The surprising result of nucleating crystals by combining hot and cold saturated solutions in a DIJ mixer motivated the theoretical analysis of the system, which showed that nucleation was enabled by the much faster energy transfer than mass transfer rates near the interface. This difference in rates enabled the temperature of the hot solution to drop to approximately the average temperature of the two solutions before its solution concentration had significantly changed, resulting in a supersaturation sufficiently high to nucleate crystals.

The theoretical analysis is supported with process simulations. The two-dimensional energy and solute mass balances are solved using COMSOL input with an analytical solution for the velocity field, to generate spatial distributions of temperature, concentration, and supersaturation near the hot-cold interface within a cooling DIJ mixer. In the most important spatial region of importance for characterization of nucleation, the two-dimensional fields are shown to be very close to analytical solutions derived from a one-dimensional approximation of the energy and molar balances. This simplification applied the derivation of design criteria to facilitate quick assessment of whether any particular pharmaceutical-solute combination will nucleate crystals in a cooling DIJ mixer, based on the physicochemical properties. These criteria could save time and material by avoiding or reducing trial-and-error experiments, which is helpful at the early stage of pharmaceutical process development. The process simulations are then expanded to include the full solution of the spatially-varying population balance model [5] to calculate the crystal size distributions throughout the spatial domain. The mathematical models and methodologies are validated for the crystallization of L-asparagine monohydrate (LAM) in aqueous solution.

References

[1] M. Jiang, Z. Zhu, E. Jimenez, C. D. Papageorgiou, J. Waetzig, A. Hardy, M. Langston, and R. D. Braatz, "Continuous-flow tubular crystallization in slugs spontaneously induced by hydrodynamics," *Cryst. Growth Des.*, vol. 14, no. 2, pp. 851–860, 2014.

[2] M. Jiang, M. H. Wong, Z. Zhu, J. Zhang, L. Zhou, K. Wang, A. N. Ford Versypt, T. Si, L. M. Hasenberg, Y. E. Li, and R. D. Braatz, "Towards achieving a flattop crystal size distribution by continuous seeding and controlled growth," *Chem. Eng. Sci.*, vol. 77, pp. 2–9, 2012.

[3] B. K. Johnson and R. K. Prud'homme, "Chemical Processing and Micromixing in Confined Impinging Jets," *AIChE J.*, vol. 49, no. 9, pp. 2264–2282, 2003.

[4] M. Midler Jr., E. L. Paul, E. F. Whittington, M. Futran, P. D. Liu, J. Hsu, and S. H. Pan, "Crystallization method to improve crystal structure and size," US5314506 A, 24-May-1994.

[5] X. Y. Woo, R. B. H. Tan, and R. D. Braatz, "Precise tailoring of the crystal size distribution by controlled growth and continuous seeding from impinging jet crystallizers," *CrystEngComm*, vol. 13, no. 6, pp. 2006–2014, 2011.

[6] T. G. D. Capstick and I. J. Clifton, "Inhaler technique and training in people with chronic obstructive pulmonary disease and asthma," *Expert Rev. Respir. Med.*, vol. 6, no. 1, pp. 91–103, Feb. 2012.

452797 Real Time PAT Monitoring of the Solvent-Mediated Transformation of Ciclesonide Acetone Solvate into Methanol or Cyclohexane Solvate

Abstract:

Due to structural differences, pseudopolymorphs of a given drug substance exhibit different physicochemical properties, which impacts drug dissolution, bioavailability, and efficacy. Therefore polymorphic transformation has been a vital aspect of pharmaceutical crystallization process engineering. In this work, three solvates of ciclesonide were characterized by Differential Scanning Calorimetry ï¼ˆDSCï¼‰, Thermogravimetry (TG), and X-ray Powder Diffraction (XRPD) analysis. The driving force of solvates transformation process was investigated via measuring solubility of ciclesonide solvates in different mixed solvent systems. To follow the crystallization process kinetics and understand the phase transformation mechanisms, , several Process Analytical Technology (PAT) techniques such as Fourier Transform Infrared Spectroscopy (FTIR), Raman Spectrum (RAMAN), Particle Video Microscope (PVM) and Focused Beam Reflectance Measurement (FBRM) were employed to monitor the phase transformation from acetone solvate into methanol or cyclohexane solvate in real time. In addition, the effects of temperature, solvent composition, and stirring rate on phase transformation process were discussed. Furthermore, it was demonstrated that real time PAT process monitoring can help to realize better process control and achieve desired crystal product quality attributes.

472470 Crystal Morphology and Size Control in Pharmaceutical Industry

Abstract:

Crystal Morphology and Size Control in Pharmaceutical Industry

Junbo Gong[1,2*], Yan Wang[1,2], Jingkang Wang[1,2]

[1] National Engineering Research Center of Industry Crystallization Technology, School of Chemical Engineering and Technology, Tianjin University, Tianjin, China;

[2] The Co-Innovation Center of Chemistry and Chemical Engineering of Tianjin, Tianjin University, Tianjin, China;

E-mail address: junbo_gong@tju.edu.cn * wangyan57@tju.edu.cn

Abstract

Crystal shape and size engineering is an integral part of pharmaceutical crystallization. It aims to design and control crystal habit and particle size distribution (PSD) of drug substance. This program is of great significance due to the strong impact of product quality, including bioavailability, dissolving rate, flowability, etc. Downstream particle handling is also affected, such as filtration, drying and compaction. The factors influencing crystal morphology and size come from two groups: the one is the operating condition, which includes supersaturation, temperature, the hydrodynamics of the crystallizer (e.g., stirring rate, mixing conditions) and external field (e.g., magnetic field, ultrasound field), and the other is chemical composition of the solution which consists of solvents and additives. In order to control crystal shape and PSD, experiments should be designed and carried out to monitor crystal growth process. The influence on crystallization thermodynamics and kinetics should also be investigated. During the research, microscopy techniques, like polarizing optical microscope, scanning electron microscopy (SEM) and atomic force microscope (AFM) will be applied to analyze crystal shape. Particle sizing techniques are also used to count for PSD, like the laser diffraction (LD) and focused beam reflectance measurement (FBRM). On the other hand, the use of additives has attracted increasing attention in the drug manufacturing process. The presence of additives could substantially affect the solid forms and morphology. High throughput screening (HTS) techniques to additives should be developed to accelerate the screening efficiency. The effects of additives on crystal shape may be resulted from the specificity of interaction between the additive and the solute molecule, charge distribution, hydrophilic-hydrophobic property, stereoselectivity as well as diffusion rate and competitive adsorption, etc. With fast advances made in computational techniques, modeling and prediction of crystal morphology can be realized in the absence or presence of additives. They not only provide powerful tools to design and screen effective additives, but lead to further understanding of crystal growth mechanisms.

Fig. 1. SEM of the drug substance without (a) with (b) the modification of the crystallization environment

472381 Liquid Liquid Phase Separation during the Cooling Crystallization of Pyraclostrobin

Abstract:

Liquid−Liquid Phase Separation during the Cooling Crystallization of Pyraclostrobin

Crystallization, as a method of isolation and purification, is widely used in fine chemicals, food and pharmaceutical industries. It is a powerful mean to improve the properties of particle, such as changing crystal form, narrowing crystal size distribution (CSD), modifying crystal habit and so on. However, sometimes droplets like oil drops are generated instead of crystals are appeared from the supersaturated solution during the cooling and/or solventing-out crystallization process. This phenomenon is always termed as liquid¨Cliquid demixing, liquid¨Cliquid-phase separation (LLPS) or oiling out.

The main purpose of this investigation was to investigate the cooling crystallization process of pyraclostrobin in isopropanol/cyclohexane solvent mixtures. It was found that liquid−liquid phase separation will occurred under some certain conditions. In order to understand the mechanism of the liquid−liquid separation and the influence of cooling rate on oiling out, the phase diagram and cloud point curves at different cooling rate were experimentally measured with the aid of in situ tools, such as PVM, FBRM, ATR-FTIR and Raman. The phase diagram includes the solubility curve, the liquid−liquid separation curve, supersolubility curves, and clear point. It was found from the phase diagram that a normal single-phase crystallization would change to a oiling-out crystallization once the concentration of pyraclostrobin exceeded a certain value during crystallization. Contrasting with cloud point curves of different cooling rate, it is obvious that the cooling rate has a significant effect on the metastable state critical oiling out point. The concentration corresponds to the metastable state for critical oiling out point decreases with the increase of cooling rate. In the cooling and heating cyclic process, the solution has experienced five stages. During cooling down the solution from high temperature, oiling out including stable LLPS and metastable LLPS occurred, then nucleation and growth of crystals happened; upon heating up, the crystals dissolved gradually, then the solution with a small amount of particles transformed into transient oil phase and finally it formed a stable and clear solution.

Considering the crystal properties of final products, some experiments designed to investigate the influence of cooling rate and initial concentration. And the result showed that the higher cooling rates, the more serious agglomeration and poorer crystal habit of final products would get. Another result is that initial concentration has little influence on the crystal properties. High quality of products can be obtained by designing route of cooling crystallization according to the phase diagram.

Figures

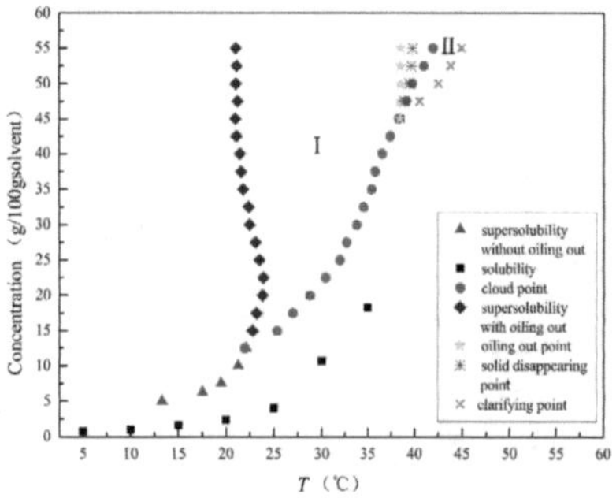

Figure 1 Phase diagram of pyraclostrobin in 10 wt % isopropanol/cyclohexane solvent mixtures.

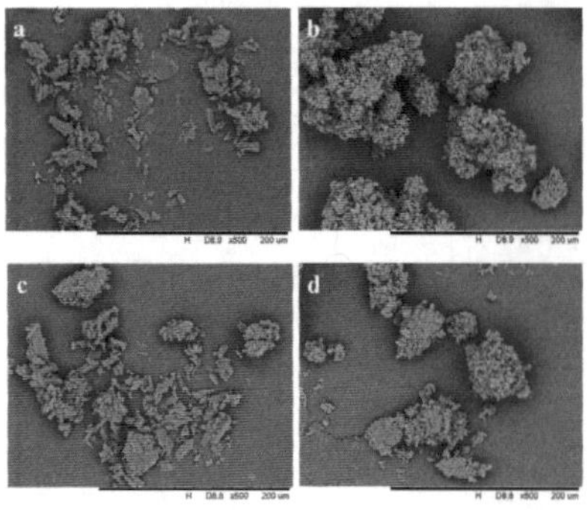

Figure 2 SEM images of pyraclostrobin product obtained from different experiments

a refer to pyraclostrobin products at the cooling rate of 0.1 K/min under concentration of 10g/100g solvent; b refer to pyraclostrobin products at the cooling rate of 0.5 K/min under concentration of 10g/100g solvent; c refer to pyraclostrobin products at the cooling rate of 0.1 K/min under concentration of 15g/100g solvent; d refer to pyraclostrobin products at the cooling rate of 0.5 K/min under concentration of 15g/100g solvent.

472527 Discriminate the Differences Between Surface Water and Channel Water in the Formation of Solvates

Abstract:

Discriminate the differences between Surface water and Channel water in the formation of solvates

Yumin Liu[1,2], Dandan Han[1,2], Jingkang Wang[1,2], Junbo Gong[1,2*]

[1] National Engineering Research Center of Industry Crystallization Technology, School of Chemical Engineering and Technology, Tianjin University, Tianjin, China;

[2] The Co-Innovation Center of Chemistry and Chemical Engineering of Tianjin, Tianjin University, Tianjin, China;

E-mail address: liuyumin@tju.edu.cn junbo_gong@tju.edu.cn *

Abstract

Solvates are defined as Active Pharmaceutical Ingredients (APIs) solids which incorporate one or more solvent molecules in the crystal lattice or channel to form new crystalline substances. Converting the drugs into hydrophilic solvates can help to improve the solubility in water and increase drug bioavailability. Meanwhile, it also can improve the stability of drugs, thus facilitating the formulation and storage of the drugs. There are four binding modes between compounds and solvents, just presented in the following Figure 1. Liquid inclusions and incorporation into the lattice are composed of solvates. With the increased incidence of solvate formation, it is essential to get a comprehensive understanding of their structure, physical stability, and the implication of de-solvation on the properties of the APIs.

Hydrate are the most abundant solvates, owing to the high propensity of water to be entrapped in the lattice of APIs, with its small size and tendency to form multi-directional hydrogen bonds. Pharmaceutical solids may come in contact with water during processing steps, such as crystallization, lyophilization, spray-drying, wet granulation, or aqueous film-coating, and they may be exposed to humid air during storage. Absorbed water molecules may reside on crystal surfaces, in crystal channel, or in crystal lattice structures which can change the packing in the unit cell. Surface water and channel water is quite different from each other. Without a doubt, we can utilize SXRPD to discriminate that differences between them. But in our laboratory, several other offline analytical techniques, such as XRPD, TGA and DVS, were also adopted to distinguish it and achieved remarkable results.

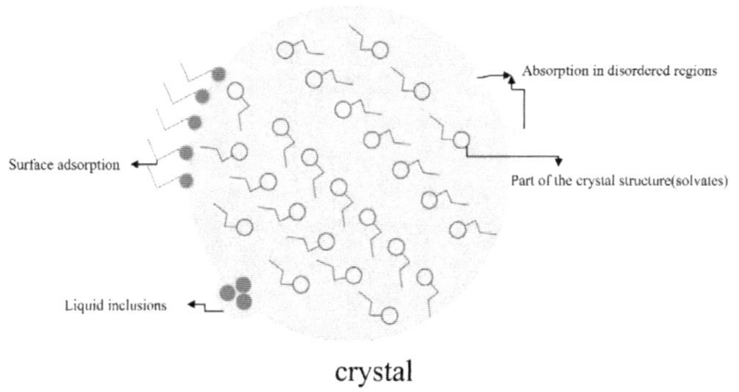

Fig.1 Different principles of solvent associations with crystalline solids

Fig.1 Different principles of solvent associations with crystalline solids

472078 Influence of Ultrasonic Wave on the Crystallization of Lysozyme

Abstract:

Due to the highly sensitivity of proteins to the crystallization conditions, a slight change in operating parameters might affect the outcome of the crystallization. In this study, the effects of ultrasonic wave on the crystallization of lysozyme were investigated by applying different levels of ultrasound during the purification process of lysozyme from the aqueous solutions of hen egg white through the solvent freeze out (SFO) technology. Tetragonal lysozyme crystals were obtained during the process, which were then verified by X-Ray Powder Diffraction (XRPD). The quality of lysozyme crystal (composition and purity, size and shape, enzyme activity) was measured using SDS-PAGE, polarized light microscopy and spectrophotometer, respectively. It was found that the purity and size of lysozyme crystal were affected significantly by ultrasonic wave, while the enzyme activity of lysozyme crystal werenâ€™t. In addition, the influence mechanism of ultrasonic wave on lysozyme crystallization was discussed. The results obtained by this work will provide essential information for the purification of other proteins.

471180 Site-Specific Conjugation of a Drug to an Antibody Via Nucleotide Binding Site

Abstract:

Antibody-drug conjugates (ADCs) are recent pinnacle of immunotherapeutic agents that enhance the antitumor effects of antibodies and reduce adverse systemic effects of potent drugs. To obtain an efficient ADC, the drug should be conjugated to the antibody with a linker while preserving antibody activity and high potency of the drug with an optimized drug-antibody ratio (DAR). Conventional ADCs are typically produced by conjugating a drug to an antibody through the side chains of either surface-exposed lysines or free cysteines generated through reduction of inter-chain disulfide bond. Many obstacles have been associated with the conventional drug conjugation methods to antibodies, such as aggregation of antibodies, high number of the DAR yielding rapid *in vivo* clearance, heterogenous modifications causing weak performance of the ADC, and requirement of harsh chemical modification resulting with damaged antibodies. In this study, we describe a site-specific method for conjugation of drugs to monoclonal therapeutic antibodies utilizing nucleotide-binding site (NBS). We used an NBS ligand â€"ring structured small moleculeâ€" that has moderate binding affinity to the NBS located between heavy and light chains on the variable region of Fab fragment to conjugate a drug with a PEG linker to the antibody. We have used previously described UV-NBS method that utilizes UV-energy to initiate a photo-cross-linking reaction between the NBS and the NBS-ligand, to obtain controlled site-specific conjugation of the drug to the antibody. We have assessed the effectiveness of the ADC generated with the UV-NBS method utilizing various drugs such as paclitaxel and bortezomib in combination with various antibodies including Rituximab, anti-CD133 and anti-CD38. We further characterized *in vitro* efficacy, toxicity, and the DAR of the ADCs as well as activity and sensitivity of the antibodies post-conjugation with the drugs.

461822 Acid-Sensitive Silylated or Acetalated Polysaccharides for Vaccine Applications

Abstract:

Acid-degradable polymers are well-suited for use as drug delivery vehicles because numerous physiological sites (e.g., intracellular endocytic pathway) are acidic. In particular, our work has concluded that acid-sensitive polymers are advantageous for vaccine and other immune modulatory therapies because once phagocytosed by immune cells, the low pH environment of the phagosome leads to triggered release of the encapsulated cargo. Here we report the synthesis of acid-sensitive silylated or acetalated polysaccharides derived from either dextran or inulin. For the silyl chemistry, various alkyl substitutions on the silicon center were performed, including trimethylsilyl dextran (TMS-DEX), ethyldimethylsilyl dextran (EDMS-DEX), and triethylsilyl dextran (TES-DEX) on dextran. Because dextran and acetalated dextran are not inherently immunogenic, we chemically modified a naturally-occurring, immunostimulatory polysaccharide, inulin, to produce tunable, acid-sensitive biopolymers acetalated inulin (Ace-IN) and trimethylsilyl inulin (TMS-IN). These polymers were then fabricated into microparticles (MPs) via emulsification followed by solvent evaporation. Our data indicates the MPs are stable at extracellular pH 7.4 and display a wide range of degradation half-lives (fifteen minutes to greater than nine days) when exposed to acidic pHs of 2.0 and 5.0. The half-lives are dependent on the extent of silylation/acetalation and steric crowding on the silicon center (trimethyl to ethyldimethyl to triethyl). Silyl-DEX Ace-IN, and Silyl-IN MPs exhibit cytocompatibility when cultured with macrophages. TES-DEX, Ace-IN, and TMS-IN MPs, composed of highly hydrophobic moieties or the parent immunostimulatory inulin, elicited substantial in vitro production of tumor necrosis factor alpha (TNF-Î±), a cytokine associated with an innate immune response. In vivo immunization with a model ovalbumin antigen encapsulated in silylated or acetalated polysaccharide MPs, without a separate adjuvant, results in a dual humoral and cellular response that is superior to an alum-adjuvanted formulation. Overall, we present Silyl-DEX, Silyl-IN, and Ace-IN as members of the acid-degradable polymer family for potential use in subunit vaccines and other drug delivery applications.

467628 The Effect of Particle Physico-Chemical Properties on Phagocytosis

Abstract:

The effect of particle physico-chemical properties on phagocytosis

Anusha Garapaty[1], Julie A.Champion[1], (1) School of Chemical and Biomolecular Engineering, Georgia Institute of Technology, Atlanta, GA

The tunability of polymeric particle properties makes them attractive for a wide range of applications in drug delivery, imaging and diagnostics. Their physico-chemical properties such as size, shape, surface chemistry and stiffness can be altered to influence their interactions with cells. It has become imperative to establish a correlation between these physico-chemical properties and biological behavior to design targeted delivery systems. During these delivery applications, particles introduced into the body interact with macrophages of the reticuloendothelial system. These macrophages rapidly clear them through a process known as phagocytosis thus limiting their potential in biomedical applications involving either phagocytic inhibition or macrophage targeting.

The influence of material parameters size, shape and mechanical stiffness on phagocytosis has been established previously [1-3]. However, the combinatorial effects of these material biophysical parameters on phagocytosis is still unknown. Our objective in this work is to understand how the combined effects of particle physico-chemical properties size, shape and stiffness affect the process of phagocytosis using particles which can be independently tuned in each of these properties while keeping the chemistry constant.

Tunable particles have been fabricated by the development of a template based layer-by-layer (LbL) assembly process. Polystyrene particles of varied sizes were chosen as the template. Shape is imparted to the polystyrene particles by heat liquefaction while stretching. The polystyrene template particles are adsorbed with polyelectrolytes poly (vinyl pyrrolidone) (PVP), poly (ethyleneimine) (PEI), and poly (acrylic acid) (PAA) by LbL self assembly. Mechanical

stiffness can be varied either by maintaining the solid template core or by making hollow capsule particles through tetrahydrofuran treatment and thiol crosslinking of the polyelectrolyte layers. To study specific Fc interaction with macrophages, the particles were coated with IgG antibody. Our results show that for a spherical shape, when the size is altered cores are internalized slightly more than corresponding capsules during phagocytosis. However, when the shape is altered to a rod for a particular size, capsules exhibit enhanced internalization compared to corresponding cores. These particles thus demonstrate that shape and stiffness play a crucial role in tuning the specificity of IgG functionalized particle interactions with macrophages. This novel polymer particle platform can also be further extended to functionalization of active ligands for phagocytic inhibiton or modulation of macrophage activity and phenotype in immunological therapeutic applications.

References

1. Beningo, K.A. and Y.-l. Wang, Fc-receptor-mediated phagocytosis is regulated by mechanical properties of the target. Journal of cell science, 2002. 115(4): p. 849-856.

2. Champion, J.A. and S. Mitragotri, Role of target geometry in phagocytosis. Proceedings of the National Academy of Sciences of the United States of America, 2006. 103(13): p. 4930-4934.

3. Champion, J.A., A. Walker, and S. Mitragotri, Role of particle size in phagocytosis of polymeric microspheres. Pharmaceutical research, 2008. 25(8): p. 1815-1821.

473372 Process Development and Scale-up of an ADC Conjugation Process

Abstract:

Antibody-drug conjugates (ADCs) are a class of pharmaceuticals that use an antibody to selectively deliver a cytotoxic drug to cancer cells. Critical quality attributes (CQAs) are controlled by means of the input quality of antibody and small molecule drug/linker components as well as execution of the conjugation within defined process parameter ranges. Fairly minor changes in ADC process parameters may have the potential to result in changes to the drug substance CQAs. Furthermore, the limitations of in-process sampling may not offer immediate detection nor the potential for corrective action. Thus, in-depth understanding of the conjugation process, determination of scale-dependent parameters, and operating knowledge of manufacturing equipment are essential elements necessary to develop a control strategy. This presentation will provide an overview of the experiences learned from a successful scale-up campaign.

469000 *In Situ* crosslinked Endosomolytic Polymeric Vesicles for Immunotherapeutic Targeting of the Sting Pathway

Abstract:

Cyclic dinucleotides (CDNs) are an emerging class of immunomodulators that have shown particular promise in cancer immunotherapy for their ability to elicit the production of type I interferons and trigger the generation antitumor leukocyte phenotypes in immunosuppressive tumor microenvironments (TMEs). The endogenous ligand of the stimulator interferon genes (STING), 2â€™3â€™ cyclic guanosine monophosphate-adenosine monophosphate (cGAMP) has recently shown promise as an immunotherapeutic capable of eliciting tumor regression and immune memory. However, a number of drug delivery barriers limit the therapeutic efficacy and translation of CDNs, including nuclease degradation, poor intracellular uptake, and inefficient cytosolic delivery. To address this challenge, we have developed polymeric vesicles that actively enhance endosomal escape of cyclic dinucleotides for efficient delivery to the cytosolic STING receptor.

Reversible addition-fragmentation chain transfer (RAFT) polymerization was used to synthesize poly[(ethylene glycol)-b-(butyl methacrylate-co-diethylamino ethyl methacrylate-co-pyridyl disulfide ethyl methacrylate)] (PEG-b-DBP) amphiphilic diblock polymers that self-assemble into endosomolytic pH-responsive vesicles with d < 100 nm in aqueous conditions with neutral surface charge. Vesicles encapsulated cGAMP with efficiencies of 38% and significantly enhanced potency *in vitro*, manifesting in multiple order of magnitude decreases in cGAMP EC50 when encapsulated. Nanoparticle cGAMP formulations demonstrated significant inhibition of tumor growth and increased mouse survival times in a murine melanoma model. Collectively, these data demonstrate that endosomolytic polymer nanoparticles significantly improve CDN potency by enhancing delivery to the cytosolic STING sensor and suggest that encapsulated cGAMP formulations may prove to be a powerful cancer immunotherapeutic.

472945 Impact of Dose, Route, and Composition on the Immunogenicity of Polyelectrolyte Multilayers Assembled on Gold Templates

Abstract:

Novel biomaterial vaccines are promising in treating challenging diseases such as cancer and infectious diseases. We recently developed a vaccine platform based on self-assembly of immune signals into immune polyelectrolyte multilayer (iPEMs). These iPEMs vaccine can be delivered on gold nanoparticle template to stimulate effector cytokines *in vitro* and expand antigen-specific T cells *in vivo*. Here we investigate the impact of vaccine dose, route, and adjuvant choice on the generation of T cell immunity and anti-tumor response. Three injection routes (i.e. intradermal, subcutaneous and intramuscular injections) and three iPEM dosing levels were employed. Intradermal injection induced the most potent antigen-specific T cell responses, and in all cases, the level of response was dose dependent. Importantly, we discovered that iPEMs generate durable memory T cells responses, indicated by potent, antigen-specific $CD8^+$ T cells responses in mice receiving a vaccine challenge 70 days after a priming immunization. In a common exogenous antigen melanoma model, iPEM vaccines inhibited tumor growth more effectively than equivalent admixed formulations consisting of antigen and one of several toll like receptor agonists used as adjuvants. This study reveals the ability of iPEMs to enhance response to several different classes of vaccine cargos using common injection routes, creating a new vaccine and adjuvant delivery platform based entirely on immune signals.

473160 Diminution of Macrophage Attachment on CD47-Functionalized and Polydopamine-Coated Surface

Abstract:

CD47 is a ubiquitously expressed protein known as "marker of self". Recently, CD47 functionalized surface has shown the ability to reduce inflammatory response and platelet adhesion. Polydopamine is an ultrathin but robust film formed by pH-induced polymerization. Polydopamine has been used as a universal surface modification agent capable of further surface functionalization. The purpose of this study is to use polydopamine as a platform to immobilize CD47 onto substratum in order to prevent macrophage attachment. The surface of 6-well cell culture plates was modified with biotin molecules through polydopamine coating. CD47-streptavidin fusion protein was then immobilized onto the biotin-functionalized surface through the high affinity between biotin and streptavidin. The effect of CD47-SIRPα interaction on cell attachment was studied by inoculating J774A.1 macrophages and human mesenchymal stem cells (hMSCs) on CD47 immobilized cell culture plates. Macrophage cells have high amount of SIRPα expressed on the cell surface, while hMSCs do not have any SIRPα. CD47 modified surface deferred macrophage attachment for up to 8 hours. However, hMSCs attached to CD47 immobilized surface within 2 hours. Our results demonstrate that CD47-functionalized surface can prevent macrophage attachment due to CD47-SIRPα interaction.

457963 De Novo Design of Skin-Penetrating Peptides for Enhanced Transdermal Delivery of Peptide Drugs

Abstract:

Skin-penetrating peptides (SPPs) are attracting increasing attention as a noninvasive strategy for transdermal delivery of therapeutics. Recent work has demonstrated their potential for dermal delivery, and experimental as well as theoretical studies have elucidated their mechanism of action. However, discovery of such peptides is largely driven by experimental screening of large libraries rather than theoretical or computational designs. Recent mechanistic studies have shown that SPPs must possess affinity for both keratin in skin and the drug of interest to be effective enhancers. Based on this finding, we computationally generated and screened virtual libraries of disulfide-bonded, cyclic peptides against keratin and Cyclosporine A (CsA) to identify SPPs capable of enhancing transdermal delivery of CsA. The sequential screening procedure was performed using the docking software HADDOCK and returned a list of sequences that were tested *in vitro* to validate their ability to bind CsA and enhance its transdermal permeation. The selected sequences were experimentally tested and were found to bind keratin and CsA, as determined by mass spectrometry and affinity chromatography, as well as to enhance CsA transport across the skin. The leading heptameric sequences were tested and yielded CsA permeation on par with previously discovered SPP (SPACE). An octameric peptide yielded significantly higher delivery of CsA compared to heptameric SPPs. Further, the safety profile of the selected sequences was validated by incubation with skin keratinocytes. This method represents an effective procedure for the de novo design of skin-penetrating peptides for the delivery of desired therapeutic or cosmetic agents.

472586 Ligand-Targeted Conjugate Systems for Delivery of siRNA to Tumors

Abstract:

Small interfering ribonucleic acids (siRNA) are a class of nucleic acid-based drugs that have the potential to knock down any gene with a high degree of specificity. They harness a natural gene regulation and antiviral mechanism for efficient destruction of any mRNA. This is useful for both the understanding of and treatment of diseases ranging from genetic disorders and viral infections to cancer. However, to be effective, the siRNA must travel from the injection/infusion site to the tissue of interest and then reach the cytoplasm of the desired cell type. Nucleic acids are highly charged and cannot cross the cell membrane unaided. Exposure to serum endonucleases also accelerates the destruction of the siRNA payload. Methods to cross these barriers include targeted delivery using vehicles ranging from lipid nanoparticles to polyplexes. Unfortunately, the delivery materials tend to be toxic, limiting the efficacy in patients, and also tend to localize to immune cells, the endothelium, or the liver. Currently, there are very few biomaterials that can successfully deliver siRNA to tissues outside of this group, and none are in Phase II clinical trials.

Recently, advances in siRNA stabilization technology have made unencapsulated siRNA conjugates viable, as chemical modifications to the RNA backbone have been used to reduce susceptibility of the siRNA to endonuclease degradation. These conjugates also have desirable treatment properties, including an in-human safety profile of up to 300 mg/kg and a clear regulatory path due to a well-defined chemical structure. Furthermore, due to their small size and thus ability to diffuse relatively quickly through extracellular matrix, the conjugates can be injected subcutaneously, a home procedure familiar to many diabetics on insulin, as opposed to nanoparticles, which must be infused intravenously at a clinic or hospital.

We have developed an azide-alkyne cycloaddition based system that allows for the delivery of ligand-targeted siRNA conjugates to models of glioblastoma (a form of brain tumor) *in vitro* and *in vivo*. The ligand used is based on chlorotoxin. *In vitro* studies using U87 (glioblastoma) cells show a similar degree of knockdown to commercially available transfection materials relative to a scrambled siRNA control. Confocal microscopy studies using derivatives of these ligands with fluorescently-labeled siRNA demonstrate a high degree of uptake relative to unconjugated siRNA in both cases. These conjugates are also able to traverse the bloodstream, deliver a nucleic acid payload into the tumor cells, and efficiently knock down expression of a housekeeping gene. Using fluorescent derivatives of our siRNA, we are able to observe localization of the siRNA to subcutaneous or intracranial tumor xenografts. Intravital imaging experiments demonstrate the ability of these conjugates to pass the blood-brain barrier and permeate the tumor. *In vivo* knockdown studies show a moderate degree of tumor-specific knockdown of a housekeeping gene in an intracranial model of glioblastoma using U87 cells.

These results demonstrate that siRNA knockdown of genes in tumors using these modular conjugate systems are feasible and present a way forward for using siRNA in the treatment of cancer. Future work involves using biologically-relevant siRNA to demonstrate anti-tumor effects in this and syngeneic tumor models.

472815 Design,Development and Optimization of pH-Responsive Hydrogels for the Oral Delivery of Human Growth Hormone

Abstract:

Protein therapeutics have the potential to treat a variety of debilitating disorders ranging from diabetes, growth hormone deficiencies, and even certain forms of cancer. However, due to their large and low stability, their administration has been traditionally limited painful and embarassing injections. An orally delivered carrier capable of protecting the delicate cargo during its transit through the gastrointestinal tract and undergoing targeting delivery to the absorptive region of the upper small intestine has the potential to overcome limits of protein stability, improve oral bioavailability and provide a more patient tolerated means of protein administration. To this end, a pH-responsive hydrogel carrier composed of methacrylic acid (MAA), N-vinyl pyrrolidone (NVP), and poly(ethylene glycol), P((MAA-co-NVP)-g-EG), was developed to deliver higher molecular weight protein therapeutics via the oral route. By varying crosslinking density and length, the swelling behavior of these gels could be modulated and adjusted to accomodate the larger protein, with loading efficiencies up to 60% for human growth hormone. Additionally, due to their pH-responsive behavior, the protein was not released under simulated gastric conditions but was preferentially released in simulated intestinal conditions. The hydrogel carriers also proved to be non-toxic in both model intestinal cell lines - Caco-2 and HT29-MTX - and in vivo upon administration to C57Bl/6 mice. Finally, when administered to the intestine, human growth hormone loaded P((MAA-co-NVP)-g-EG) hydrogels successfully delivered their protein payload with oral bioavailabilities of 0.5 - 5%, which is similar to other reported oral delivery stratgies for protein therapeutics.

460422 Design of Biodegradable Layer-By-Layer Films Capable of Localized, Efficient, and Sequential Gene Delivery

Abstract:

Layer-by-layer (LbL) films containing cationic polyelectrolyte and anionic bioactive molecules such as DNA are promising biomaterials for controlled and localized gene delivery for a number of biomedical applications such as cancer DNA vaccine delivery and tissue engineering. The LbL technique allows a wide variety of biological molecules including DNA, RNA, protein, and peptides to be incorporated into surface coating with exact amount and spatial distribution. Additionally, the release time and dosage are controllable by the LbL film-based gene delivery vehicle. Although nonviral gene delivery vectors generally show lower cytotoxicity than their viral counterpart, their low transfection efficiency prevents their clinical translation. In order to improve the transfection efficiency, we synthesized a novel bioreducible poly(amido amine) (PAA). We incorporated a highly transfecting monomer, 5-amino-1-pentanol (APOL) into the PAA molecular structure. We also modified this new type PAA with different amount of disulfide bond. The disulfide bond here can be reduced in target cell thus enabling DNA release from the polymer/DNA complex. The buffering capacity of the new APOL-containing PAA in the relevant intracellular pH range was higher than poly(ethyleneimine) (PEI) as well as the PAA without APOL. The high transfection efficiency of the APOL-PAA was confirmed by in vitro experiments using HEK 293, NIH 3T3, and MC 3T3 cells. We chose the highest transfecting APOL-PAA and applied it into our LbL film design. We observed increased transfection in the APOL-PAA containing LbL films. We also modified the LbL interior structure by incorporating non-biodegradable PEI as a barrier layer to achieve sequential disassembly and sustained DNA release. Moreover, by using two different DNA plasmids embedded in different layers of the LbL film, the two types of DNA were shown to be released at different times. Our study contributes new molecular and LbL film designs for improved DNA delivery based on the LbL film technique.

466875 Aminoglycoside-Derived Liposomes for Drug Delivery

Abstract:

Breast cancer is one of the leading cancers in women, and is responsible for an estimated 50,000 deaths every year in the US. Triple-negative breast cancer (TNBC) is diagnosed in 15-30% of all breast cancer cases, and dvelopment targeted therapeutics are challenging due to lack of estrogen, progesterone, and HER2 receptors in this disease. Aminoglycosides contain multiple sugar groups with with amines and hydroxyls, and have been investigated as antibiotics in the clinic. Aminoglycoside-derived lipid-containing polymers were formulated into nanoparticles (liposomes) for encapsulation and delivery of mitoxantrone, an FDA-approved DNA damaging drug, to different cancer cells. Five different liposomes were synthesized, formulated and characterized for their size, surface charge, and stability over time. The efficacy of liposomal mitoxantrone for ablation of TNBC, bladder, and prostate cancer cells was evaluated both as a single-agent treatment and in combination with TRAIL (Tumor Necrosis Factor-apha Related Apoptosis-Inducing Ligand), which is a protein that has been shown to selectively kill cancer cells compared to normal cells. In addition the combination of liposomal mitoxantrone with PARP (poly ADP ribose polymerase enzyme) inhibitors was also evaluated in cancer cells. Liposomal mitoxantrone demonstrated 6-10 fold higher efficacy for ablation of cancer cells compared to unencapsulated drug in vitro. In vivo biodistribution and efficacy of liposomal mitoxontrone is being studied using TNBC tumor-bearing mice.

462047 Water Soluble Enteric Polymer

Abstract:

Water Soluble Enteric Polymer

Jin Zhao[1], Oliver Petermann[2], Matthias Knarr[2], Josh Imperial[1], Debbie Holbrook[1], Harold Bernthal[1]

(1) Pharmaceutical Excipients R&D, Dow Pharma & Food Solutions, Midland, MI 48674

(2) Pharmaceutical Excipients R&D, Dow Pharma & Food Solutions, Bomlitz, ND 29699, Germany

Enteric dosage forms have been developed to protect either acid sensitive active pharmaceutical ingredients (API) from gastric conditions or the gastric tissue from an irritating APIs. Current technology typically involves coating multi-particulates, capsules or tablets with a known enteric polymer. Such coating processes usually involve organic solvents that are flammable, toxic, and explosive, or involve aqueous dispersions that are hard to handle, tend to block coating nozzles and have poor stability ADDIN EN.CITE ADDIN EN.CITE.DATA (*1-3*). To overcome these drawbacks, a new class of enteric polymer has been developed that can be directly dissolved in water and used to manufacture enteric dosage forms such as intrinsic soft and hard shell capsules, as well as enteric films and coatings.

Films and capsule shells were prepared using water soluble (WS) hypromellose acetate succinate (HPMCAS) aqueous solutions. The rheology of aqueous solutions of WS HPMCAS showed a reversible thermal gelation behavior which can be adjusted to fit various film forming manufacturing requirements by varying the polymer type, molecular weight, polymer concentration and the usage of plasticizers. The resulting films demonstrated comparable mechanical strength to films made from organic solutions. The API diffusion across the film also demonstrated acid resistance and release performance that satisfied USP enteric requirement.

Water soluble HPMCAS capsules were produced by a dipping process where the cold aqueous solution can be dipped with relatively low pin temperatures and less aggressive drying process (Figure 1).

The findings from this research can help pharmaceutical scientists manufacture intrinsic enteric dosage forms such as soft and hard capsules, chewable films, as well as apply enteric coatings with cost effective process. The enteric capsules can be manufactured without additional coating steps. This will not only enable controlled oral delivery of acid sensitive biomolecules, vaccines, and live organisms, but also enable rapid prototyping of New Chemical Entities (NCEs) and New Biological Entities (NBEs) that require enteric protection. Such enteric dosage form can provide not only enteric release but also delayed or sustained release in both pharmaceutical and nutraceutical applications.

Drying

Figure 1. Enteric capsules made from WS HPMCAS aqueous solutions.

Acknowledgements

The authors wish to thank Robert Schmitt and Shrikant Khot for their insights and suggestions throughout this research.

References

ADDIN EN.REFLIST 1. M. A. Osman, R. B. Patel, D. S. Irwin, P. G. Welling, Absorption of theophylline from enteric coated and sustained release formulations in fasted and non-fasted subjects. *Biopharmaceutics & drug disposition* **4**, 63-72 (1983).

2. M. A. Osman, R. B. Patel, D. S. Irwin, P. G. Welling, Absorption of theophylline from enteric coated and sustained release formulations in fasted and non-fasted subjects. *Biopharmaceutics & drug disposition* **4**, 63-72 (1983)

3. P. Mura, F. Maestrelli, M. Cirri, M. L. Gonzalez Rodriguez, A. M. Rabasco Alvarez, Development of enteric-coated pectin-based matrix tablets for colonic delivery of theophylline. *Journal of drug targeting* **11**, 365-371 (2003).

4. D. Cade, H. straub. (Capsugel France SAS, WO2013/164121).

468054 Preparation of a Drug Delivery System (DDS) of Hydropylic and Hydrophobic Drugs: In Vitro Validation in Human Lung Cancer Cells

Abstract:

PREPARATION OF A DRUG DELIVERY SYSTEM (DDS) OF HYDROPYLIC AND HYDROPHOBIC DRUGS: IN VITRO VALIDATION IN HUMAN LUNG CANCER CELLS

Fernando Mesías-Recamán, Jesús A. Rodríguez-Rodríguez, Miguel A. Galán, Eva Martin Del Valle

It is well known that current therapies present many side effects due to their low tumor cells-recognition capacity (low specificity), low solubility, low specificity, low bioavailability, prolonged circulation time in the body (1, 2) and high toxicity (3). In this regard, developing new medical strategies which could solve these disadvantages is necessary. Therefore, in this work, two different drugs with different hydrophobicity were combined to obtain an important synergic effect on the tumor cell line H460 (human lung cancer).

Our strategy is based on the combination of the carboxymethyl-β-cyclodextrin (CMCD) covalently attached to cisplatin as paclitaxel (PTX) carrier. In this manner, cisplatin and PTX could be delivered at the same time obtaining a synergic effect. CMCD was selected due to it presents low toxicity, small size, non-immunogenicity, excellent biocompatibility and high conjugation capacity with a variety of lipophilic compounds (4, 5, 6).

Furthermore, this molecule presents a hydrophilic ring and a hydrophobic cavity (7) that allows the link of cisplatin in its surface and forms the inclusion complexes with PTX. Cisplatin was covalently attached to the outside oligosaccharide ring through activation of the carboxylate functional groups of the CMCD, while PTX was included inside the oligosaccharide ring forming an inclusion complex. In addition, the solubility of both drugs was increased in aqueous conditions when drugs were bind to/formed the inclusion complex with

CMCD.

This system was characterized by NRM and FTIR, and thermodynamic studies. The PTX inclusion was determined by HPLC.

Later, their stability over time and the paclitaxel and cisplatin release kinetics were also studied. Finally, the efficacy of the both drugs delivery system has been validated *in vitro* with the H460 cell line by MTT assays and promising results have been obtained. Thus, this system could be a novel strategy to avoid the side effects that current antitumor therapies generate.

References

[1] Ganta S., Devalapally H., Shahiwala A., Amiji M. A review of stimuli-responsive nanocarriers for drug and gene delivery. J. Control Release. 2008, 126(3)187-204.

[2] Barreto J. A., Malley W. O., Kubeil M., Graham B., Stephan H., Spiccia L. Nanomaterials: Applications in Cancer Imaging and Therapy. Adv. Mater. 2011, 23(12):H18-40.

[3] Kesharwani P., Jain K., Jain N. K. Dendrimer as nanocarrier for drug delivery. Prog. Polym. Sci. 2014, 39(2):268–307.

[4] Zhang J. and Ma P. X. Host-Guest Interaction Mediated Polymeric Core-Shell Assemblies: Versatile Nanocarriers for Drug Delivery. Angew. Chem. Int. Ed. Engl. 2009, 48(5):964-968.

[5] Davis M. E., Brewster M. E. Cyclodextrin-based pharmaceutics: past, present and future. Nat. Rev. Drug. Discov. 2004, 3(12):1023-1035.

[6] Zhang J. X., Ma P. X. Host-guest interactions mediated nano-assemblies using cyclodextrin-containing hydrophilic polymers and their biomedical applications. Nanotoday. 2010, 5(4):337-350.

[7] Álvarez-Parrilla E., De la Rosa L. A., Torres-Rivas F., Rodrigo-García J., González-Aguilar G. A. Complexation of apple antioxidants: chlorogenic acid, quercetin and rutin by β-cyclodextrin (β-CD). J. Inclusion Phenom. Macrocycl. Chem. 2005, 53:121-129.

472207 A QbD Approach for Identifying Critical Process Paramerters in Drug Substance Manufacturing

Abstract:

The Vertex approach to critical process parameter identification via an effect threshold concept will be discussed and accompanied by examples applying the method to drug substance development and manufacture.

481150 QbD and Continuous: A Step Change in Performance

Abstract:

Traditionally the majority of active pharmaceutical compounds have been manufactured at large plant scales using batch methodologies. Continuous manufacturing is becoming an increasing attractive alternative however due to apparent benefits such as smaller manufacturing footprints, increased process safety, lower manufacturing costs and increased supply-chain flexibility. Realising all of the above benefits using conventional continuous equipment is however quite challenging and sometimes not achievable. In this project, a new approach to the design of continuous manufacturing platforms is being presented which is not restricted to using such conventional equipment but which is based around the bespoke design and manufacture of all unit operations, optimised to fit the specific process being examined. Processes are quantified in detail using methods such as computational fluid dynamics (CFD) leading to tailored equipment designs which can be manufactured with high precision using metallic 3D printing.

In order to demonstrate the suitability of applying this design approach to the entire manufacturing process, a case study is also presented in which two reactions are coupled together, followed by an extraction and wash step and a liquid-liquid separation. All unit operations within this platform are designed using fundamental engineering principles combined with numerical analyses (based upon the requirements of the specific process), prior to being 3D printed and combined to create a bespoke, sealed, hard-piped manufacturing system. The optimised design of the units enables extremely high throughput rates to be achieved corresponding to 580,000 kg/L/day (compared to an approximate batch productivity of 525kg/L/day).

472550 A QbD Approach to Experimental Designs with Kinetic Parameters

Abstract:

Kinetic parameters are commonly ignored when applying traditional DoE approaches to chemiscal reactions, resulting in a failure to accurately capture and describe the time-dependent nature of the system. An alternative method utilizes such parameters as responses in the traditional DoE, allowing the impact of process parameters on the chemical kinetics to be evaluated and communicated within a standardized QbD framework.

455311 New Trends in Materials-Sparing and Modelling Approaches Towards Pharmaceutical Development

Abstract:

Rising development costs and longer development times that cut into exclusivity periods have pushed the pharmaceutical industry into re-examining current development processes. Historically, the development of pharmaceutical products relied heavily on trial-and-error approaches. Over the last couple of decades, the concept of QbD (Quality by Design) has been incorporated to build robustness and consistent quality into the final product. More recently, there has been a significant push to develop materials-sparing techniques and predictive modelling approaches to develop a sound framework of product and process knowledge at small scales in order to detect risks early in development and facilitate efficient diagnosis and mitigation strategies for challenges at the commercial scale. Such knowledge promotes both the development of robust products and also enables an acceleration path for development activities when costly API (active pharmaceutical ingredient) is limited.

In this talk, examples of how materials-sparing approaches have been used for the development of both an oral solids drug product (OSD) and a biologic drug product will be presented. For the OSD case, plastic/elastic material properties were leveraged to develop an empirical heuristic used to predict the feasibility of two materials to successfully yield a bilayer tablet. The primary tool used to develop this assessment was a compaction simulator fitted with an instrumented die. The materials-sparing work-flow efficiently guided the formulation development of bilayer tablets (using <50g of material); the approach was validated at pilot and commercial-scales and during accelerated stability testing. For the biologic drug product, an FEM-based model was used to optimize commercial lyophilization cycles by leveraging knowledge of heat and mass transfer principles. Lyophilization, which involves freezing a product and then reducing the surrounding pressure to remove the frozen solvent via sublimation, is used to increase the shelf-life of biologic products by enabling products to be stored as dry powders instead of in solution. Thus, heat and mass transfer rates during both freezing and sublimation processes impact product quality and stability. However,

technology transfer and scaling approaches typically utilize large DOEs (design of experiment) to determine cycle parameter ranges through a trial-and-error approach, which is costly both in terms of material needs (API) and resources (time, utility costs). By developing an understanding of how heat and mass transfer rates change with scale, the transfer of a product from lab to commercial scale was successfully completed using only two engineering runs. The presentation will focus primarily on key decisions made during the development process that were enabled by the use of materials-sparing laboratory techniques or modelling predictions and discuss how the propagation and acceptance of these capabilities can streamline the development of future products.

461149 The Journey of Mechanistic Models from Design to Optimization/ Launch..... and Beyond

Abstract:

This work describes a diversity of technical needs that led to the development of several mechanistic models to guide the design of specific drug substance processes at early phases of development. Some of these former models were also used for process optimization to define design spaces, predict process acceptable ranges and normal operating ranges and define conditions for process validation. Considerations on the requirements for models classified as medium impact and the approach to include mathematical expressions in the common technical document will be discussed. Finally, some details on the planned integration of mechanistic models to the manufacturing environment post-validation will be provided.

471359 Scale-up of Reaction and Crystallization Steps of a Commercial Manufacturing Process in a Multi-Purpose Plant Using Mechanistic Modeling

Abstract:

This work presents an approach on modeling a manufacturing process in a multi-purpose plant, that comprises multiple reaction and crystallization steps, from the laboratory to production scale. In a multi-purpose plant, more often than not, the equipment used might not be the most suited for a particular process (e.g., mass transfer limitations due to poor mixing capabilities of an impeller) and thus it becomes crucial to use modeling tools and first principle methodologies to increase process understanding and further optimize the process to highest performance possible.
The case-study to be shown comprises a manufacturing multi-step, multi-phase process already in commercial phase, in which the reaction and crystallization unit operations are both modeled in laboratory and large scales. It is assessed if the conditions used in today's manufacturing process are optimized and if there is room for improvement either in yield of final product (reduce impurity formation), or particle size distribution (narrower interval and more suited range for further particle engineering). The data source for the modeling is both offline (HPLC analysis) and online (FBRM) which can be used as-is in modeling tools such as Dynochem (no data processing needed) to adjust the kinetics and crystallization constants. Both models are derived from laboratory scale experiments and subsequently are used to predict the behavior of the large scale process. After having the models validated in large scale batches, they are used to propose new conditions for the large scale process, using narrower ranges inside the Design Space of the current manufacturing process, for either yield of final product or particle size distribution.
In parallel to the kinetics and crystallization modeling, potential mass transfer limitations are assessed with Dynochem and CFD (Computational Fluid Dynamics) tools, depending on the complexity of reactor and impeller designs (lab and large scale designs). The mixing efficiency and mass transfer coefficients are derived from this study and are added to the kinetics model to increase its robustness and improve the overall understanding of the process.

469184 Case Study: QbD Approach to Design Space Verification

Abstract:

Design Space Verification was completed for a recent regulatory filing using the Quality by Design paradigm for a drug substance process. A risk analysis approach was utilized to identify the key issues to address in order to verify that the design spaces derived from laboratory scale experiments were appropriate for commercial scale production.

467964 Development of Injectable Microgels for Galns Enzyme Replacement Therapy

Abstract:

Introduction: In this study, we developed injectable microgels comprised of multiarm polyethylene glycol acrylate (PEG-Ac) with tunable degradation and controlled release features. We employed the microgels for the encapsulation and delivery of N-acetylgalactosamine 6 sulfate sulfatase (GALNS) enzyme with the broader goal of developing an effective enzyme replacement therapy. Deficiency of GALNS causes accumulation of glycosoaminoglycans and ultimately Morquio A disease. Although effective in alleviating some disease symptoms, current enzyme replacement therapy (ERT) has many limitations, such as a need for recurrent infusions and occurrence of immune responses.[1] Delivery of GALNS encapsulated in PEG microgels could offer an improvement over existing therapies, as it preserves enzyme activity, leads to longer circulation times and attenuates immune responses, while requiring a single injection.

The PEG microgels in this study were prepared by electrospraying by Michael's addition reaction. We were able to obtain microgels of size range 70-300 μm by varying the electrospraying set-up parameters. The microgels could be injected through a needle, while still preserving their integrity upon injection. Further, the degradation of the microgels could be tuned to control release of encapsulated proteins. Release kinetics for several model proteins varying in size and hydrodynamic radii was found to be dependent on the protein size, as well as mesh size and degradation rate of the hydrogel. Furthermore, we observed preservation of GALNS enzyme activity in the upon encapsulation in the microgels. The nanoporous microgel mesh size precluded protein release until sufficient microgel degradation occurred.

Conclusions: The developed drug delivery approach has potential as a modified-ERT system to improve treatment outcomes and quality of life of Morquio A patients.

References: 1. Montaño, A. M.; Tomatsu, S.; Gottesman, G. S.; Smith, M.; Orii, T. International Morquio A

Registry: clinical manifestation and natural course of Morquio A disease. *Journal of inherited metabolic disease* **2007,** 30, (2), 165-74.

471705 Computationally-Driven Design of a Heterobivalent and Gate to Improve Targeting Agent Selectivity

Abstract:

Targeted molecular imaging agents have greatly advanced the sensitivity and specificity of detection methods and increased the ability of clinicians to detect and characterize cancers. However, these imaging agents often target biomarkers that are not exclusively expressed on cancerous cells, resulting in false positives and high signal background. Many cancer types do not have a single biomarker that can consistently distinguish them from the normal tissue. Yet the overexpression of two biomarkers is a more unique phenomenon that could enable selective targeting for a diverse array of patients. The goal of this study is to develop a modular molecular targeting system, labeled a biological AND gate, that can discriminate between cells that have a moderate overexpression of two biomarkers over cells that have moderate or even high expression of only one.

The AND gate molecule is a heterobivalent ligand comprising two binding domains linked via a flexible peptide. The two binding domains target two disease-associated biomarkers with a weak affinity that precludes monovalent binding to singly overexpressing cells while enabling robust bivalent binding to dually overexpressing cells. Optimal AND gate design – including individual affinities of the binding domains, linker length, and linker structure – depends on several properties of the both the tumor and the biomarker, such as average cell surface expression, internalization rate, homogeneity across cell membrane, and potential agonistic/antagonistic effects. To guide quantitative design, a geometric, cell-surface binding model was developed and integrated with a literature biodistribution model (Thurber and Wittrup *J. Theor. Biol.* 2012). The hybrid model indicates that, given a surface expression of 100,000 biomarkers per cell (each), a 100-fold increase in specificity can be achieved over cells expressing 100,000 of only one biomarker with individual ligand affinities of 125-575 nM. Stronger affinities hinder selectivity while weaker affinities preclude even bivalent binding. The optimal flexible linker length, to maximize local ligand concentration after binding of the first domain, is 22 amino acids for a receptor radius of 20 Å. Differential results are

observed for flexible and rigid linkers.

These optimal ligand properties were then engineered into an experimental heterobivalent ligand. In our model system, the binding domains are two non-antibody protein scaffolds (10th type III domain of fibronectin (Fn3) domains) that have been previously engineered to bind either epidermal growth factor receptor (EGFR, clone A') or carcinoembryonic antigen (CEA, clone C7) with initial affinities of 0.42 ± 0.06nM and 15 ± 15 nM, respectively. To obtain ligands with affinities within the 125-575 nM range, yeast surface display libraries, created by mutagenesis of each parental gene, were sorted by flow cytometry for decreased binding to the proper target. Mutants from the sorted sub-populations were produced and titrated on cancer cells, which revealed almost no correlation between yeast-displayed affinity to recombinant target and soluble affinity to the natively-expressed target. Nevertheless, several mutants for each target were identified; C7 mutants ranged from 2.5 to 5,600 nM and A' mutants from 15 to 3,000 nM. To determine scaffold modularity, *i.e.*, whether their affinities were affected by the fusion of a second Fn3 domain, the parental ligands A' and C7 were fused on either the N- or C- termini with a non-binding Fn3 clone. The fusion of a non-binding Fn3 to the C-termini (distal to the binding site) decreased the affinity of A' 50-fold to 21 ± 12 nM and C7 affinity held at 16 ± 10 nM. When the non-binding Fn3 was fused to the N-termini (near the binding site), A' affinity decrease was similar (18 ± 6 nM), while C7 showed a decrease in affinity to 89 ± 97 nM. Two mutants of A' were also tested in the fusion context; A' mutants A'.1 and A'.2, which have affinities of 28 ± 3 nM and 17 ± 4 nM, respectively, as single domains, displayed affinities of 28 ± 12 nM and 125 ± 93 nM, respectively, when fused to the non-binding Fn3 via their C-termini. An additional high-affinity Fn3 binder to EGFR (clone D) demonstrated no effect when fused via its C-terminus (0.4 ± 0.1 nM) compared to the monovalent affinity of 0.25 nM. These variations in fusion modularity demonstrate the importance of quantifying the *in vitro* affinity of fusion components – in the fusion context – and highlight dependence on unknown binding paratope, receptor binding epitope, binding orientation, and structural folding variations. Ongoing efforts include identifying the structural effect of the fusion via circular dichroism and SEC analysis.

For initial AND gate testing, three heterobivalent ligands were synthesized: A'-C7, A'.1-C7, and A'.2-C7, where the N-terminus of C7 was fused to the C-terminus of the three EGFR binding Fn3 via a glycine-rich linker to test their selectivity on native and modified cell lines of varying EGFR and CEA expression. Preliminary data of the A'-C7, A'.1-C7, and A'.2-C7 fusions on A431 cells (<2,000 CEA, 4,000,000 EGFR per cell) show that the fusion of C7 via the C-terminus further decreases the EGFR affinity of the fusion compared to the non-binding Fn3 fusion to 102 ± 31 nM, 443 ± 238 nM, and 1200 ± 1000 nM, respectively. However, titration of A'.1-C7 on lentiviral-transfected HEK-293T cells (250,000 CEA, <2,000 EGFR receptors per cell) results in an affinity of 69 ± 10 nM, showing that the C7 affinity is indistinguishable when fused to either A'.1 or the non-binding Fn3. This data shows that the effect of the fusion can be dependent on the clone you are fusing, which is crucial when accurately measuring the properties of the system, and the importance of cell lines with accurately characterized expression to quantify individual domain affinities within the heterobivalent ligand. Ongoing studies, evaluating binding selectivity and sensitivity on cell lines with varying levels of CEA and EGFR, will be discussed. Additional results to be shared include evaluation of ligands of varying affinities, additional scaffolds, and alternative fusion methods to create a more modular system.

461410 Supramolecular Pegylation of Biopharmaceuticals

Abstract:

Introduction: The practice of medicine has been transformed by the use of protein drugs, but these also introduce an array of attendant complications arising from poor chemical and/or structural stability, thus requiring extensive effort in the development of stabilizing excipients and formulation methods. Direct chemical modification with poly(ethylene glycol) (PEG) has demonstrated improvements in protein stability and circulation half-life, a practice that has been used clinically for a number of protein therapies. Herein, we demonstrate for the first time a platform approach to PEGylation of biopharmaceuticals using only supramolecular interactions by leveraging high-affinity host-guest interactions in order to endow PEG-like functionality on proteins non-covalently.

Materials and Methods: A supramolecular macrocycle with high binding affinity was synthesized and functionalized with PEG chains of controlled molecular weight *via* "click" chemistry. Leveraging host-guest binding of this macrocycle with native residues on proteins, this PEGylated compound was added in formulation with insulin, glucagon, and a therapeutic antibody. Functional performance of all three biopharmaceuticals was assayed both *in vitro* and *in vivo.*

Results and Discussion: The supramolecular PEGylation (SM-PEG) of insulin extends *in vitro* stability and preserves potency when aged under agitating conditions, extending protein life from approximately 14 hours alone to over 100 days when in formulation with SM-PEG. Moreover, in a diabetic mouse model the duration of insulin activity can be tuned by selection of PEG molecular weight. Extending this approach to glucagon and a therapeutic anti-CD20 antibody, the SM-PEG lends significant stability and preservation of activity to these proteins as well. This approach to supramolecular modification of therapeutic proteins provides an excipient-only strategy to modulate protein activity and stability without the need for covalent modification of the protein drug molecule, and has broad possible use as a platform for biopharmaceutical formulation.

Conclusions: In biopharmaceutical practice, assuring stability and modulating pharmacokinetics are two primary areas of interest, and covalent PEGylation has been among the more successful strategies evaluated in this regard. Using high binding affinities afforded by specific supramolecular host-guest interactions, proteins can be modified non-covalently with PEG chains of various molecular weights. This supramolecular PEGylation strategy dramatically extended the stability of three therapeutic proteins, and also prolonged the activity of insulin in a modular fashion dependent on molecular weight. This approach may be broadly applicable as a platform of "smart" excipients for formulation and shelf-life extension of biopharmaceuticals.

462185 *Non-Invasive Screening of Diseases Using an Orally Available Near-Infrared Imaging Agent*

Abstract:

Early detection of diseases is crucial to improving the chances of successful treatment, such as curative surgery of cancer or early intervention with disease-modifying treatments to prevent irreversible damage in rheumatoid arthritis. Current methods of detection involve either minimally invasive blood tests or medical imaging modalities. Blood tests can provide large amounts of molecular information but lack spatial information about the site of the disease, and imaging modalities can provide spatial information but only offer anatomical data. Molecular imaging modalities can bridge the gap between the aforementioned techniques, but the drawbacks include safety concerns with ionizing radiation and the associated high cost.

We aimed at developing a near-infrared orally available imaging agent to detect diseases non-invasively. This technique employs near-infrared fluorescence as a method of detection that makes the process safe (non-ionizing radiation) but limits the detection of diseases within 1 to 2 cm from the surface of the skin. This method of detection is easily portable, cheap and the use of an orally available imaging agent makes it completely non-invasive. Importantly, oral delivery avoids associated risks with parenteral delivery and could improve compliance in a screening setting.

As proof of concept, we chose a single-digit nanomolar integrin binding peptidomimetic as a model system. The small molecule ligand has the highest affinity for Î±$_v$Î²$_3$ integrin, which is highly expressed on certain forms on cancer and on macrophages present at sites of inflammation such as arthritic joints. The binder was conjugated to several near-infrared dyes in order to make a series of agents with varying physicochemical and optical properties to ascertain what properties are important for orally available imaging agents. *In vitro* and preliminary *in vivo* experiments were conducted to find the agent that had the best mix of oral bioavailability, *in vivo* optical properties and specificity. Based on these results the IRDye 800CW conjugate was used in an orthotopic breast cancer model to show that this method of detection is effective following oral delivery. We were able to produce target-to-background ratios (TBR) of 4.7 and tumor-to-muscle ratios over 20 using our method of delivery and detection. A mechanistic model based on partial differential equations was also made to correlate the properties of the agents to TBR, in order to highlight the most important properties from a quantitative pharmacokinetic and transport perspective.

467547 Investigation of the Cellular Uptake and Cytotoxicity of Cell-Penetrating Peptides in *Candida* Fungal Pathogens

Abstract:

Cell-penetrating peptides (CPPs) have been widely studied as tools for delivering a variety of molecular cargo into cells, including DNA, RNA, proteins, and nanoparticles. Previous work with CPPs has mainly focused on their uptake in mammalian cells, but CPPs also have potential as drug delivery and research tools in other organisms, including fungi. We have investigated 12 different peptides for their ability to function as CPPs for the important human fungal pathogens *Candida albicans* and *C. glabrata*. The peptides were commercially synthesized with an N-terminal 5-carboxyfluorescein label serving as small-molecule cargo and as a fluorescent probe to report cellular uptake. We have used flow cytometry to screen the CPPs for their potential to efficiently enter the *Candida* cells. Among the 12 CPPs, we have identified several promising CPPs, including pVEC and SynB, that can enter both *C. albicans* and *C. glabrata* cells, and the overall uptake process depends on both the salt concentration in the buffer and on the concentration of the peptides. We have also identified CPPs with some specificity; for example, Hst-5 can only translocate into *C. albicans,* whereas TP-10 has higher uptake in *C. glabrata*. We found that generally the CPPs with higher uptake also have higher toxicity toward both *Candida* species, but SynB showed almost no antimicrobial activity at a concentration where nearly 100% of cells have high levels of peptide uptake. Furthermore, we explored the subcellular localization of CPPs at different concentrations with fluorescent microscopy and flow cytometry. Most of the promising CPPs showed a 3-stage behavior, where CPPs evenly distributed in the cytosol at high concentrations, localized inside vacuoles at moderate concentrations, and gathered around the membranes at lower concentration. Cecropin B, in contrast, shows â€œall-or-nothingâ€ behavior, with the peptide either present in the cytosol or exhibiting no apparent association or uptake. In addition, we have also investigated the uptake at different temperatures to aid understanding of the mechanism of uptake for the peptides. Some peptides (e.g., MPG and $(KFF)_3K$) are significantly affected by temperature, indicating a potential ATP-dependent uptake mechanism, whereas temperature has minimal impact on

other peptides (e.g., pVEC and Cecropin B). Based on these results, we are now exploring the structure-function relationship for translocation of CPPs into fungal cells.

470913 Catalase-Laden Microdevices for Cell-Mediated Enzyme Delivery

Abstract:

Catalase-Laden Microdevices for Cell-Mediated Enzyme Delivery

Junfei Xia[1], Zhibin Wang[1], Yuanwei Yan[1], Zhijian Cheng[2], Li Sun[2], Yan Li[1], Yi Ren[2], Jingjiao Guan[1,*]

[1] Department of Chemical and Biomedical Engineering, FAMU-FSU College of Engineering, Florida State University

[2] Department of Biomedical Sciences, College of Medicine, Florida State University

Abstract

Enzymes have been used as drugs to treat a wide variety of human diseases and traumas. However, therapeutic utility of free enzymes is generally impeded by short circulation time, lack of targeting ability, immunogenicity, and inability to cross certain biological barriers. Various strategies have been developed to overcome these limitations. Among them is cell-mediated drug delivery, which is featured by integrating drugs with live cells and taking advantage of the unique capabilities of the cells to achieve controlled drug delivery. We developed simple and inexpensive methods based on soft lithography and layer-by-layer (LbL) assembly for fabricating disk-shaped microparticles termed microdevices and generating the cell-microdevice complexes. [1, 2] We herein applied these methods to the fabrication of enzyme-laden microdevices and corresponding cell-microdevices complexes for the first time. We used catalase as a model enzyme in this study. It is a water soluble protein with an isoelectric point of 5.8. It is thus negatively charged at pH 7. Poly(diallyldimethyl ammonium chloride) (PDAC) was chosen here as a polycation to form a bilayer with catalase. Catalase in the microdevices was catalytically active and active catalase was slowly released from the microdevices. The catalase-laden microdevices were attached to the external surface of live suspension and adhesion cells respectively to form cell-microdevice complexes. This technique is applicable to other therapeutic enzymes and therapeutic cells, and thus promises to find clinical applications for treating various human diseases and traumas.

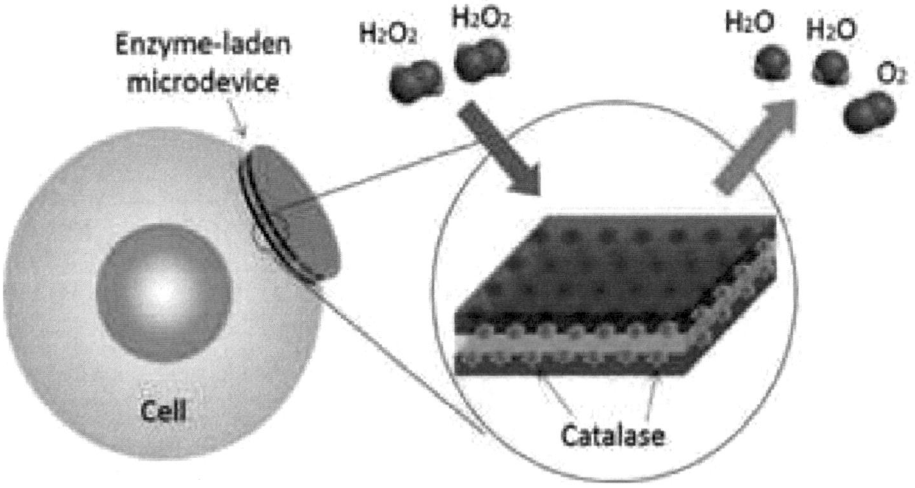

References

1. Junfei Xia, Zhibin Wang, Danting Huang, Yuanwei Yan, Yan Li, Jingjiao Guan*,*Asymmetric Biodegradable Microdevices for Cell-Borne Drug Delivery*, **ACS Applied Materials & Interfaces** (2015) 7: 6293-6299.

2. Zhibin Wang, Junfei Xia, Yuanwei Yan, Ang-Chen Tsai, Yan Li, Teng Ma, Jingjiao Guan*,*Facile functionalization and assembly of live cells with microcontact-printed polymeric biomaterials*, **Acta Biomaterialia** (2015) 11: 80-87.

472660 Ragweed Pollen-Based Oral Vaccine Delivery

Abstract:

Oral vaccine delivery although highly desirable, continues to be a difficult target. This is because the harsh acidic and enzymatic environment together with physical and biological barriers in the gastrointestinal (GI) tract pose unique challenges to deliver vaccines orally. We have developed pollen grains (PGs) as a novel method to deliver antigens across the GI mucosa. In this study ragweed pollens were used to deliver ovalbumin (OVA) orally.

Ragweed pollens were chemically processed to remove native biomolecules and to obtain clean pollen particles. Mice were then immunized with a formulation of OVA and ragweed pollens. Immune responses were characterized by determining the presence of serum anti-OVA IgG, IgG1, IgG2a, IgE, fecal IgA, salivary IgA, and vaginal IgA antibodies using enzyme-linked immunosorbent assay (ELISA). Our study shows that vaccine formulations with ragweed pollens stimulated significantly higher antibodies compared to control mice that received only OVA without ragweed pollens. Using ragweed pollens, high OVA-specific systemic immune response was induced compared to control mice. Furthermore, anti-OVA IgA antibodies were stimulated not only in the intestinal mucosa, but at the distal vaginal mucosal site. Overall, our study demonstrates the potential of ragweed pollen for a simple method of oral vaccination that can induce systemic and mucosal immunity. Further study is required to optimize the antigen dose and frequency of the delivery system.

472952 the Roles of Competing Dissolution, Diffusion and Transient Marangoni Convection Fluxes in Surfactant-Enhanced Spreading of Aerosols for Pulmonary Delivery

Abstract:

We are formulating aerosols for pulmonary delivery with surfactants to overcome aerodynamic limitations when treating obstructive lung diseases such as cystic fibrosis. The added surfactant is designed to establish transient Maragoni flows which can convect drug laterally after aerosol droplets are deposited onto airway surface liquid (ASL) and disease-associated mucus plaques. We have used a combination of aerosol droplet deposition experiments with mock systems (entangled (bio)polymer solution subphases mimicking the ASL, fluorophores mimicking drugs, multiple surfactant species) and numerical transport simulations (via COMSOL) to examine the competing roles of surfactant dissolution, diffusion and transient Marangoni convection fluxes in determining the extent of drug spreading. Large dissolution and diffusion fluxes lead to significant net drug transport, but poor spreading while small dissolution and diffusion fluxes lead to poor drug transport and poor spreading; extensive drug spreading occurs when dissolution and diffusion fluxes are similar in magnitude to the transient Marangoni convection flux. Several combinations of mock subphases, surfactants, and mock drugs will be discussed along with implications for delivery via both liquid and solid aerosols.

472776 Improving Single Use Bioreactor Design and Process Development, New Research Towards Intensifying Seed-Train and Scale-up Methods Using 5:1 Turn-Down

Abstract:

Operating bioreactor vessels at low working volumes (low turn down ratio) is often desirable but brings about challenges in regard to mixing, mass transfer, and process control. Research done towards optimizing cell culture has provided methods to improve performance and control when operating under these conditions. Implementing bottom heat exchange, making changes to impeller angle and height, taking advantage of the unique Thermo Fisher drilled hole sparge design, and implementing a new cross flow sparge into the headspace have yielded reliable mass transfer and cell culture results with the additional benefit of lower turn down mixing when draining the reactor.

462181 Characterization and Design of a 3L Single-Use Bioreactor to Support Process Development and Tech Transfer

Abstract:

The design and characterization of a 3L rigid single-use bioreactor (SUB), to replace existing 3L glass bioreactors, was pursued to further enhance the efficiency of development operations in Merckâ€™s Kenilworth biologics facility.

A myriad of bench-scale, rigid, single use bioreactors are commercially available at this time. We required one to replace our current 3L glass bioreactor vessels and meet the diverse requirements of our biologics facility, which supports a continually increasing number of cell culture projects. The new SUB needed to perform like our current bioreactors, so that process development activities would not be impacted. Overall, the SUB was expected to be versatile, robust, and simple to use.

A multi-faceted, iterative approach was used in the development of a new custom 3L SUB, with the following requisites:

- Similar configuration to larger vessels (e.g. drilled tube sparger and rigid baffles)
- Magnetic bottom drive agitation system
- Perfusion ready
- Expanded number of sensor ports and addition lines for flexibility
- Compatible with Automated sampling systems

To characterize the designs, we performed gas transfer (k_La) studies, computational fluid dynamics (CFD) modeling, and cell growth studies with both fed-batch and perfusion processes. The data generated from these studies was used to refine the design and confirm comparability to the glass vessel controls.

Implementation of the customized 3L SUBs has enabled enhanced process development capability by introducing technology to improve operational efficiency.

461864 Foam Control in the Mammalian Cell Culture Process through Extrinsic and Intrinsic Methods

Abstract:

During biopharmaceutical manufacturing, the formation of foam within bioreactors has been an industry-wide issue for decades. As the product throughput has increased over the years, higher viable cell densities have increased oxygen demand. To support this higher uptake, agitation and aeration rates have also increased, leading to more foam generation. This results in a myriad of problems, including cell entrapment within foam, cell damage caused by bubbles bursting at the surface, interruption of gas transfer across the air/liquid interface, and equipment failure due to clogged vent filters. To optimize bioreactor conditions, Biogen has been working on different methods of monitoring, removing, and controlling foam in bioreactors to dramatically reduce antifoam addition and prevent the regeneration of foam. While focusing on extrinsic methods of monitoring and removing foam through automation, the intrinsic characteristics of foam generation and stability are also being studied to find alternative methods of eliminating the foam in the cell culture process. New methods for monitoring and controlling the foam in bioreactors include using a digital camera to image the culture surface within the vessel. Antifoam is then programmed to be fed automatically based on the percentage of foam coverage. Studies of the delivery of antifoam have also shown that spraying antifoam as a mist reduces the required dose to remove the foam layer, while not significantly interrupting the mass transfer. It not only reduces the foam regeneration rate, but also protects cells from shear damage by allowing for a lower gas entrance velocity from the sparger. The foam generated during cell culture manufacturing processes has been characterized as stable Pickering foam, with slower foam decay. The foam stability was correlated to the accumulation of biomass within the culture. This stability reduces the efficacy of the traditional antifoam remediation strategy over time. Therefore, alternative methods of foam removal are now being investigated to improve the cell culture processes. Overall, the systematic ways of controlling foam have notably reduced the likelihood of equipment failure and improved bioreactor success rates.

460729 Morphological and Life Cycle Changes Exhibited at the up-Scaling of a *Saccharomyces Cerevisiae* High-Cell Density Fed-Batch Biopharmaceutical Process

Abstract:

In a previous study (Fu et al, 2014), we observed a clear difference in yeast cell volumes (determined by wet cell weight (WCW)) grown at 10L laboratory scale versus 10,000L manufacturing scale. That study also included metabolite analysis which suggested hypoxia during scale up. We hypothesized that hypoxia weakens the yeast cell wall during the scale up, leading to changes in cell permeability, and/or cell mechanical resistance.Â Cells treated with the zymolyase showed a more sensitive cell wall later in the production process at manufacturing scale, although cell viability from samples obtained at both scales were similar. In the current study we further examine the relationship between relative volumes of yeast cells to cell viability, size, morphology, and life cycle at laboratory scale (10L) and manufacturing scale (10,000L). Vi-cell and flow-cam methodologies demonstrate a larger sub-population of cells with aspect ratio 0.5 and â‰¥6 micron-diameter at 10,000L scale, corresponding to more one- and two-budded mothers in the larger scale process. In addition, results from a cell cycle study using flow cytometry and DAPI fluorescence microscopy demonstrated multi-budded cells having increased chromosomal content with a long residence in the M phase without completing cytokinesis. In summary, these results collectively suggest a possible relationship between the level of cell wall compromise with cell volume and the presence of a major sub-population of larger budding cells reflecting a longer residence in M phase.

456526 Accelerating Manufacturing in a Clinical Drug Substance Plant through Application of Plant Simulation

Abstract:

Given the major investment required to build or expand biopharmaceutical manufacturing facilities, intensification of existing assets is a highly valuable proposition. However, drug substance processes have many complex operational constraints. Increasing the run rate of a facility should only be undertaken when a thorough evaluation of all operational constraints has been performed.

In this case study, plant simulation was applied to an existing clinical drug substance manufacturing facility executing a wide variety of processes to evaluate opportunities to increase facility run rate. The modeling effort identified an opportunity to increase capacity by 40% with minimal capital investment. At current time, with partial implementation of the identified changes, a 20% increase in run rate has been realized.

461029 Model Predictive Control: A Process Development Tool for Mammalian Cell Culture Processes

Abstract:

In recent years, there has been a growing necessity to increase the speed, efficacy and information content associated to the development and scale-up of biopharmaceutical processes, particularly given the regulatory authorities current shift in philosophy towards quality-by-design (QbD). This QbD requirement for increased process understanding intensifies the work involved in the earlier phases of the drug-product lifecycle. The benefit, however, is that the increased process knowledge can speed the technical transfer from development into manufacturing, deliver a more optimised, robust process with higher titres and greater reproducibility and aid in troubleshooting and root-cause analysis of deviations during production.

A model predictive control (MPC) technology platform incorporating Raman spectroscopy was developed as a tool to streamline upstream bioprocess development and applied to a fed-batch CHO culture. In this talk, the core elements of the MPC will be discussed, namely the development of the Raman PAT method for the online quantification of cell density, substrate and metabolite concentrations, the process model and the MPC control algorithm.

The application of the MPC to process development will be demonstrated through a case study of its application to a CHO fed-batch culture. The MPC enabled the transition from traditional bolus to a PAT-enabled continuous feeding strategy in order to prevent nutrient depletion and deliver a stable macro-environment for the cells. Overall, application of the glucose set-point control strategy resulted in a 1.5 fold increase in peak viable cell density (VCD) and the integral of the viable cell density (IVC) which is directly related to increased titre.

466474 Accurate Determination of B_{22} Data Multivariate Analysis to Improve B_{22} As a Predictive Tool

Abstract:

ABSTRACT

Protein-protein molecular interactions are known to be involved in protein solution aggregation behaviour and are a common issue for the manufacturing of therapeutic proteins such as mAbs. Much effort has been employed to gain a better understanding of aggregation, however the mechanisms leading to protein aggregation are still not fully understood. The osmotic second virial coefficient (B_{22}) is a fundamental physiochemical property that describes protein-protein interactions in solution, which can be a useful tool for not only predicting the aggregation propensity of proteins but also protein solubility, crystallisation and phase behaviour.

A rapid and high-throughput way of determining B_{22} is self-interaction chromatography (SIC). However, the method is still critiqued to be less accurate than its predecessors such as static light scattering. This paper is evaluating the impact of the different experimental strategies and theoretical estimations on the parameters involved in the second virial coefficient determined by SIC. This has been done to help focus on the small variations and combinations that have the largest impact on B_{22} and can directly determine a critically different value.

A number of journal papers recently published describe important experimental considerations for more accurate B_{22} estimations. Many of these small differences in the experimental procedure could result in wrongly estimated B_{22} values and the interpretation that the protein-protein interactions are repulsive when they are actually attractive and an instable condition for the protein. To easily identify what initial parameters that have the greatest impact an extensive multivariate and sensitivity analysis has been performed to report the impact of a single or multiple parameters on B_{22}. The values that were found to be the most accurate representations of B_{22} were compared to aggregation data to identity that B_{22} was a useful and powerful predictor of aggregation propensity.

Apart from proving to be an accurate predictor for stability in pre-formulation and formulation studies, this tool has also shown to be a good indicator for suitable conditions that can be used for purification and manufacturing of mAbs to enable better product stability while not compromising recovery and quality.

In summary, this work is providing a good insight into important considerations when determining protein-protein interactions and virial coefficients to obtain accurate values that can be used as a powerful tool for biopharmaceutical development.

REFERENCES

Hedberg, S. H. M., Heng, J. Y. Y., Williams, D. R. & Liddell, J. M. (2015). Self-Interaction Chromatography of MAbs: Accurate Measurement of Dead Volumes. *Pharmaceutical Research,* 32, 3975-3985.

Quigley, A., Heng, J., Liddell, J. & Williams, D. (2013). The Accurate Measurement of Second Virial Coefficients Using Self Interaction Chromatography: Experimental Considerations. *European Journal of Pharmaceutics and Biopharmaceutics,* 85, 1103-1111.

458703 Multivariate Data Analysis (MVDA) for Cell Culture Process Development

Abstract:

Multivariate Data Analysis (MVDA) for Cell Culture Process Development

John Bowers, Maria Khouzam, Louis Obando, John Higgings, and Balrina Gupta.

Analyzing mammalian cell culture results has traditionally been done after batch completion by creating and reviewing univariate plots of each variable separately. This approach results in a large amount of data that is not easily interpreted. MVDA Multivariate Data Analysis (MVDA) resolves these because it analyzes parameters and their covariance of large data sets to summarize the performance of the process. Real-time MVDA allows for online process monitoring, and provides immediate understanding.

To apply MVDA during cell culture development, several hurdles were overcome. The number and variety of bioreactors generated our first challenge. Experiments were performed in systems ranging from an automated set of twenty four 250 mL mini-bioreactors to a manufacturing skid with two 500 L stainless steel bioreactors. Since batch context was not available in a consistent format across the bioreactor controllers, special software was developed to enter it into PI. Furthermore each reactor had its own configuration, so a universal configuration structure was develop and the actual configuration of each bioreactor entered into it.

In addition, continuous process data were collected into the data archiving system while off-line results were collected into a separate electronic notebook (ELN) system, making it harder to use combined results to holistically analyze new batches. To combine offline measurements with continuous bioreactor data, SIMCA-Online was configured to extract the daily offline measurements from E-Notebook BioAssayÂ® (PerkinElmer) database. Batch information and the continuous data were taken from a PI data historian. A â€œgrab and holdâ€ of the last entered value was sufficiently representative of daily offline measurements to account for the sampling rate difference between the daily and continuous measurements in the model. Models based on these combined modes significantly better described culture behavior, and were much better predictors of batch quality measures, such as titer or charge variants.

Furthermore, modeling approaches were developed to account for scale changes and intentional parameter adjustments. Â Most parameters used were intrinsic, such cell and metabolite concentrations, pH, DO, temperature etc. To account for scale effects in useful extrinsic parameters, such as gas flows or feed amounts, a volume based scaling was introduced. With these, we were able to use results from automated 250 mL mini-bioreactor system Ambr® 250 (Sartorius Stedim Biotech) to build models for 3-L and larger scale bioreactors.

In order to account for intentional changes in parameters, such as pH or temperature settings, variables were tracked as deviations from setpoints. The deviation from setpoint allowed for diagnostics of equipment performance, without the multivariate statistics being dominated by an intentional setpoint change. We applied this approach to process ranging studies to determine the impact of changing a parameter on overall process performance live, giving immediate feedback on what a parameter change impacted. When needed, the setpoints were then included in the models to quantify the magnitude of the change to performance or quality attributes.

We successfully applied MVDA using SIMCA-13 and SIMCA-Online® (Umetrics) to several development activities, such as assessing the impact of the parameter changes on the process performance, confirming similarity between master and working cell banks, and demonstrating process comparability across scales. Applying the techniques described here, as much information as possible was obtained from development activities, even in the context of continuing process changes. Six to eight initial similar batches can be used to create models for future work. These models, in turn, can be used to improve process understanding to guide future work. The new results can then be included in models to further improve understanding and model performance.

472155 Gold-Layersome Nanoshells: Preparation and Characterization for Photothermal Therapy

Abstract:

Gold nanoshells have attracted great attention in medicine due to their surface plasmon resonance (SPR) bands, nontoxic properties and ease of surface functionalization. A unique property of gold nanoshells is the ability to tune the optical resonances within the near-infrared (NIR) region. The high absorption efficiency of gold nanoshells and their ability to efficiently convert absorbed light into heat has enabled them to be an effective candidate for photothermal therapy for cancer treatment. In this work we present a new approach to synthesize gold nanoshells with variable shell morphologies using soft hollow templates; anionic liposomes coated with the cationic polyelectrolyte poly-l-lysine (PLL) (ca. 100 nm hydrodynamic diameter). PLL was used to enrich the template surface by adsorbing negatively-charged gold species in solution.

Two different approaches to create gold nanoshells were compared. In both approaches the nanoshells were characterized by transmission electron microscopy, field-emission scanning electron microscopy and UV-vis/NIR spectroscopy. The first method was in situ reduction of chloroauric acid in presence of layersomes, which has been reported for gold nanoshell formation on liposomes. This approach led to a packed structure of gold nanoparticles on the layersome surface with little control over nanoparticle or shell morphology. To gain this control and achieve a more continuous gold shell, a second approach based on a seeded growth method was used. Small gold nanoparticles (1-3 nm diameter) were synthesized and anchored onto the surfaces of layersome via electrostatic interactions to nucleate the growth of the gold overlayer. The resulting gold nanoshells exhibited controllable shell thickness by adjusting the gold ions concentration. Upon changing the shell thickness, the absorption wavelength corresponding to these plasmonic layersomes could be tuned to the biologically safe near-infrared wavelength. Changing chloroauric acid concentration also led to gold nanoshells with rough structures (branched, star or flake-like) that may have been due to the shape directing effect of PLL. It is hypothesized that the key role of PLL as a structure directing agent is originated from the electrostatic interaction between positive amine groups in PLL and the negatively charged AuNPs. The prepared gold nanoshells have the potential to be used for photothermal therapy and photothermal mediated drug delivery.

471453 Chiral Nanoparticles for Enantioslective Adsorption and Separations

Abstract:

The surfaces of chemically synthesized Au nanoparticles have been modified with *D*- or *L*-cysteine to render them chiral and enantioselective for adsorption of chiral molecules. Their enantioselective interaction with chiral compounds and pharmaceuticals has been probed by optical rotation measurements during exposure to mixtures with varying enantiomeric excess of chiral compounds such as *R*- and *S*-propylene oxide, *R*- and *S*-2-butanol, or *R*- and *S*-propranolol. The ability of optical rotation to detect enantiospecific adsorption arises from the fact that the specific rotation of polarized light by *R*- and *S*-propylene oxide is enhanced by interaction with Au nanoparticles. This effect is related to previous observations of enhanced circular dichroism by Au nanoparticles modified by chiral adsorbates. More importantly, chiral Au nanoparticles modified with either *D*- or *L*-cysteine selectively adsorb one enantiomer of the probe molecules such as propylene oxide from a racemic solution, thus leaving an enantiomeric excess in the solution phase. Au nanoparticles modified with *L*-cysteine (*D*-cysteine) selectively adsorb the *R*-propylene oxide (*S*-propylene oxide).

A simple adsorption model and accompanying experimental protocol have been developed to enable optical rotation measurements to be analyzed for quantitative determination of the ratios of the enantiospecific adsorption equilibrium constants of chiral species on the surfaces of chiral nanoparticles, ${K_L}^S / {K_D}^S = {K_D}^R / {K_L}^R$. This analysis is robust in the sense that it obviates the need to measure the absolute surface area of the absorbent nanoparticles, a quantity that is somewhat difficult to obtain accurately. This analysis has been applied to optical rotation data obtained from solutions of *R*- and *S*-propylene oxide, *R*- and *S*-2-butanol, or *R*- and *S*-propranolol, in varying concentration ratios, with *D*- and *L*-cysteine coated Au nanoparticles, in varying concentration ratios.

Recently, tetrahexahedral (THH, 24-sided) Au nanoparticles modified with *D*- or *L*-cysteine (Cys) have been used as enantioselective separators of the chiral pharmaceutical propranolol in solution phase. Polarimetry has been used to measure the rotation of linearly polarized light by solutions containing mixtures of propranalol and Cys/THH-Au NPs with varying enantiomeric excesses of each. Polarimetry yields clear evidence of enantiospecific adsorption of propranalol onto the Cys/THH-Au NPs. This extends prior work using propylene oxide as a test chiral probe, by using the crystalline THH Au NPs with well-defined, crystal facets to separate a real pharmaceutical. This work suggests that chiral nanoparticles, coupled with a density separation method such as centrifugation, could be used for enantiomeric purification of real pharmaceuticals. The model developed earlier has also been used to extract the enantiospecific equilibrium constants for *R*- and *S*-propranalol adsorption onto the *D*- and *L*-Cys/THH-Au NPs.

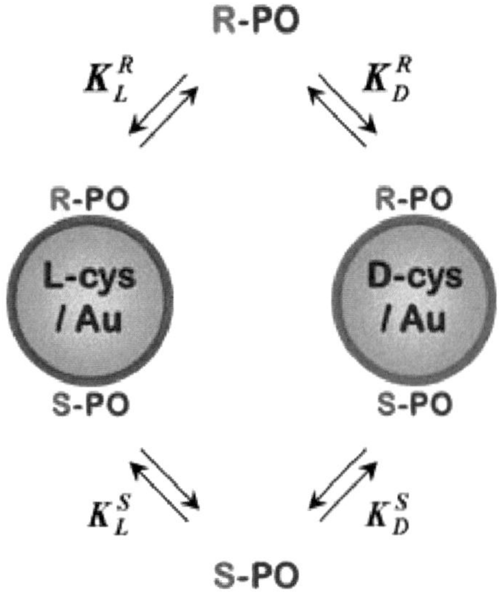

The figure illustrates the reversible equilibrium adsorption of *R*- and *S*-propylene oxide (PO) on Au nanoparticles modified with *L*- or *D*-cysteine. The equilibrium constants are enantiospecific, ${K_L}^S = {K_D}^R$ ' ${K_D}^S = {K_L}^R$.

460345 Triggered Ligand Clustering on Lipid Nanoparticles Enables Selective Targeting and Killing of Untargetable Cancer Cells: The Case for 'sticky Patches'

Abstract:

Objective: HER2-targeted nanoparticles encapsulating chemotherapeutics have shown promise for reducing the severity of side effects associated with traditional chemotherapeutics. However, the ability of HER2-targeted nanoparticles for effective targeting stops to hold on cancer cells expressing less than 200,000 HER2 receptors per cell (<1+ HER2 by immunohistochemistry). For cases of cancer with HER2 expression below this threshold, there is a lack of targeted treatment options because tumors are still considered untargetable. To address this issue, we have developed pH-responsive lipid nanocarriers (liposomes) that present a high local multivalency (â€˜very sticky patchesâ€™) on the surface of the liposome containing HER2-targeting lipopeptides. Sticky patches are formed by inducing preferential partitioning of lipopeptides into lipid raft-like domains that are triggered to form on the liposome membrane at acidic pH values ($6.5<pH<6.7$) matching the tumor interstitial pH.

Methods: Using HER2-targeting sticky liposomes loaded with doxorubicin, on a variety of cells expressing a wide range of HER2 (>1M copies per cell to 54,000 copies per cell), we evaluate the: 1) targeting selectivity, 2) killing efficacy, 3) endocytic pathway, 4) intracellular localization, 5) mechanism of action, and 6) potential toxicity to human cardiomyocytes and normal human mammary epithelial cells. In all studies, we compare to liposomes uniformly functionalized with HER2-targeting ligands via (or not) a PEG tether, and to non-targeted liposomes.

Results: Our studies show that sticky liposomes selectively and effectively kill otherwise untargetable breast cancer cells with as few as 54,000 HER2 receptors per cell utilizing both clathrin- and caveolar-mediated endocytosis, and are nontoxic to normal breast cells and cardiomyocytes which minimally express HER2. Sticky liposomes also demonstrate K_D values which are independent of HER2 expression, and long enough residence times on the cell surface ($t_{1/2}$ of dissociation ranged from 82 to 140 minutes) to allow for internalization of vesicles ($t_{1/2}$ of internalization ranged from 27 to 29 minutes).

Conclusions: These studies demonstrate the potential of sticky liposomes to selectively target and effectively kill cancer cells considered otherwise untargetable.

Acknowledgements: NJCCR, American Cancer Society RSG-12-044-01, NSF DMR-1207022, NSF CBET-1510015, GAANN.

470220 Superparamagnetic Nanoparticles for Triggered Drug Release from Alginate Hydrogels

Abstract:

While magnetic nanomaterials have already been used in clinics for contrast enhancement in magnetic resonance imaging (MRI) [1], there has been no clinical approval for a drug delivery system containing such nanoparticles, yet. Magnetic nanoparticles currently investigated for their possible application in biomedicine are predominantly different crystalline polymorphs of iron oxides. For small enough crystal sizes, iron oxides exhibit the so-called "superparamagnetic" behavior, a feature combining high magnetization with very low coercive forces. Such superparamagnetic nanoparticles show great potential in therapeutic applications due to their ability to transform the energy of an alternating magnetic field to thermal energy in what is often called magnetic fluid hyperthermia [2]. These particles dissipate thermal energy by magnetic relaxation through the Brownian and Neel mechanisms. Therefore, aqueous suspensions at relatively higher nanoparticle concentrations (g/L), the so-called "ferrofluids", also increase their temperature in the presence of an alternating magnetic field [3]. In this work, a composite multi-scale structure consisting of the biopolymer alginate, functional nanoparticles and a model drug is fabricated and analyzed. We examine the potential of flame-made SiO_2-coated Fe_2O_3 nanoparticles [4] as the stimuli-responsive material in the multi-scale composite structure. We perform detailed physicochemical and magnetic characterization on the hybrid alginate hydrogel beads and evaluate their potential in magnetic fluid hyperthermia and enhanced biomolecule release in the presence of an external AMF. The possibility to externally stimulate drug release will open up new possibilities in intelligent, on-demand drug administration [5].

References

[1] A. Singh, S. K. Sahoo, *Drug Discov. Today*, **19**, 474-481 (2014).

[2] A. Jordan, R. Scholz, P. Wust, H. Fahling, R. Felix, *J. Magn. Magn. Mater.*, **201**, 413-419 (1999).

[3] G. A. Sotiriou, M. A. Visbal-Onufrak, A. Teleki, E. J. Juan, A. M. Hirt, S. E. Pratsinis, C. Rinaldi, *Chem. Mater.*, **25**, 4603-4612 (2013).

[4] A. Teleki, M. Suter, P. R. Kidambi, O. Ergeneman, F. Krumeich, B. J. Nelson, S. E. Pratsinis, *Chem. Mater.*, **21**, 2094-2100 (2009).

[5] A. Teleki, F. L. Haufe, A. M. Hirt, S. E. Pratsinis, G. A. Sotiriou, *RSC Adv.* **6**, 21503-21510 (2016).

472317 Dual Drug-Loaded Janus Particles for Co-Delivery of Hydrophobic and Hydrophilic Compounds

Abstract:

The desire for drug delivery systems with tunable properties and multiple functionalities has spawned a new generation of particulate carriers, including Janus particles. Biocompatible Janus particles with internal compartmentalization are attractive as drug delivery vehicles because they offer a platform for co-delivery of two drugs with decoupled release kinetics.

Curcumin (CUR) and quercetin (QCT) were chosen as the model hydrophobic compounds for this study, while acetaminophen (APAP) and naproxen (NPX) were chosen as the model hydrophilic-hydrophobic drug pair. Curcumin and quercetin are bioflavonoids with anti-inflammatory, anti-proliferative, and immunosuppressive properties used to treat a host of diseases, including multi-drug resistant cancer and arthritis. Acetaminophen and naproxen are often used in combination due to their additive effects in pain management and treatment of rheumatoid arthritis.

The goal of this work is to obtain dual-loaded Janus particles with a different drug in each compartment, and to determine their encapsulation efficiency and drug content. Hydrophobic drugs are encapsulated by the single O/W emulsion technique. Hydrophilic compounds require special modifications for co-encapsulation with hydrophobic drugs due to their poor oil solubility and tendency to escape to the outer aqueous phase during the emulsification and solvent evaporation steps. Three different strategies for incorporating hydrophilic drugs were employed: 1) O/W emulsion with partially-water miscible solvent, 2) O/W emulsion with co-solvent (i.e., acetone, methanol, ethanol), or 3) W/O/W double emulsion. The encapsulation efficiencies and drug loading percentages are measured using UV/Vis spectroscopy and compared for the different synthesis methods.

461094 Preparation of Colloidal Microcapsules for Controlled Drug Delivery Systems

Abstract:

Nanoparticles (NPs) are one of the most used and studied materials in nanotechnology. Among the other usages, nanoparticles can also be used as a drug delivery agent in biomedical applications. They can be exploited as a sensor in the diagnosis and a delivery agent in the treatment of diseases. NPs can be synthesized by using a variety of constituents, primarily metallic, polymer, composite and lipid.. Solid lipid nanoparticles (SLNs), prepared using various physiological lipids, emulsifiers and water, have been used since the 1990s as colloidal nanocarrier systems, either in combination with or as an alternative to liposomes, emulsions and polymeric nanoparticles [1]. Possessing adjustable release kinetic, determined by the method of synthesis used, its level of biocompatibility, its stability and the degree to which active materials may be encapsulated within its structure, SLNs are ideal for use within the body and may be administered via different routes (parenteral, topical, ocular, etc.). There are several methods that could be utilized to synthesize SLNs, including high pressure homogenization, high shear homogenization, etc. However, if the incorporation of selected compounds is intended, particularly drugs, genes, proteins and other biomolecules, methods such as high pressure or high shear homogenization which create harsh environments for the material, need to be avoided. By using microemulsion method, the nanoparticle formulations of interest can be prepared in vials rapidly, reproducibly and cost effectively in a two-step process involving only mild operating temperatures. Moreover, the uniform nanoparticles produced become more biocompatible due to the elimination of organic solvent use during its preparation. Furthermore, the solubilization properties of hot microemulsions enable SLNs to be loaded with various drugs and is particularly useful for drugs with poor water solubility.

As well as NPs are being used in several areas including medical and pharmaceutical individually, more versatile systems can be established via self-assembly of NPs. Colloidal microcapsules are one of the most interested part of these self-assembled structures whose shell side are composed of nanoparticles [2]. Layer-by-layer method (LbL) is

one of the most used methods in the production of self-assembled structures. These self-assembled structures, independent from their single form, possess attractive properties such as high stability, enhanced functionality and increased loading capacity. Additionally, using pharmaceutically active ingredients such as drug crystals or drug loaded nanoparticles as the core material of colloidal microcapsules provides controlled release of the encapsulated material with the help of shell layers [3]. In this study, novel self-assembled colloidal microcapsules by using Ibuprofen crystals or $CaCO_3$ nanoparticles as core material and SLNs as shell material were prepared by using LbL method. The performance of establishing the layers was monitored by measuring the change of zeta-potential values. Also the self-assembled structures were visualized with SEM. Controlled drug release from the microcapsules was monitored by using UV-visible spectroscopy. Due to several advantageous properties such as being composed of physiologically related lipids, possessing properties like small and adjustable structure, high surface area, enhanced stability, stimuli-responsivity and increased encapsulation efficiency, these unique systems will be an ideal alternative to the conventional nanocarriers, which are especially used in pharmaceutical and biomedical applications as controlled drug delivery agents.

[1] Mehnert, W., Mader, K., 2001, Advanced Drug Delivery Reviews, 47, 165-196.

[2] Bae, J. et al. 2013, Current Organic Chemistry, 17, 3-13.

[3] Sukhorukov, G. B. et al. 2004, Langmuir, 20, 3398-3406.

461183 Stimuli Responsive Lipid Nanotubes for Drug and Protein Delivery

Abstract:

Lipid nanotubes (LNT) are one of the most attention drawing systems due to their superiorities provided by their asymmetric geometry such as high interior volume, high drug loading capacity, easy surface modifications and high circulation persistence as well as their high biocompatibility. Thus, they have high potential to be used in many different applications ranging from electronics to bionanotechnology. One of the most important applications of LNTs is their usage as drug delivery agents. Due to their exceptionally high interior volume compared with most of the other carrier systems they are expected to effectively entrap both small drugs and biomacromolecules.

Recently, we designed a molecule called as AQUA (AQ-NH-$(CH_2)_{10}$COOH, where AQ is anthraquinone group) that can form self-assembled lipid nanotubes. The molecular structure has been designed very carefully so that each group in the structure has key properties for both nanotube formation and its functions. AQUA LNTs have dual responsive character provided by COOH and AQ groups and its morphology can be controlled reversibly by external stimuli (pH and redox reaction). In the design step of AQUA molecules, we especially took into consideration to use them as delivery agents for some selected pharmaceuticals such as Doxorubicin (DOX), Hemoglobin (Hb) and BSA.

Total Hb entrapment capacity is determined by UV-vis spectroscopy from the absorbance value of the samples which were treated with Drabkinâ€™s reagent at 420 nm. The entrapment capacity is significantly affected by the solution pH and the loaded Hb amounts are constant and at its maximum, in the pH range of 5.0-7.0. AQUA nanotubes can hold Hb approximately two times that of their own weight. (i.e loading ratio is ~2) However, loading ratio decreases to 1.1 as the pH is increased to 9.0 due to the reduced electrostatic attraction. However, even at the high pH values AQUA nanotubes can carry significant amount of Hb. The capability to make Ï€-Ï€ stacking interactions of the AQ group in the AQUA increase the loading capacity and helps to obtain high loading ratios in a wide pH range.

Hb release profile to PBS buffer at pH 7.0 changes with loading pH and release temperature however, even at the maximum measured release conditions, AQUA nanotubes keep more than 80 % of the initially loaded Hb after three days and from this time the Hb release almost vanish. This such a low Hb release at the physiological conditions along with the high loading capacity, is very important and advantageous for the artificial blood application potential of AQUA nanotubes.

In the BSA loading and release studies Biuret method is used and the BSA amount is determined from absorbance value at 540 nm by UV-vis spectroscopy. Maximum BSA loading is achieved at pH 3.0 and pH 6.0 with the loading ratio of 0.8. When the BSA loading ratios of AQUA nanotubes are compared with the systems in the literature, it is seen that AQUA nanotubes are loaded with 10-50 times more BSA than most of them systems in the literature. At room temperature, the highest release is observed at pH 7.0 and totally 59 % ±0.5 of the loaded BSA is released after 48 hours which is determined as 18 %±0.5 for pH 5.0 and 20 %±0.5 for pH 6.0. When the temperature is increased to 37 ^{o}C at pH 7.0, the total released amount in the 48±0.5 hours increases to 68 %±0.5 due to the increase of the water solubility of BSA with temperature [1]. Unlikely to Hb, high release rates can be reached for BSA. The in vitro release profiles from AQua nanotubes shows opposite behaviors for Hb and BSA and both of those profiles are advantageous for different purposes. The very low release percentages for the Hb even when it is at its maximum, is a desired result for the artificial blood applications while the very high and pH and temperature dependent release profiles for BSA are pleasing for the controlled release applications.

The entrapment and release studies carried out with two different model proteins showed that the special functional groups of the AQUA molecule are quite effective to obtain strong attraction interactions with the guest molecules and it is possible to control these interactions on demand with the help of the environmental triggers due to the stimuli responsive character of our nanotubes. After these promising results the entrapment and release studies of the small drugs was carried out by using doxorubicin (DOX) as the model drug.

DOX is an anthraquinone based antibiotic which is commonly used for cancer treatment. However, it has various important side effects such as cardiotoxicity, supressing blood cell production, low therapeutic index, etc. So, developing effective carrier systems for DOX is a widely interested topic [2]. One of the most common cancer types for which DOX is used as the therapeutic agent is the colorectal cancer (CRC). Although parenteral ways are generally preferred for the treatment of the colorectal cancer, oral or intracolonic administrations can also be used.

AQUA and DOX have some similarities in their molecular structure and they both contain anthraquinone groups which we believe affects the affinity of DOX to AQUA LNTs via Ï€-Ï€ stacking interactions. In addition, the presence of carboxylic acid group in the molecular structure of AQUA helps us to encapsulate significant amount of DOX due to the hydrogen bonding and electrostatic interactions. There are some studies in the literature in which DOX is added to molecule structure/carrier via a covalent bond. These kinds of approaches will require more complex synthesis procedures which will prevent their use in large scale applications. Here we present a drug delivery system formed by relatively simple, but smartly tailored molecules.

The pH sensitive character of the AQUA LNTs is advantageous for increasing their efficiency on drug entrapment and release as well as providing selective release of the DOX to the tumor cells. DOX loading capacity is effected by the parameters such as drug concentration, pH, drug to carrier mass ratio. The maximum loading capacity is reached at pH 9 when the DOX/LNT ratio is 1/1 (g/g). In this condition AQUA LNTs can entrap almost as much DOX as their own mass. This loading capacity is significantly higher than the ones that could be reached in the literature [3]. Also it has been found that the release of DOX from the AQUA LNTs changes depending on the temperature and pH. The AQUA LNTs releases significantly more DOX at pH 5.5 which is the pH of the many cancer tissues than physiological pH (i.e. 7.0) This shows that AQUA LNTs selectively releases the DOX to the cancer tissue which is very advantageous for their usage as the carrier system in the cancer treatment.

The abovementioned results show that the rationally designed new stimuli responsive nanotube system is versatile as a highly functional carrier system for various different guest molecules changing from small drugs to biomacromolecules.

[1] M. A. Masuelli, Advances in Physical Chemistry, 2013, **2013**, 360239

[2] B.D. Weinberg, H. Ai, E. Blanco, J.M. Anderson, J. Gao, J Biomed Mater Res., 2007, **81**, 161.

[3] S. Ilbasmis-Tamer, H. Unsal, F. Tugcu-Demiroz, G.D. Kalaycioglu, I.T. Degim, N. Aydogan, Colloids Surf B Biointerfaces, 2016, **143,** 406.

466051 Uncertainty Analysis As Essential Step in the Establishment of the Dynamic Design Space of Primary Drying during Freeze-Drying

Abstract:

Since large molecules are considered the key driver for growth of the pharmaceutical industry, the focus is shifting from small molecules to biopharmaceuticals. The list of approved biopharmaceutical drug products from the Food and Drug Administration (FDA) and European Medicines Agency (EMA) consists of approximately 50% freeze-dried products (Constantino and Pikal, 2004). Therefore, freeze-drying is the preferred way to stabilise biopharmaceutical drug products which are unstable in aqueous solution. However, freeze-drying requires a lot of energy and has a long processing time, making it a costly process (Burns, 2009; Gieseler, 2012). The largest fraction of freeze-dried biopharmaceutical products are therapeutic protein formulations and vaccines (Pikal, 2002).

Conventional pharmaceutical freeze-drying (lyophilisation) is a batch process that consists of three consecutive steps: freezing, primary drying (ice sublimation) and secondary drying (desorption of unfrozen water). During the primary drying step the temperature at the sublimation front is critical, and should not exceed a critical value (dependent on the freeze-dried product) to prevent cake collapse. In this project, the focus is on the modelling of the primary drying step. The purpose of the model is to use it to support future optimisation of the process parameters during operation of the freeze-dryer in a dynamic way. During primary drying only two process parameters are adaptable, i.e. the shelf temperature and the chamber pressure. In conventional pharmaceutical manufacturing, both process parameters are fixed, resulting in a conservative cycle. However, by dynamically updating both variables (i.e. resulting in a time variant Design Space), a substantial decrease in processing time can be obtained.

The modelling approach can benefit from incorporating model parameter uncertainty to prevent that the critical temperature at the sublimation front is exceeded. Once the time variant Design Space is created for a certain set-up (i.e. a combination of a specific product and a piece of freeze-drying equipment), the values for both process variables can be fed to the equipment and a significant decrease in processing time can be obtained. As a result, the cost of processing will also be decreasing. Moreover, end product quality is guaranteed.

Burns, L. (2009). The biopharmaceutical sectorâ€™s impact on the US economy: analysis at the national, state and local levels.

Constantino HR, Pikal MJ. Lyophilization of Biopharmaceuticals. Arlington (VA): AAPS Press; 2004.

Gieseler, H. (2012). Insights in lyophilization 2012. Current best practices & research trends. Antwerp, 2012.

Pikal, M. J. (2002). Freeze Drying. Encyclopedia of pharmaceutical technology, 1299-1326.

472886 Tunable Atomization of Viscous Liquids in the Production of Amorphous Solid Dispersions

Abstract:

Atomization of bulk liquids into droplets underlies several critical properties of amorphous dispersions of pharmaceuticals such as dry particle size, morphology, bulk density, physical stability, and dissolution behavior. Because these properties must often be maintained throughout the life cycle of a compound, a comprehensive understanding and control of atomization is required over a range of process scales. Droplet sizes of low viscosity Newtonian fluids are relatively well predicted by several empirical and semi-empirical correlations in the literature. However, viscous and / or non-Newtonian fluids, such as the polymer-rich solutions used to create amorphous solid dispersions, deviate significantly from these historical correlations. This work will present an atomization model for high viscosity liquids that derives its functional form from first principles of fluid dynamics for a pressure-swirl atomizer. The model is subsequently tuned to match experimental data using a small number of empirically determined free parameters. This methodology allows droplet size to be reliably predicted from controllable process parameters such as atomization nozzle geometry, solution flow rate, and liquid properties.

468023 Secondary Drying Scale-up Methodology: Eliminating a Bottleneck with a Lean Development Approach

Abstract:

Secondary drying scale-up methodology: eliminating a bottleneck with a lean development approach

T. Porfirio[1,2*], P. Valente[1], I. Matos[1], J. Moreira[1], J. Vicente[1], M. Temtem[1], V. Semiao[2]

[1] Hovione Farmaciencia SA, Sete Casas, 2674-506 Loures, Portugal; *tporfirio@hovione.com

[2]LAETA, IDMEC, Mechanical Engineering Department, Instituto Superior Tecnico, Universidade de Lisboa, Av. Rovisco Pais 1, 1049-001 Lisbon, Portugal

Owing to the short residence time of the droplets/particles within the spray drying chamber and process constraints to maintain product quality, most spray drying processes include a secondary drying (post-drying stage) to reduce residual solvent content to values below ICH guidelines (International Conference on Harmonization of technical requirements for registration of pharmaceuticals for human use). In spite of the widespread use of secondary drying process the physical mechanisms governing it are not yet fully understood [1, 2]. When using vacuum dryers, this complementary step is often the bottleneck of spray drying process and thus there is an increasing need to optimize process conditions to maximize throughput and ensure product quality. The process optimization for spray dried dispersions (SDD) is normally challenging due to the presence of an amorphous polymeric matrix with a glass transition temperature that limits the operating conditions in order to avoid the crystallization of the product.

A systematic methodology is proposed to allow an early recognition of scale-up and drying issues before process implementation on industrial scale to support the SDD development. The methodology includes a model that describes the drying mechanism of free-flowing powders under-vacuum considering the penetration theory [3,4]. This theory assumes a continuous mixing process with an alternate fictitious sequences of static periods and perfect mixing periods [5]. The understanding of the drying mechanism is crucial to optimize the process where the fundamental drying mechanism and mass transfer principles play an important role. The proposed methodology was created under the Quality by Design and Development by Design paradigms where the scale-independent relations (such as drying rate curve) and product/process understanding are found on lab-scale trials in order to achieve a lean development (Figure 1).

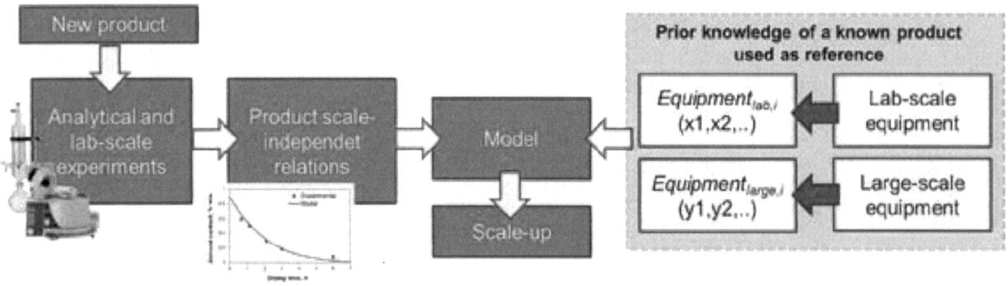

Figure 1: Proposed methodology framework

The method was constructed including the prior knowledge and the data of a known product used as reference. Several testes at lab- and large-scale units were performed for this reference case in order to establish the relevant equipment's parameters.

With that, when a new product is introduced, the proposed methodology provides an analytical assessment to evaluate a glass transition temperature, density, particle size and other important quality attributes that are relevant for the drying performance and understanding, e.g. the plasticization curve of the SDD to assure the amorphous state during the process since the risk of crystallization grows when the drying temperature is too close to the glass transition temperature. Succeeding, the product and process behavior and understanding are attained on lab-scale experiments to predict scale-independent relation. The lab-scale experiments are also important to understand the product behavior in terms of agglomeration and accumulation.

With the obtained scale-independent relation, it is possible to fit the model parameters and extrapolate for the large-scale unit where the model parameters of the equipment were previous stablished.

The intended benefits of this method are:

- Reduce secondary drying process time through process optimization;
- Support secondary drying development and scale-up;
- Tech transfer between secondary drying units.

A case study comprising the secondary drying process development of a SDD will be presented by following the proposed methodology. In this case study, operating conditions (namely the temperature and rotation speed) were optimized by the model without compromising the amorphous state of the temperature in order to reduce the drying time.

[1] Gianfrancesco, 2012. New method to assess water diffusion in amorphous matrices during storage and drying. Food Chemistry, 132, 1664-1670

[2] Kougoulos, 2011. Impact of agitated drying on the powder properties of an active pharmaceutical ingredient. Powder Technology, 210, pp. 308-314

[3] Schlunder and Mollekopf, 1984. Vacuum contact drying of free flowing mechanically agitated particulate material. Chemical Engineering and Processing, 18, 93-111

[4] Tsotsas and Schlunder, 1986. Contact drying of mechanically agitated particulate material in the presence of inert gas. Chemical Engineering and Processing, 20, 277-285

[5] Shani and Chaudhuri, 2012. Contact drying: a review of experimental and mechanistic modeling approaches. International Journal of Pharmaceutics, 434, 33-348

471328 Scientific and Regulatory Considerations on Particle Engineering for Dry Powder Inhaler (DPI)

Abstract:

Dry Powder Inhaler (DPI) is a leading pulmonary drug delivery technology for the treatment of asthma, chronic obstructive pulmonary disease (COPD), and respiratory infections. For DPI drugs, due to significant impact of the dry powder properties on the inter-particle interactions as well as aerosol performance, particle engineering provides an effective approach for development and optimization of particulate formulation to ensure efficient drug delivery as well as optimal therapeutic outcomes. An advanced particle engineering technique can further improve the pharmaceutical manufacturing process to minimize all possible sources of the product variability and to ensure manufacturing consistency as well as drug product quality.

Recently, a variety of particle engineering techniques have been adopted for formulation development and commercial manufacturing of DPI drugs. For each particle engineering technique, in-depth understanding and control of the particulate and solid-state properties for dry powders is critical. This presentation will discuss various particle engineering techniques available for manufacture of DPI and will focus on scientific and regulatory considerations for characterization and control of dry powder properties for DPI drugs. This presentation will also include a discussion of regulatory submission expectations with regard to characterization and control of the particulate and solid-state properties for dry powders based on commonly seen deficiencies in new drug applications (NDAs) and abbreviated new drug applications (ANDAs). The objective of this presentation is to illustrate how to use a risk based approach as advocated by the Quality by Design (QbD) paradigm for developing particle characterization methods as well as establishing control limits for key dry powder properties of DPI drugs to ensure reproducible commercial manufacturing for consistent product quality.

469151 Production of Pharmaceutical Tablets By FDM 3D Printing

Abstract:

Production of pharmaceutical tablets by FDM 3D printing

Production of tablets using 3D printers is a novel method of pharmaceutical formulation. Although still in the research&development phase, since recently it already has its first commercial application. The main advantage of this approach is that every single tablet can be printed to meet the needs of a specific patient, the dosage and dissolution rate can be controlled by the composition and porosity of the tablet. Furthermore, two or more drugs can be present in one tablet by using multiple printer nozzles with varying feed materials. The topic of this work is fabrication of pharmaceutical tablets, using Fused Deposition Modeling (FDM), a commercially available 3D printing technology, which utilizes thermoplastic polymer filaments as feed material. Biodegradable filaments were produced from mixtures of pharmaceutical polymers with additions of plasticizers and defined contents of a model drug, processed by hot-melt extrusion. For this purpose, the DIY hot-melt extruder Filastruder v1.5 was employed, which surpassed standard pharmaceutical extruders in terms of price, maintenance time and replaceability of its components, making it ideal for the purposes of an extensive parametric study of filament composition. This study resulted in production of drug-laden filaments with mechanical properties suitable for 3D printing, which were then successfully used as feed material for FDM 3D printer to produce tablets of varying shape and porosity. Furthermore, the effect of various printing parameters on the quality of the printed tablets was documented. Composition analysis by HPLC has confirmed that the drug contents in the tablets correspond with the initial drug loadings of the filaments. XRPD and DSC analyzes have shown, that the drug in the tablets exists in the form of an amorphous solid solution. Dissolution analysis has illustrated the effect of tablet shape, porosity and printing settings on the dissolution profile.

Although some cases of 3D printed tablets through Fused Deposition Modeling were already reported in the past, to the best knowledge of the author, this work presents the only case, where a conventional drug (with a relatively low melting point), which did not contribute to the printability of the material, was processed. Also, the extrusion and printing temperature is the lowest achieved so far in the field of FDM tablet printing. This makes the documented method more universal (applicable for a wider range of drugs) and presents potential for future research, which will involve incorporation of multiple drugs into one tablet, increasing resolution of the printing and expanding control over dissolution behavior.

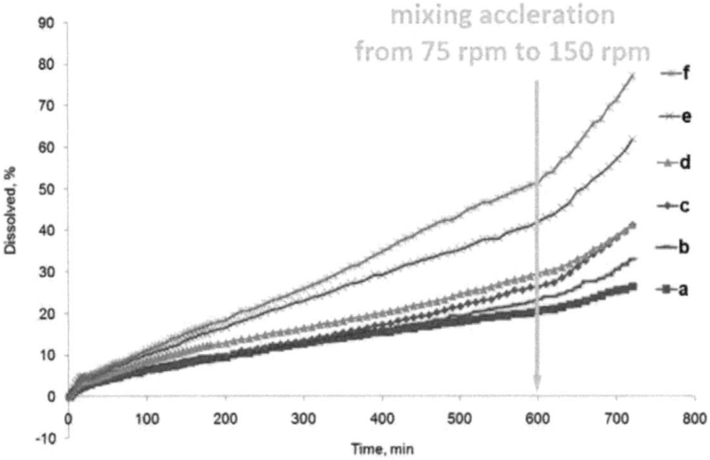

a) A disc-shaped tablet with deformations, resulting from using an unsuitable model, **b)** A disc-shaped tablet from a more suitable model, printed with a slight drop-shaped deformation on one side, **c)** A disc-shaped tablet from the same model as b), but with no deformation (due to better printing settings), **d)** A caplet-shaped tablet, **e)** A caplet-shaped tablet with micropores (pore height of 0.6 mm), which were most probably at least partially clogged during dissolution due to the slight swelling of the tablet, **f)** A caplet-shaped tablet with larger pores (pore height of 1.2 mm), which most probably stayed at least partially open during dissolution, which explains the highest dissolution rate of this tablet

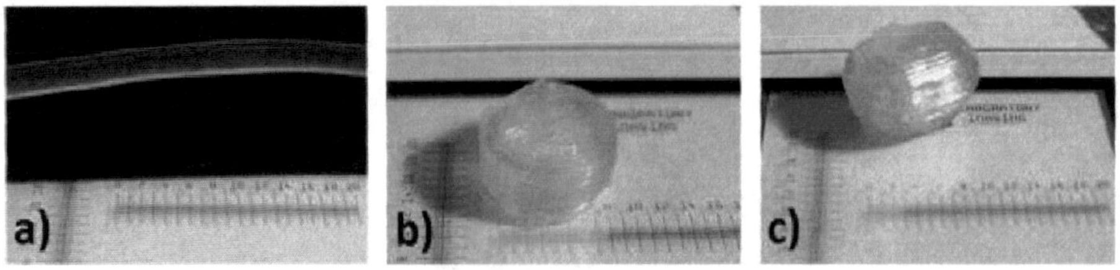

a) A biodegradable drug-laden filament, suitable for 3D printing purposes, **b)** A cylinder-shaped tablet,

c) A caplet-shaped tablet with predefined partially visible micropores

453052 Experimental Analysis of Pharmaceutical Twin-Screw Wet Granulation Using Hydrophilic and Hydrophobic Formulations

Abstract:

Introduction

Traditional pharmaceutical solid-dosage manufacturing exists of batch-wise processes which are rather inefficient and expensive. Today, efforts are made to shift from batch to continuous processing. Twin-screw granulation is a promising continuous wet granulation technique for solid dosage manufacturing. Nevertheless, little is known about how process and equipment settings affect granule formation mechanisms and kinetics.

Objective

The aim of this study was to evaluate the twin-screw granulation process for both hydrophilic and hydrophobic formulations to understand the role of the individual screw modules and process settings, as well as their interaction upon granule formation. This involved investigating the influence of process settings on granule properties at different locations in the screw barrel.

Results from this experimental study will be used for the development, calibration and validation of a two-dimensional Population Balance Model (PBM) involving particle properties (granule size and liquid content or density/porosity) relevant for twin screw granulation. The 2D PBM will incorporate the rate processes which are considered dominant in the wetting and mixing zones of the granulator, such as aggregation and breakage, based on an inflow and outflow granule size distribution in each zone. The experimental data from different locations inside the granulator will be used to obtain the unknown model parameters for individual screw modules.

Material and methods

Materials

Two hydrophobic model drug formulations were used. Therefore, two poorly water soluble API's (60% w/w) were separately blended with 16% (w/w) lactose monohydrate (Lactochem® Regular, DFE Pharma, Goch, Germany), 16% (w/w) microcrystalline cellulose (Avicel® PH 101, FMC, Philadelphia, USA), 3% (w/w) hydroxypropylcellulose (Klucel® EXF, Ashland, Covington, Kentucky, USA) and 5% (w/w) crosscarmellose sodium (Ac-Di-Sol®, FMC, Philadelphia, USA).

Because a poor flowability was observed for one model formulation, 0.5 % (w/w) colloidal silicon dioxide (CAB-O-SIL® M-5, Cabot Corporation, Alpharetta Georgia, USA) was added to that formulation in order to ensure adequate dosing of the dry ingredients into the granulation module. Additionally, a hydrophilic formulation existing of 97.5% (w/w) lactose monohydrate (Lactochem® Regular, DFE Pharma, Goch, Germany) and 2.5% (w/w) Polyvinylpyrrolidone (Kollidon®30, BASF, Ludwigshafen, Germany) was studied. All formulations were processed using demineralized water as granulation liquid.

Methods

Granules were produced using the high shear wet granulation module of the ConsiGma^{TM}-25 system (GEA Pharma systems, Collette, Wommelgem, Belgium).

To develop and calibrate the 2D PBM model, a 3-level full factorial Design of Experiments (DoE) (MODDE 11, MKS Umetrics, Umeå, Sweden) was performed for three formulations by varying the following parameters: screw speed (450-900 rpm), material throughput (5-25 kg/h) and liquid-to-solid ratio (formulation dependent). The temperature of the cooling water around the granulator barrel was kept constant at 25°C. This DoE was executed for one screw arrangement existing of 2 kneading zones of each six kneading elements ordered at a stagger angle of 60°. Several responses were assessed and are described below.

In order to gain insight in the granulation progress along the length of the granulation barrel, samples were collected in well-defined zones of the granulator unit:

- Location 1: Before the first kneading zone, where the granulation liquid is added (wetting)
- Location 3: After the first kneading zone
- Location 5: After the second kneading zone
- Location 6: After the chopper section being a kneading block at the end of the screw (i.e. granulator outlet).

Granules collected at these locations were oven dried (24h, 40°C, 25% RH) and granule size distribution (QICPIC®, Sympatec GmbH, Clausthal-Zellerfeld, Germany) was measured. Furthermore, all samples were sieved in different fractions (i.e. >2000, 1000-2000, 850-500, 150-500, 75-150 and <150 μm) using a Retsch VE 1000 sieve shaker (Haan, Germany). True density (Helium-pycnometry, Accupyc 1330 Pycnometer, Micrometrics, Norcross, USA) and specific surface area (Brunauer–Emmett–Teller (BET)-theory) was measured for every fraction.

Moreover, distribution of moisture among the different sieve fractions was assessed using a colored tracer (0.5% w/w cochenillered) dissolved in the granulation liquid. UV spectroscopy was used to measure tracer content in every sieve fraction, allowing to visualize the moisture content per sieve fraction at the different locations of the granulator. This experimental data about granule properties collected at intermediate locations inside the screw barrel will help to describe the dynamics during twin-screw granulation to be included in the aimed 2D PBM.

The granules collected at the end of the barrel (location 6) were oven dried (24h, 40°C, 25% RH) and analyzed for several responses: torque, granule size distribution (QICPIC®, Sympatec GmbH, Clausthal-Zellerfeld, Germany), friability (Friabilator, PTFE Pharma Test, Hainburg, Germany), flowability (Ring Shear

Tester RST-XS, Dietmar Schulze, Wolfenbüttel, Germany), mean residence time and Peclet number (determined with residence time distribution – SentroPAT FO system, Sentronic, Dresden, Germany).

Conclusion:

This experimental study proved the ability to capture the stepwise formation and transition of granules at the different locations inside the granulator in terms of granule properties (e.g. granule size, moisture, density/porosity).

It was affirmed that process parameters like liquid-to-solid ratio can dictate the formation of granules and their corresponding quality attributes. Furthermore it should be emphasized that differences are observed between hydrophilic and hydrophobic formulations. This could be derived from the measured moisture content in the different sieve fractions along the length of the granulator barrel. Results showed a more heterogeneous moisture distribution among the different size fractions in the hydrophobic formulations where moisture is less likely to be found in the smaller size regions. This could be explained by a significant fraction of un-wetted particles.

472245 Alginate/Chitosan Microparticles for Gastric Passage and Intestinal Release of Therapeutic Protein Nanoparticles

Abstract:

Proteins have significant potential to treat diseases due to their specific mechanisms of action compared to small molecule drugs. Protein nanoparticles offer a means to turn soluble protein, which has very low cellular uptake, into a colloidal protein carrier capable of enhanced uptake and subsequent intracellular activity. Previous work produced a stable therapeutic nanoparticle formulation encapsulating a bacterial effector protein, AvrA, which is capable of modulating inflammatory signals in an in vivo inflammatory bowel disease (IBD) mouse model. A major drawback is that these nanoparticles are instilled transrectally, limiting delivery to lower regions of the colon. To implement our protein nanoparticles as a clinically viable treatment for IBD with access to the entire intestinal tract and improved patient compliance, an oral formulation must be designed.

Localized intestinal delivery of protein therapeutics via the oral route is a highly attractive approach, however, the harsh acidic and enzymatic conditions in the gastrointestinal tract challenge retention of protein function. The approach is to deliver protein nanoparticles within gastro-protective microparticles made from alginate and chitosan that would release the nanoparticles in the intestine. A flow focusing microfluidic device was used to control alginate droplet diameters by tuning the continuous and dispersed phase flow rates. Alginate droplets encapsulating protein nanoparticles were then simultaneously crosslinked with calcium and layered with chitosan to form uniform, coated microparticles with high loading efficiency. Protein nanoparticles encapsulated within crosslinked alginate/chitosan microparticles retained approximately 50% of their activity after 1 hr incubation in simulated gastric fluid (SGF). Subsequent incubation in simulated intestinal fluid (SIF) released all remaining active protein within 3 hrs. Alginate vs. chitosan droplets, chitosan concentration, alginate concentration, number of layers, and crosslinker type were tuned to optimize protein protection in SGF and controlled release in SIF. Altogether, these hydrogel microparticles could serve as a potential oral delivery vehicle for protein therapeutics.

458095 Photodynamic Enhancement of Drug Delivery to Tumors

Abstract:

Pancreatic cancer is associated with one of the poorest prognosis of all malignancies, with an unchanged five- year survival rate of 5% or less. The complex desmoplastic microenvironment impeding effective drug delivery to pancreatic tumors and the compensatory ability of cancer cells to adapt alternative survival pathways suggest that treatments will need to attack several targets. Photodynamic therapy (PDT) may have a specific role in enhancing drug delivery to pancreatic tumors and ameliorating drug resistance mechanisms by the targeted destruction of specific pathways. We have developed a bi-directionally cooperative combination regimen of PDT and irinotecan for pancreatic cancer using two clinically approved, nanotechnology-enabled therapies with a clear molecular basis. In an orthotopic pancreatic cancer mouse model, a combination of photodynamic therapy (PDT) and FDA-approved liposomal irinotecan (nal-IRI, MM-398) synergistically reduced the primary and metastatic tumor burden, and prolonged animal survival. This presentation will discuss how PDT increases tumor permeability and irinotecan transport by several mechanisms including disruption of efflux pumps, vascular and stromal integrity. Our findings provide a framework for the translation of mechanistic insights into rational new therapies for several cancers. This study is supported by the following NIH grants: 1K99CA194269 (HH); 5K99CA175292/ 4R00CA175292 (IR); 4R01CA15841505, 5R01CA16099805 5P01CA08420312 (TH). JL was supported by the Herchel Smith fellowship and Harvard College Research Program.

471199 Highly Efficient Encapsulation of Small-Molecule N-Acetylcysteine within PLGA Nanoparticles to Restore Redox Balance in Oxidant-Stressed Environments

Abstract:

Introduction: Overproduction of reactive oxygen species (ROS) is a major factor in the pathogenesis of multiple sclerosis[1]. Localized antioxidant delivery is therapeutically relevant to counteract and neutralize ROS in such oxidant-stressed systems. Poly(lactic-co-glycolic acid) (PLGA) nanoparticles can provide a localized, controlled release of encapsulant, though reported encapsulation efficiencies of small-molecule therapeutics have been low: 12%[2] and 16%[3] for dopamine and N-acetylcysteine (NAC), respectively. Here we investigate how modifying the traditional double-emulsion, solvent evaporation method by doping external water phases with the small-molecule drug affects encapsulation efficiency, temporal release profiles, released drug activity, and released drug protective effect on oligodendrocyte progenitor cell (OPC) viability after H_2O_2 stress. We aim to deliver NAC, a potent, cheap, FDA-approved antioxidant, with high efficiency and temporal control.

Materials and Methods: NAC-loaded PLGA nanoparticles with varying lactide:glycolide (L:G) ratios and inherent viscosities were prepared by the double-emulsion, solvent evaporation method. The solvent for all water phases was deionized water. Briefly, the internal water phase (10% w/v NAC) was dispersed via sonication within the oil phase (15% PLGA in dichloromethane). The primary emulsion was dispersed in the outer water phase (1% PVA with or without 2.5% NAC) via sonication. The secondary emulsion was stirred in 0.3% PVA with or without 5% NAC for 4 hours. Particles were washed 3x in either DI water or 2.5% NAC and lyophilized. Release was determined by incubating particles at 37°C in phosphate-buffered saline and measuring supernatant NAC via Ellman's Reagent. LAP photoinitiator radical concentrations were measured via luminol to determine the dose response of NAC radical scavenging. OPC viability in response to oxidative conditions with and without the NAC antioxidant was measured via the CellGlo reagent.

Results and Discussion: Encapsulation efficiencies (relative to theoretical load without doped external phases) for NAC-loaded PLGA nanoparticles were greatest with external phase doping – reading greater than 300% of the case when undoped. With the doped particles, NAC was released for 14 days. NAC ranging from 0.5-1000 µM (the particle supernatant concentration range for each sample time) scavenged photoinitiated LAP radicals in a dose-dependent manner, with 0.5 µM and 1000 µM NAC scavenging 6.82% and 95.5% of radicals, respectively. Ongoing experiments show NAC-loaded nanoparticle antioxidant activity and protective effect on OPC viability.

Conclusions: Drug-doping of external water phases during the double emulsion, solvent evaporation method gave NAC-loaded PLGA nanoparticles with significantly higher encapsulation efficiencies than without doping, and up to 20-fold higher than previously reported values. NAC scavenged LAP radicals in a dose-dependent manner and was released from PLGA nanoparticles for at least two weeks. This indicates that these particles will provide a larger and more continuous small molecule dose longer than previously reported methods. Efficiently loaded antioxidant nanoparticles capable of controlled and localized delivery could have therapeutic potential in oxidant-stressed systems and bias progenitor cells such as OPCs toward a self-renewing fate.

References: [1]Lehnardt S. ***J. Neurosci,*** 2002. 22:2478-86. [2]Pahuja R. ***ACS Nano,*** 2015. 9:4850-71. [3]Lancheros R. ***Univ. Sci.,*** 2014. 19:161-168

472474 Fabrication and Characterization of PLGA Nanoparticle-Bacteria Conjugate Nanobeads for Increased Tumor Penetration

Abstract:

Cancer is the second cause of overall mortality in the world and systemic chemotherapy is a major therapeutic approach for nearly all types and stages of cancer. Grand challenges of limited efficacy and systemic toxicity of existing chemotherapeutic treatments can be addressed through better drug carrier targeting and enhanced penetration of drug carriers. Attenuated strains *Salmonella* Typhimurium have previously been shown to have the ability to preferentially colonize tumor tissues. However, clinical success in immunocompetent hosts has been rare due to insufficient tumor colonization by bacteria[1-3]. We hypothesize that a combinatorial therapy approach based on integrating bacteria with chemotherapeutics-loaded nanoparticles will amplify both forms of therapy. Thus, we have developed a Nanoscale Bacteria-Enabled Drug Delivery System (NanoBEADS)[4] in which the functional capabilities of bacteria are interfaced with and augmented by chemotherapeutic-loaded nanoparticles. Each NanoBEADS agent is constructed by interfacing a *Salmonella* Typhimurium VNP20009 (i.e. a live engineered tumor-targeting bacterium) with an ensemble of poly(lactic-co-glycolic acid) nanoparticles loaded with chemotherapeutics for the purpose of using the bacterium to transport the particle deep into the tumor. To create the NanoBEADS, *S.* Typhimurium VNP20009 bacteria were cultured in Luria-Bertani broth (1% tryptone, 0.5% yeast extract and 0.5% sodium chloride) and coated with biotin-conjugated rabbit polyclonal anti-*S.* Typhimurium antibody. The nanoparticles were fabricated using a modified nanoprecipitation method[5] and fluorescently tagged with 6,13-Bis(triisopropylsylylethynyl)pentacene (TIPS pentacene). With a size of approximately 110 nm in diameter and a zeta potential of about -30 mV, the carboxyl end groups allow for EDC coupling to streptavidin. The biotin-coated bacteria were then mixed with streptavidin coated nanoparticles for 30 minutes at room temperature to promote self-assembly of the particles to bacteria. Scanning electron microscopy (SEM) images of NanoBEADS show the average number of nanoparticles

attached to each bacterium to be 35. In-vitro assays were carried out to compare the penetration of passively diffusing PLGA nanoparticles and NanoNBEADS in multicellular tumor spheroids.

1. Heimann, D. M. & Rosenberg, S. A. Continuous intravenous administration of live genetically modified salmonella typhimurium in patients with metastatic melanoma. *J. Immunother.* **26,** 179–80 (2003).

2. Toso Gill, V., Hwu, P., Marincola, F., Restifo, N, Schwartzentruber, D, Sherry, R, Topalian, S, Yang, J., Stock, F., Freezer, L., Morton, K., Seipp, C., Haworth, L., Mavroukakis, S., White, D., MacDonald, S., Mao, J., Sznol, M., and Rosenberg, S., J. & Toso, J. Gill, V., Hwu, P., Marincola, F., Restifo, N, Schwartzentruber, D, Sherry, R, Topalian, S, Yang, J., Stock, F., Freezer, L., Morton, K., Seipp, C., Haworth, L., Mavroukakis, S., White, D., MacDonald, S., Mao, J., Sznol, M., and Rosenberg, S. Phase I study of the intravenous administration of attenuated Salmonella typhimurium to patients with metastatic melanoma. *J Clin Oncol.* **20,** 142–152 (2002).

3. Pawelek, J. M., Low, K. B., Bermudes, D. & Pawelek, J. M., Low, K. B., Bermudes, D. Bacteria as tumour-targeting vectors. *Lancet Oncol.* **4,** 548–556 (2003).

4. Traore, M. A., Damico, C. M. & Behkam, B. Biomanufacturing and self-propulsion dynamics of nanoscale bacteria-enabled autonomous delivery systems. *Appl. Phys. Lett.* **105,** 173702 (2014).

5. Niu, X., Zou, W., Liu, C., Zhang, N. & Fu, C. Modified nanoprecipitation method to fabricate DNA-loaded PLGA nanoparticles. *Drug Dev. Ind. Pharm.* **35,** 1375–83 (2009).

460403 Interstitial Release of Cisplatin from Triggerable Liposomes Enhances Efficacy Against Triple Negative Breast Cancer Solid Tumor Analogues

Abstract:

In solid tumors, including Triple Negative Breast Cancer (TNBC), limited penetration of therapeutics leads to heterogeneous intratumoral distributions reducing efficacy. Liposomal cisplatin decreases toxicities associated with free cisplatin. However, the heterogeneous intratumoral distributions of liposomes combined with limited release of cisplatin from liposomes may also reduce efficacy. This study uses pH-responsive liposomes designed to release cisplatin within the acidic tumor interstitium (7.0 < pH â‰¤ 6.0) with the aim (1) to improve the penetration of the free drug within tumors on the assumption of greater diffusivities based on the free drug's much smaller size and (2) to increase the availability of the free agent near cancer cells.

Introduction: In advanced solid tumors, including Triple Negative Breast Cancer (TNBC) solid tumors, the limited penetration of therapeutics results in limited therapeutic effects. TNBC is a subgroup of breast cancer associated with a high chance of cancer reoccurrence outside the breast and has been found to exhibit sensitivity to DNA-damaging platinum-based compounds, which is the basis for increased clinical use of platinum-derived agents. To enable selective and effective treatment of TNBC solid tumors, we study a drug delivery carrier of CDDP that is expected to result in potentially deep penetration and homogeneous distribution of the drug within the tumor interstitium. The carrier is a tunable (pH-releasing) liposome encapsulating CDDP. Tumor selectivity by these liposomes is mediated by the EPR effect. Efficacy is mediated by the pH-triggered release of contents within the tumor interstitium (7.0 < pH < 6.5) by extravasated liposomes. The small size of CDDP relative to its nanometer-sized carrier should exhibit greater diffusivities and deeper tumor penetration.

Methods: To study the efficacy of these liposomes on TNBC cells two different TNBC cell lines were studied: MDA-MB-231 and MDA-MB-468. The efficacy of pH-releasing liposomes (LD_{50}) loaded with CDDP was compared to the efficacy of other constructs including conventional (non-p-releasing) liposomes, empty liposomes, and free cisplatin. Following incubation of 6 and 24 hours, in addition to cell viability, cisplatin per cell, cisplatin intercalated to DNA and (if applicable) retention of cisplatin within the liposomes, were measured. In multicellular spheroids exhibiting intracellular pH gradients measured in situ by confocal microscopy using a membrane impermeant pH-indicator, we evaluated the spatiotemporal distributions of liposomes and contents. In dose-response studies, upon completion of incubation with all constructs, spheroids were washed and their volume change (volume at time t/volume at initiation of treatment x100%) was monitored in time followed by an outgrowth assay.

Results: On MDA-MB-231 and MDA-MB-468 cell monolayers, extracellular acidification resulted in lower LD_{50} values only for pH-releasing liposomal cisplatin (101.9 ± 24.1 and 14.5 ± 3.6 μg/mL, respectively) which were significantly less than for non-pH-releasing liposomal cisplatin (374.2 ± 20.0 and 33.7 ± 5.0 μg/mL, respectively). In contrast, at neutral pH, the LD_{50}values of both liposome types were comparable. Killing efficacies correlated strongly with the nuclear DNA-associated cisplatin independent of the delivery modality or the extracellular pH. pH-releasing liposomal cisplatin decreased the volumes of multicellular spheroids by 46.4-70.0 % more than non-pH-releasing liposomal cisplatin. pH-releasing liposomes resulted in improved penetration and more homogeneous profiles within spheroids of a cisplatin-surrogate cell-permeant fluorophore compared to non-pH-releasing liposomes.

Conclusions: These findings demonstrate the potential of intratumorally CDDP-releasing nanometer-sized carriers to selectively and effectively control TNBC solid tumors.

Funding Sources:

American Cancer Society (ACS): RSG-12-044-01

National Science Foundation (NSF): DMR-1207022

AAUP-AFT Professional Development Fund

457602 Bioresponsive Polymer Coating on Targeted Drug Nanorods

Abstract:

Non-specific delivery and ineffective drug release at the target site are among the top challenges to effective treatment. The first challenge reflects the fact that interactions with carriers (*i.e.*, nanoparticles) can denature the active ingredients of drugs by shear stress. The second challenge is low affinity for specific receptor proteins overexpressed on the target cell surface. The third challenge of ineffective drug release can be overcome by exploiting the disease microenvironment. Here, we report a one-step process to prepare bioresponsive polymer coated drug nanorods (NRs) from liquid precursors using the solvent diffusion method. A thin layer (10.3 ± 1.4 nm) of poly(e-caprolactone) (PCL) polymer coating was deposited on the surface of camptothecin (CPT) anti-cancer drug NRs. The mean size of PCL-coated CPT NRs was 500.9 ± 91.3 nm length x 122.7 ± 10.1 nm width. The PCL polymer coating was biodegradable at acidic pH 6 as determined by Fourier Transform Infrared (FT-IR) spectroscopy. CPT drugs were released up to 51.5% when PCL coating dissolved into non-toxic carboxyl and hydroxyl groups. Trastuzumab (TTZ), a humanized IgG monoclonal antibody, was conjugated to the NR surface for breast cancer cell targeting. Combination treatments using CPT and TTZ decreased the HER-2 positive BT-474 breast cancer cells by 66.9 ± 5.3 % *in vitro*. These results suggest highly effective combination treatments of breast cancer cells using bioresponsive polymer coated specific drug delivery.

466655 Internalization of Nanoparticles Functionalized with Low Molecular Weight Protamine into Erythrocytes

Abstract:

Erythrocytes have been exploited extensively for their potential applications as delivery vehicles due to their remarkable biocompatibility, biodegradability and prolonged life-span in circulation. The mature human erythrocyte is a terminally differentiated cell that does not possess an endocytic uptake mechanism. As a result, the efficient migration of bioactive molecular cargos such as large macromolecular drugs, genetic material, or proteins through the plasma membrane of erythrocytes remains as a major challenge for intracellular delivery applications. A variety of techniques have been successfully developed to entrap proteins, drugs, and enzymes within erythrocytes. The majority of these techniques, however, require disruption of the cell membrane to create large pores or perturbations through which biomolecules can diffuse. These techniques include chemical methods such as drug-induced endocytosis, mechanical methods such as hypotonic osmosis/dialysis, and electrical methods such as electroporation. The main disadvantages associated with these methods are the loss of structural integrity and cellular components of the cells due to weakened cell membranes. As a consequence, these modified erythrocytes are easily recognized by the host immune system. To achieve the benefits associated with erythrocytes as a drug delivery vehicle it is vital to maintain complete structural and functional integrity of erythrocytes.

Toward this goal, we introduced a novel method for transporting nanoparticles (NPs) into intact erythrocytes by combining the advantages of both cell-penetrating and targeting peptides. The method employs an ERY1 (WMVLPWLPGTLDGGSGCR) peptide that has a high affinity for the erythrocyte membrane protein glycophorin A and a low molecular weight protamine (LMWP) peptide that has been shown to transport proteins and small molecules into erythrocytes. These peptides are conjugated to a NP formed between fluorescently-labeled bovine serum albumin and a grafted copolymer of poly(ethylene) glycol and poly-L-lysine.

The formation of nanoparticles was observed using SEM, TEM, and dynamic light scattering. The nanoparticles had a core/shell structure and their size ranged from approximately 20-100 nm. Confocal microscopy and flow cytometry data confirmed that uptake of LMWP functionalized NPs was significantly higher (approximately > 3-5 fold) than that of other NPs at 6 h. Under these conditions, synthesized NPs were not toxic to the erythrocytes and no significant hemolysis was observed. Catalytic activity of the encapsulated protein was not negatively affected upon encapsulation or crosslinking. In conclusion, NPs were efficiently internalized into RBCs without adversely affecting cellular function and this method has the advantage over existing methods in that it does not appear to alter the structural integrity of the cell. The above approach of utilizing a surface modified NP architecture to improve internalization into erythrocytes may serve as a universal platform to deliver therapeutic proteins/drugs into RBCs in future. Furthermore, this nanoparticle system capable of entering red blood cells and extending circulation time has enormously broad potential, including improved treatment of blood-specific diseases such as malaria as well as treatment of acute lymphoblastic leukemia.

466720 Enzyme Replacement Therapy Extended to the Brain through Nano-Polymersomes

Abstract:

Approximately 1 in 5,000 to 8,000 children are born annually with a lysosomal storage disease (LSD). Current treatment of LSDs, called enzyme replacement therapy (ERT), involves weekly to monthly intravenous(IV) infusions of missing endogenous enzymes. While ERT has proven effective in LSDs without central nervous system (CNS) involvement, the brain has remained untreatable due to the presence of the blood-brain barrier (BBB). The BBB is made up of endothelial cells with tight junctions, possessing a width between connections of 0.2 nm in size, which prevents passage of 98% of small molecule drugs, including enzymes, from the blood into the brain. Patients with neuropathic LSDs, which account for 50-70% of diseases in this category, present with severe CNS degeneration, ataxia, and premature death, with no treatment available on the market. The disease of focus, GM1 gangliosidosis, causes premature death between the ages of two and four with no hope available for parents of patients. Encapsulating enzyme into polymeric vesicles, called polymersomes, facilitates the transport of this missing enzyme into the brain through IV injections, effectively extending ERT therapy to the brain for the first time.

Polyethylene glycol-b-poly(lactic acid) (PEGPLA) has been proven to self-assemble into vesicles via both passive mixing and the solvent injection, using dimethyl sulfoxide (DMSO) and water. The mixing enzyme in GM1 gangliosidosis, β-galactosidase (βgal), is fluorescently tagged with reactive Alexa Fluor 488 at an efficiency of 13 dye molecules per mole of protein, with a finally protein concentration of 0.35 mg/mL attainable. Fluorescently tagged βgal is dissolved in stirring water prior to injection of DMSO and PEGPLA. Apolipoprotein E (ApoE) and CF350 attachment is facilitated through the introduction of homobifunctional PEG during injection. Polymersome diameter is characterized by dynamic light scattering (DLS). Self-assembly was stopped by freezing prior to lyophilization, with mannitol as a lyoprotectant. βgal loaded polymersomes were released *in vitro* in buffers of pH 7.4 and 4.8. Attachment was confirmed with DLS and Lowry assays. An immortalized feline fibroblast line, GM1/SV3, was used in *in vitro* studies to confirm

particle uptake and treatment efficacy.

Previous studies indicated that incorporating 2 wt%/v mannitol during formation led to PSDs 54 ± 0.8% smaller after lyophilization. Therefore, 2 wt%/v mannitol was used to aid in the formation of brain-deliverable polymersomes encapsulated with βgal. PEGPLA polymersomes loaded βgal at 0.08 ± 0.03 mg βgal/mg polymersome with activity maintained at a value of 452 ± 82 a.u. DLS data indicates that polymersomes with βgal had diameters statistically similar to control. Release of βgal was doubled over 6 hours when loaded polymersomes are placed into citrate buffer (pH=4.1) when compared to tris buffer (pH=7.4), indicating increased enzyme release expected in the lysosome. First order kinetics of release were followed for three hours when βgal was released from polymersomes in acidic buffer prior to a plateau in mass released, indicating the potential for sustained long term release from polymeresomes. Activity of βgal was maintained after release. CF350 Amine was attached to polymersome surface, with an increase in diameter from 147.2 ± 24 nm to 210.5 ± 8.6 nm (Fig 1A). Fluorescence microscopy confirmed attachment (Fig 1C), with 79 ± 21% of polymersomes labeled with blue fluorescence. Furthermore, our carriers were capable of both encapsulating Alexa Fluor 488 and attaching CF350 at a high efficiency, with 87 ±12% of polymersomes presenting with both colors in flow cytometry. Attachment of ApoE was confirmed with an increase in polymersome diameter to 194.4 ± 3.5 nm and an increase in protein content from 0.06 ± 0.11 mg/mL to 1.03 ± 0.71 mg/mL. GM1SV3 cells, an immortalized cell line from felines with GM1 gangliosidosis, present low density lipoprotein receptors (LDLR) on the surface, as shown in confocal microscopy, indicating that they are a good model for ApoE mediated delivery through the BBB. ApoE increases the uptake of polymersomes into GM1/SV3 fibroblasts when compared to unlabeled control polymersomes.

This work highlights our ability to create polymersomes capable of protecting βgal in the blood stream, shown by limited release in a pH 7.4 environment, targeting the nanocarrier to the BBB, shown by the facilitated attachment of ApoE, and after transcytosis, releasing βgal in the lysosome of neural cells, shown by extended release in a pH 4.1 enviornment and cell studies in GM1SV3 cells. This combination of ERTand nanotechnology, demonstrating the capability of our carrier to transport enzymes to the brain while maintaining their activity, will create a paradigm shift in the treatment of CNS disease, providing treatment for currently untreatable and fatal diseases like GM1 gangliosidosis.

473905 Regulatory Considerations for Introducing New Technologies during Drug Development

Abstract:

This presentation will cover regulatory aspects to consider when introducing new technologies during the development phase, with emphasis on changes made during the latter stage of product development. Some case studies of new technologies will be also be discussed.

466568 Viscoelasticity and Molecular Mobility of Celecoxib/Pvp/Tpgs Melt: A New Approach for Amorphous Drug Dispersion Assessment during Hot Melt Extrusion Product Design

Abstract:

Innovations at the Product Design and Discovery Interface

Abstract # 466568

Title: Viscoelasticity and molecular mobility of celecoxib/PVP/TPGS melt: A new approach for amorphous drug formulation assessment during hot melt extrusion (HME) product design.

Introduction/Motivation/Objectives:

Formulating drugs (active pharmaceutical ingredients) as amorphous solid dispersions is an alternative for improving the solubility of poorly water soluble drugs. Hot melt extrusion (HME) has demonstrated to be the adequate manufacturing unit process for compounding such amorphous solid dispersions. However, the risk of a potential event of inadequate amorphous content in the extrudate is always present, which is of critical significance when assessing the risk to the quality of such formulations during developmental product design phases.

Celecoxib is a poorly water soluble (BCS class II) non-steroidal anti-inflammatory drug. When compared to other BCS class II compounds, it has relatively low molecular mobility (high viscosity) and crystal growth rate upon cooling from the undercooled melt state [1,2]. The presence of surfactants or polymers has demonstrated to significantly increase or inhibit the radial crystal growth rate of Celecoxib, respectively [3].

We have previously studied the impact on the Celecoxib crystallization by PVP (polymer) and TPGS (surfactant) in terms of molecular mobility. Over all it was demonstrated that the polymer decreases the mobility of the drug, whereas the surfactant acts as a plasticizer. The dynamic viscosity of the Celecoxib with polymer and/or surfactant follows Arrhenius temperature dependence near the melting point (10°C of undercooling), however the VTF model results in higher agreement as the undercooling degree increases.

Even though, viscosity has been assessed for HME processability of melt formulations, it has not been associated to the amorphous content in the extrudate when processed using HME. Herein we present the link between viscoelasticity and amorphous state and content in the extrudate of mixtures of celecoxib, PVP and/or TPGS. We expect these results to complement the knowledge available for when assessing a drug as a candidate for a solid dispersion formulation during developmental stages of a drug product.

Sample Preparation:

Physical samples were prepared by mixing the drug with the corresponding amount of polymer and/or surfactant. Samples were mixed manually following a mixing protocol for homogeneous material distribution, and hence reproducibility.

Binary and ternary mixtures consisted of 50 wt.% TPGS or 50 wt.% PVP dispersed in a 50 wt.% Celecoxib, and ternary mixtures of 25 wt.% PVP, 25 wt.% TPGS and 50 wt.% Celecoxib. Samples where extruded using a Thermo Scientific MiniLab II HAAKE Rheomex CTW5 II HME. Extruded amorphous content and state was characterized using X-Ray Powder Diffraction (XRD), and Raman spectroscopy combined with a microscope for area of sample selection to obtain the spectra. The latter allowed for testing of drug homogeneous content throughout the extruded material.

Methodology/Data Analysis:

Extruded drug samples where analyzed in terms of amorphous content and amorphous state. The results where compared with our previous data on viscoelasticity and molecular mobility of the melt. Together we linked the molecular mobility (viscoelasticity) of the melt with the amorphous content in the extrudate of celecoxib mixed with PVP and/or TPGS.

Results and Analysis:

The samples were extruded using a bench-top HME (Thermo Haake MiniLab II) at a temperature of 140°C and a speed of 25 1/min (rpm). Raman spectroscopy of the extrudate samples resulted in a significantly less amorphous content when celecoxib was combined with PVP only, than when combined with both PVP and TPGS. This is evidenced with a noticeable larger Raman intensity of the celecoxib/PVP/TPGS mixture of each of the most important functional groups of celecoxib (CF_3, SO_2 and NH_2) compared to celecoxib mixed with PVP.

Powder diffraction characterization (XRD) showed a significant reduction in the crystalline state of the drug when mixed with PVP compared to PVP/TPGS, evidenced by the disappearance of the characteristic peaks of Celecoxib between 13 and 25 degrees (2Θ). On the other hand, TPGS increased the intensity of these peaks, and hence the crystalline state.

SEQ Table * ARABIC 1. Comparison between the off-line rheological and Raman data. A decrease in molecular mobility of the melt results in a higher amount of amorphous content in extrudate.

Samples	Off-line rheology			**HME**	Raman
	Molecular mobility (viscosity)*	Viscoelastic behavior*	Viscoelastic behavior* (140°C)		Amorphous content in extrudate*
Celecoxib/PVP	Decreased (viscosity increase)	Liquid to Elastic	Liquid	**140°C, 25 min^{-1}.**	Increased
Celecoxib/PVP/TPGS	Decreased *(Decreased much less than with PVP alone)*	Liquid to Elastic	Liquid		Increased *(Increased much less than with PVP alone)*
Celecoxib/TPGS	Increased (viscosity decrease)	Liquid to Elastic	Liquid		Decreased

*Compared each to pure Celecoxib.

The table summarizes the molecular mobility behavior and the amorphous content of the extrudate. When analyzed together, scientifically justifies the inclusion of viscoelasticity when designing a drug product formulation for HME. A decrease in viscosity results in an increase of the amorphous content of the drug, whereas an increase in molecular mobility results in a decrease in amorphous content in the extrudate. However, it must be highlighted that this correlation is conditional to the materials present in the melt, polymers (PVP) or surfactants (TPGS) as when these are combined less amount of amorphous content may result.

Conclusions:

We have elucidated the relationship between the molecular mobility of celecoxib when in presence of PVP (polymer) and/or TPGS (surfactant), the elastic/liquid (viscoelasticity) behavior, and the resulting amorphous content of the celecoxib in the extrudate. This work provides the proof-of-concept that viscoelasticity characterization combined with analytical techniques in HME could be used to tailor an amorphous drug with reproducible and adequate amorphous content manufactured using HME, being this correlation currently unknown and highly beneficial for when designing an amorphous drug dispersion.

References

[1] Baird, J. A.; Santiago-Quinonez, D.; Rinaldi, C.; Taylor, L. S. Role of Viscosity in Influencing the Glass-Forming Ability of Organic Molecules From the Undercooled Melt State. *Pharm Res* **2011**, *29*, 271–284.

[2] Baird, J.A.; Van Errdenbrugh, B.; Taylor, L.S. A Classification Tendency of Organic Molecules from Undercooled Melts. *J Pharm Sci* **2010**, 99, 3787-3806.

[3] Mosquera-Giraldo, L. I.; Trasi, N. S.; Taylor, L. S. Impact of Surfactants on the Crystal Growth of Amorphous Celecoxib. *International Journal of Pharmaceutics* **2014**, *461*, 251–257.

[4] Trachenko, K. The Vogel-Fulcher-Tammann Law in the Elastic Theory of Glass Transition. *Journal of Non-Crystalline Solids* **2008**, 354, 3903-3906.

[5] Gupta, P.; Chawla, G.; Bansal, A.K. Physical Stability and Solubility Advantage from Amorphous Celecoxib: The Role of Thermodynamic Quantities and Molecular Mobility. *Molecular Pharmaceutics* **2004**, 1, 406-413.

472436 Miniaturized Purification Platform for Automated Screening of Flow Synthesis Using Extraction and Crystallization

Abstract:

R&D for continuous flow chemistry (e.g. end-to-end continuous manufacturing and automated synthesis lab) necessitates a modular and reconfigurable chemical synthesis platform capable of screening of a variety of chemical pathways. Flow microreactors are now commercially available, and more recently, drop-on-demand formulation device, such as ink- and droplet-based pharmaceutical printers, have made it to market. While these devices can be naturally combined to construct a modular end-to-end continuous flow platform, the platform cannot be fully realized without its purification counterpart. Nevertheless, microscale purification platform is not yet well established; Although a number of innovative separation techniques have been reported in the microfluidics literature, including evaporation, extraction, crystallization, filtration and chromatography, most of these processes have been designed as a standalone for lab-on-a-chip applications. An automated and reconfigurable purification device which can seamlessly integrate with other parts of the flow synthesis platform in a flexible and scalable manner is therefore desirable. In this work, a microscale purification platform which can virtually be constructed using off-the-shelf parts is described. Specifically, our device consists of a combination of commercially available parts, such as capillaries, chips, mixers, fittings, etc. as well as custom parts, including tubing heaters and zero-dead volume Y-junctions. In addition, the device support multiple reservoirs for use with aqueous and organic solvents, a Peltier-based heating and cooling system, a couple of process analytical tools, including UV and non-contact Raman spectroscopy and an auto-sampler for sample collection of offline analysis using techniques such as MS and UPLC. The device operation is controlled using a LabView software which integrates, controls and monitors electronic actuators, including syringe pumps, valves, pressure sensors, etc., and executes a sequence of instructions to perform priming, loading and cleaning. The capability of the device is demonstrated via automated screening of continuous and segmented (air and liquid) flow extraction and crystallization under various favorable and unfavorable conditions for the purification of model APIs including diphenhydramine and atropine.

464011 Novel Validation Method Development for a Drug Delivery Device Used during Breastfeeding

Abstract:

4.5 million infants die every year. Many of these deaths can easily be prevented, with existing medications, if hygienic and effective paediatric formulations were available. In low-resource settings, the need for medications, which do not require refrigeration, can easily be accurately dosed, and can be hygienically administered is especially high, due to electricity, potable water, and literacy limitations. Breastfeeding infants are especially in need of appropriate drug and delivery device combinations, considering few drug delivery systems are commercially available specifically for this population, and infant compliance can be low.

We are responding by developing a novel device for administering medications or nutrients to infants, and *in vitro* validation methods for characterising itâ€™s efficacy. The device, when worn by a mother during breastfeeding, releases active pharmaceutical ingredients contained in the device into the milk directly consumed by the infant.Â It therefore has many potential advantages. For example, by suspending the medication or nutrient into the breast milk directly, and since it could be disposable, it overcomes the need for potable water as a suspension or cleaning fluid. It contains a solid dosage form, and therefore does not require refrigeration. Since the device is pre-loaded with medication a mother does not need to be literate to measure out a dose. Furthermore, since it is used during breastfeeding, infants may find it to be a more natural means of receiving medicine than conventional spoons or syringes.

Several validation methods have been used and developed in optimising the device design and potential corresponding dosage forms. These include a breastfeeding simulation apparatus, a novel infant tongue peristalsis simulation apparatus, and a novel tablet disintegration characterisation method using texture analysis.

The breastfeeding simulation apparatus was used to assess the release from the device of several dosage forms containing zinc, used in the treatment for diarrhoeal disease. The apparatus mimics maternal lactation, by flowing human milk through a silicone breast mimic. The milk then passes through the device loaded with the therapeutic insert, and then is sucked by a modified breast pump, at pressures physiologically representative of breastfeeding, into an exit line from which samples are collected with a fraction collector. The generated breast milk samples supplemented with zinc, collected over a simulated breastfeed are then analysed via inductively coupled optical emission spectrometry to test drug release rates from the drug delivery system. Zinc release over the course of 20 minutes, more than the time typically used during breastfeeding, in the simulations for four rapidly disintegrating tablet formulations ranged from 32-51%, indicating further formulation optimisation is needed.

An infant tongue mimic system was developed to simulate *in vitro* the peristaltic motions and associated pressures exerted from an infant tongue during breastfeeding. The system was used to characterise the potential effects of tongue movement on the disintegration of model tablets from the drug delivery system. The tongue mimic system includes a modified peristaltic pump equipped with a novel tongue mimic, a piezoresistive sensor for pressure measurements, a tube for holding the tablet under model breastfeeding flow conditions, and an adjustable palate mimic, allowing for pressure to the tablet to be modified based on clinical data. The release profiles of the model compound, sulforhodamine B, measured at two rotational values possible during breastfeeding for two potential compression values, representative of two potential tongue strengths of infants were statistically different ($P=0.0013$ and $P<0.001$, using 2-way ANOVA with matching for the lower and higher compression values respectively). For breastfeeding rotational rates representing non-nutritive breastfeeding, compression amount was found to be a significant factor ($P=0.0162$). In addition to being useful for characterising dosage forms for the device under the influence of tongue motion during suckling, this system could be useful in characterising other oral dosage forms or products.

Disintegration characterisation in human milk of potential tablet formulations is critical for screening dosage form appropriateness for the device. Therefore, a texture analysis disintegration method was also developed to analytically characterise tablet disintegration in opaque media. This method was developed to overcome visibility challenges which would be associated with use of the USP disintegration test with human milk. In this method, a texture analyser applies a load to a rapidly disintegrating tablet immersed in opaque media, and load data is collected over time as the tablet disintegrates, with fluctuations in the load indicating tablet fracturing. An in-house tablet formulation of a rapidly disintegrating tablet, and a commercially available tablet formulation were characterised in water, bovine, and human milk. These were found by one-way ANOVA to have statistically different disintegration times in each media ($P<0.0001$). This system is useful in characterising other tablet formulations for the device, and may be useful in the characterisation of other materials in opaque media.

Further work aims to continue testing various designs of the drug delivery system, as well as other formulations of corresponding dosage forms, to optimise the release of active pharmaceutical agent from the device.

470420 Photothermal Therapy of Bladder Cancer and Melanoma Via Targeted Single-Walled Carbon Nanotubes

Abstract:

Single-walled carbon nanotubes (SWNTs) are unique in that they strongly absorb near-infrared (NIR) light, while biological systems have very low levels of absorption of NIR light. This project focuses on targeting the SWNTs using conjugation with the human annexin A5 protein. Annexin A5 (molecular weight of 36,000) is a monomeric protein that binds with high affinity to phophatidylserine (PS) in phospholipid bilayers. PS is the most abundant anionic phospholipid of the plasma membrane and is tightly segregated to the internal side of the plasma membrane in most cell types. Anionic phospholipids are largely absent from the external surfaces of resting mammalian cells under normal conditions. PS is expressed on the external surface of tumor cells and also of endothelial cells that line the blood vessels in tumors, but it is not expressed on the outside surface of the vascular endothelium in normal organs. The SWNTs serve as an NIR target resulting in tumor heating and ablation.

This project focuses on the photothermal therapy treatment of two types of cancer that initially occur on surfacesâ €"bladder cancer and melanoma. Approximately 80% of bladder cancer patients have recurrent tumors starting with stage I superficial and quickly progressing towards more muscle invasive and metastatic malignancies with each recurrence. The high incidence of reappearance is believed to be due to the residual tumor left behind in a significant number of patients. The current form of treatment is a transurethral resection for lower grade tumors and cystectomy for higher grade masses. Novel treatments are needed for eliminating all bladder tumor cells with a minimum number of treatments, ideally one. One type of skin cancer, malignant metastatic melanoma (CMM), has a high mortality rate. When CCM becomes metastatic, patient survival rates fall as low as 5%. Over the last half-decade rapid advances in biologic therapeutics and nanotechnology have spawned a new generation of treatment options proving a rich opportunity to investigate combinatorial therapies.

In the study of this therapy for treating bladder cancer, *in vitro* binding tests confirmed a strong binding affinity of SWNT-annexin A5 to MB49 mouse and J82 human bladder cancer cells. Combining SWNT-annexin A5 heating via NIR light confirmed significant cell death as compared to untreated controls for both cell lines. The *in vitro* tests provided statistically significant validation for the potential of this targeted ablation therapy. *In vivo* testing on immune competent mice was conducted to confirm the efficacy of this treatment even further. A biodistribution study followed by FT-Raman analysis confirmed no non-specific accumulation of SWNT-annexin A5 in any organs. Using a 360° diffuser to deliver NIR light to the inside of mouse bladders, NIR power tolerance tests confirmed that no healthy tissue damage occurred at 50 J/cm^2. *In vivo* therapy testing of MB49 cells implanted in the bladder of immune competent mice confirmed a promising treatment modality which is currently being further investigated with more testing using bioluminescence imaging of the cells in combination with fluorescence microscopy.

The work on treating melanoma used B16F10-luciferase mouse melanoma cells injected into immune competent mice. Plasma levels of SWNT-annexin A5 were monitored for up 24 hours after i.v. injection in B6 immune competent albino mice; plasma levels of the conjugate were negligible 1 hour after injection. The minimum injected dose of SWNT-annexin A5 used in combination with photothermal therapy using NIR irradiation was found for obtaining complete eradication of primary tumors (implanted subcutaneously into the flank). Having established the necessary photothermal therapy treatment conditions, we are currently undertaking an evaluation of the synergism between this photothermal therapy and the immunostimulants anti-PD1, anti-CTLA-4, anti-CD73, and cyclophosphamide to ascertain possible mechanisms of harnessing photothermal antigen production to eradicate distant metastases following primary tumor ablation.

471289 Bioproduct Design through Market Assessment: NSF Innovation Corps (I-Corps™) Participation Guides the Development of New *Escherichia coli* Strains

Abstract:

Host cell proteins (HCPs) complicate the recovery of biologically-derived products due to their presence as the target product is isolated by various unit operations that include chromatography, precipitation, and other common bioseparation techniques. While the technology exists to rewrite the genome of the host cell used for expression, few have attempted to use these technologies to reduce HCP burden on the bioseparation steps. Such a space to investigate is quite large because of the number of different techniques and conditions used for bioseparation. Nevertheless, our team set out to elucidate a product design that could gain traction with academia and industry. Participation in the I-Corps™ program defined the characteristics of a minimum viable product through a combination of hypothesis-driven discovery and potential end-user comments.

To wit, a target bioseparation technique (diethylaminoethyl or DEAE ion exchange chromatography) and improvement threshold was identified based on extensive discussions with potential end-users, and a minimum viable product was constructed. An engineered *Escherichia coli* cell line (Lotus®) was built via homologous recombination that displays excellent fermentation and expression characteristics. Moreover, DEAE column capacity was improved by 37% based on chromatography breakthrough data. This presentation will describe the fundamental research efforts, I-Corps™ participation, minimum viable product development, and finally the assessment of the effort.

451290 Recombinant Snake Antivenom Peptide

Abstract:

Tens of thousands of people die by snakebite every year, primarily in developing countries. Many more survivors continue to suffer permanent physical and mental disabilities from a bite. Cytotoxic snake venoms are present in many countries including from American rattlesnakes. Current Fab-based treatment for US snakes costs over $1K per vial but is covered by some insurance policies for humans but not normally for animals. A small ribosomal peptide was reported to neutralize snake venom that was identified as the N-terminus of a serum protein in the North American Opossum (*Didelphis virginiana*). We have found that the chemically synthesized peptide is able to neutralize the hemorrhagic toxins in different species of snake venoms when tested in mice. In addition to activity studies, we have explored the possibility of making the peptide recombinantly by a process that would be amenable to a large scale production. The goal of our work is to propose a large scale, low cost process for the peptide that can be implemented in developing countries where snakes are prevalent and the population cannot afford expensive horse serum-based antivenom treatments. Our process obviates the use of snake venoms that are currently required to inject horses for the production of antibodies. We have successfully produced the peptide in *E. coli* and purified sufficient quantity to demonstrate activity in mice. Process optimization is currently underway, in addition to expanded activity studies. The presentation will cover details of our process studies and the activity of the *E. coli*-produced peptide.

457484 CFD MODEL Based Near-Infrared Spectroscopy Implementation for in-LINE ASSAY Monitoring of a Pharmaceutical Suspension

Abstract:

INTRODUCTION AND PURPOSE

During bottle filling of a pharmaceutical suspension formulation, a decreasing trend in assay is observed at the end of the batch. This leads to approximately 17% of the batch content which cannot be commercialized. The root cause can most likely be found in foam present in the top layer of the suspension in the holding tank which contains less API. The aim in this study is to implement a PAT solution for establishing a batch-dependent endpoint (I.e. assay â‰¤ 95%) determination while reducing in-process sampling drastically.

MATERIALS AND METHODS

Materials

Tests were performed with a high-dose (i.e. 100 mg/mL) pharmaceutical suspension formulation. Lab scale samples were prepared in a 10L mixing vessel. Full scale batches are manufactured in a 150L compounding vessel following a one-pan process. After the transfer to a holding tank, the full scale batches are gradually discharged to a specific bottle filling system with the help of a peristaltic pump. During this process step, the suspension is stirred continuously.

Methods

NIR spectra were recorded using a Fourier-transform Near-Infrared (NIR) spectrometer (Bruker Matrix-F Duplex) equipped with a TE-InGaAs detector, a quartz halogen lamp and a fiber-optic non-contact probe. Every 30

seconds, a spectrum was acquired in the 12000-4000 cm^{-1} region with a resolution of 8 cm^{-1}, averaged over 64 scans. Ultra Performance Liquid Chromatography (UPLC) was selected as reference technique for API content determination. All UPLC measurements were performed at Janssen Pharmaceutica, Beerse, Belgium, using their developed and validated method (confidential).

RESULTS AND DISCUSSION

Samples in a range of 85-115% target concentration were analyzed with both NIR spectroscopy and UPLC. A lab-scale calibration model was developed which shows acceptable feasibility for NIR spectroscopy as potential PAT tool for the API assay determination in suspension.

For approaching the production measurement setup as good as possible, the calibration model was optimized by using a circular system with a similar flow as in production (i.e. 800 mL/min) with the help of a customized T-piece. Samples with concentrations varying between 90-103% were measured and included in the prediction model.

At production scale however, the developed model showed a weak performance with an unexpected variation in NIR assay results. It was assumed that the signal variation was created by disturbed flow patterns/turbulence in front of the lens due to possible back pulse formation at the system output. A Computational Fluid Dynamics (CFD) simulation of the flow pattern of the suspension inside the measuring interface confirmed this assumption.

CFD simulations demonstrated that an extension of the original T-piece could optimize the measurements by transferring the back pulse formation further away from the lens. This mechanical solution allowed a constant sample volume per time unit in front of the NIR laser. This was confirmed with additional lab-scale tests and the follow-up of several production scale batches.

CONCLUSIONS

Despite the dominantly present water absorption bands, NIR spectroscopy is feasible to establish good API assay predictions of this high solid-load suspension formulation.

Experiments indicated that a fine-tuning of the interfacing device is crucial to achieve adequate in-line NIR measurements. Therefore, sufficient process knowledge, including fluidum mechanics, is desired to rationalize the choice for an optimal measurement setup.

CHALLENGES AND FUTURE WORK

In the future it is aimed to investigate if NIR spectroscopy in combination with this specific interfacing device is applicable for other liquid pharmaceutical formulations.

466854 Using Optical Coherence Tomography (OCT) to Analyze in Real-Time the Tablet Coating Variability of a Continuous Drum Coating Process

Abstract:

This study demonstrates an in-line measurement system used in a continuous drum coater (Driaconti-T lab, Driam Anlagenbau GmbH). Process monitoring is performed by an optical coherence tomography (OCT) sensor (OCT Pharma 1D, Phyllon GmbH) providing cross-sectional images of film-coatings in a non-destructive and contactless manner.

The continuous coater has three chambers, each with a product volume of around 2 liters of tablets. In continuous operation mode, the tablets are inserted in chamber 1 and transported to chamber 3, while being coated in each of them. Thus, the product is handled in small batches, without exchange of product between the chambers, resulting in a defined residence time for each tablet.

OCT measurement and subsequent automatic data analysis is a method to determine the coating thickness from OCT images without the need of chemometric calibration models, based on a known refractive index of the coating materials. OCT was used to obtain the temporal evolution of the quality parameters, i.e., mean coating thickness, inter-tablet coating variability and intra-tablet coating variability during the coating process. Thereby, the influence of spray rate, air temperature and spray pattern pressure on the quality parameters was studied based on a full-factorial DoE. Additionally, the variability of single drum segments of an extended period of time was studied.

Continuous coating, here in the form of small batches, is an important enabler for continuous manufacturing of solid dosage forms. In combination with an in-line OCT system the coating process can be monitored and actively

controlled.

467352 Advanced Flexible Control System Implementation into Direct Compaction Continuous Pharmaceutical Manufacturing Pilot-Plant

Abstract:

Currently, pharmaceutical companies are going under paradigm shift from batch to continuous manufacturing. Real time monitoring and supervisory control system are necessary part of efficient continuous manufacturing. However, the minimum level of control necessary to produce the product with desired critical quality attributes depends on several factors including drug content (low dose, high dose), formulation properties (e.g. flowability, segregation), and product properties. Nevertheless, different control strategies such as feedforward (FF), feedback (FB), or combined FF/FB[1] as well as different control algorithms such as PID, MPC, or hybrid MPC-PID can be potentially used. Selection of a right control strategy and algorithm is mandatory task both for efficient manufacturing as well as regulatory point of view. However, because of different levels of complexities and unavailability of systematic method and tools, the identification of required level of control, type of control strategy and control algorithm are still a challenging task.

In this work, a flexible control system has been implemented into a direct compaction continuous tablet manufacturing pilot-plant. The flexible control system consists of three control strategies (FF, FB, combined FF/FB[1]) and three control algorithms (PID, MPC, hybrid MPC-PID) giving options for multiple control schemes. The flexible control system is designed in such a way that any control scheme can be easily selected. The control system is also modular in nature that can be easily transferred to any other control platform and manufacturing plant. The continuous tablet manufacturing pilot-plant runs under each control scheme and their performance are evaluated using standardized performance indicators. Based on the results obtained, a systematic methodology is proposed to assist the selection of control strategy and algorithm for continuous pharmaceutical manufacturing process.

Direct compaction continuous tablet manufacturing process is considered to demonstrate the implementation of flexible control strategy. The flexible control strategy consist of the five supervisory feedback control loops and three supervisory forward control loops among which the required number of control loops can be selected. Feedback loops are for drug concentration, powder level in chute, main compression force, tablet weight and hardenss control while the feedforward loops are for proactively mitigate the effects of process and raw materials variabilities. For example, measured powder bulk density is considered in real time in tablet press to proactively mitigate the effect of powder bulk density on tablet weight and hardness via changing the fill depth. Similarly, the feed forward loop to mitigate the effects of lubricant concentration variability on tablet hardness and thereby on tablet dissolution is added. As well as a feed forward control loop to mitigate the effects variation on blend composition on tablet potency. All feedforward control loops can not be used simultaneously because of limitations of number of actuators and therefore the methodology plays an important role to identify the appropriate control loops for a particular pharmaceutical product. The flexible control system are implemented using industry standard control platform (e.g. DeltaV), PAT data management tool (synTQ), OPC communication protocol, fieldbus devises, chemometric tools and real time monitoring tools.

Reference

1. Singh, R., Muzzio, F., Ierapetritou, M., Ramachandran, R. **(2015)**. A combined feed-forward/feed-back control system for a QbD based continuous tablet manufacturing process. PROCESSES Journal, 3, 339-356.

468862 Raman Spectroscopy- Towards the Prediction of Quality Attributes and Application in Cell Culture Process Development

Abstract:

In recent years, the United States Food and Drug Administration (FDA) and the European Medicines Agency (EMA) have stressed the need of improved process understanding and control in the area of pharmaceutical development, manufacturing and quality assurance, summarized in the Process Analytical Technology (PAT) initiative. Lately, Raman spectroscopy has emerged as a promising online monitoring and control tool in bioprocesses. The advantages of this technology are the invasive and nondestructive operations, the simultaneously detection of multiple process variables, as well as the online operation mode and real-time information release.

Raman spectroscopy is mainly used for monitoring purposes at production stage, where variations are small and the process is already well defined. However, the exploitation of Raman technology for process development, where different process conditions, media, products or cell lines are tested resulting in a large variability and few experiments at similar conditions, could bring very important benefits, in terms of process understanding, speed and cost reduction. This work presents several ideas, experimental results and recommendations to overcome those challenges in process development.

Different fed-batch cultures, varying in cell lines, products and platform media, were monitored online. In addition, diverse spiking strategies were performed to artificially increase the calibration range as well as the quality of the prediction models. The evaluation of advanced modeling techniques to derive optimal results from the obtained spectral data plays a central role in this work. Those techniques included an in-depth analysis of the pretreatment procedure, wavelength selection and outlier elimination with regards to robust calibration and prediction of the components of interest.

Besides classical parameters such as glucose, lactate, titer and viable cell density, all 20 amino acids could be very well predicted. More than that, even the prediction of quality attributes like aggregates and glycans showed promising results. In addition, the potential of generic models, which can be applied independently of the cell line, process platform and medium, could be proven. This enables the Raman spectroscopy as an online monitoring tool of multiple process variables and its implementation into a supervisory control of a continuous integrated bioprocess to ensure consistent process efficiency and shaping the product qualities.

470108 Hydrodynamics and Mass Transfer Effects in Mini Vessel Dissolution Testing Systems

Abstract:

In the pharmaceutical industry, dissolution testing is routinely conducted to evaluate the *in vitro*performance of solid dosage forms in pharmaceutical development, determine the dissolution behavior of formulations and provide an optimization of drug release dosage forms, and as a quality control tool to insure that solid dosage forms have consistent dissolution properties. The USP Dissolution Testing Apparatus 2 is the device most commonly used for this purpose. In a typical test, a tablet is added to the vessel and samples are manually withdrawn from the vessels over time and analyzed for their drug content.

The USP 2 drug dissolution testing method is not always appropriate, especially for dissolution testing involving very small tablets or when not enough drug substance may be available for proper dissolution testing in the USP 2 Apparatus, especially during the initial stage of drug development. Small-volume dissolution testing apparatuses (mini vessel systems) can be advantageously used instead of the standard USP 2 system to overcome these limitations. The most common of these mini vessel systems is similar in shape and operation to the USP 2 apparatus, but it only requires a small volume vessel (200 mL) and a mini paddle impeller.

Before mini vessel systems can gain wide acceptance in industry, their hydrodynamics needs to be studied, and their operating conditions must be defined in order to compare the velocity distribution and mass transfer coefficient in a mini vessel system with those in the standard USP2 system. Very few studies on mini vessel systems are available. Therefore, the objective of this work was to quantify the hydrodynamics and drug mass transfer rates in mini vessel systems, and determine the operating conditions under which mini vessel systems can be reliably used instead of USP 2 system for dissolution testing purposes.

The velocity distribution in a minivessel was obtained here using Computational Fluid Dynamics (CFD) at different agitation speeds. The predictions were compared to the velocity vector maps obtained using Particle Image Velocimetry (PIV). Four horizontal iso-surfaces were selected to be studied in detail. The CFC predicted and PIV measured flow patterns were found to be in good agreement. As expected, the 3D flow velocity magnitude increased as the agitation speed increased. However, the velocity in the inner core region below the impeller always remained very low.

The tablet-medium mass transfer process was additionally modeled, and the mass transfer coefficients at four different agitation speeds were obtained and compared to the standard system. The mini vessel system has a flow pattern similar to that of the standard USP 2 system. It was found that similar velocities at the tablet surface (on the iso-surface at y=7mm, i.e., at the vessel bottom) in the mini vessel system and in the USP 2 system were obtained when the mini vessel was operated a 100 mL, 125 rpm and the USP 2 system was operated at 900rpm and 100rpm, indicating that these two systems should result in similar dissolution profiles when operated under these conditions.

471592 3D Raman Spectroscopy – a Method for Mapping Ingredient Distribution throughout a Tablet Volume

Abstract:

Blending of powders can often result in agglomeration when one or more raw materials in a blend are cohesive, hydrophilic or have the tendency of acquiring electrostatic charges. Agglomeration often leads to failure modes in pharmaceutical tablets, particularly "low dose" formulation tablets where the active pharmaceutical ingredient (API) is present in small amounts. When powders agglomerate, the uniformity of the final tablet is adversely affected, leading to variation in the tablet potency and dissolution profiles, which are usually undesirable. New technology enables the quantitative examination of the microstructure within tablets to determine the extent of agglomeration and to assess its effects on product performance. This can be useful as a means for understanding how the process causes agglomeration. If process parameters can be correlated to the tablet microstructure, this knowledge can assist in designing tablets of desirable quality thereby reducing product failure drastically.

Raman imaging is a powerful tool for studying tablet structure. It has higher chemical specificity than NIR and is also less affected by physical characteristics of the sample. The main disadvantage of Raman, its low quantum yield, requiring long integration times, is vanishing quickly, due to vast advances made recently in laser and detector technology. This is causing a significant renewed interest in using chemical mapping to examine the microstructure of pharmaceutical tablets.

This talk will focus on the use of a new Raman imaging platform, the mPAT system from H_2Optx, which is equipped with a 10x10 micron 2D spatial resolution along with the ability to shave layers off a tablet for depth analysis. This level of spatial resolution enables the user to extract detailed structural data that reveals the relationships between process, product structure, and product performance. This technique has the ability to do a quantitative content determination within a sample and, combining this for each pixel, generate 3D maps w.r.t. the individual constituents.

Tablets manufactured from traditional batch processing methods and continuous manufacturing methods were examined, including tablets made via Direct Compaction (DC), Wet Granulation (WG) and Roller Compaction (RC). Tablets using several formulations for each of the three methods were made with various levels of induced agglomeration and were analyzed using the mPAT system. The effects of the different processing conditions on agglomeration were characterized and their effect on tablet microstructure and dissolution were quantified. These observations will be discussed in this presentation.

463044 Real-Time Monitoring and Modeling of Milling in a Continuous Tableting Line Via Dry Granulation

Abstract:

In the Pharmaceutical Industry, batch operation is the traditional mode of manufacture. During batch production, the quality of the batch usually is subject to batch-to-batch variability and thus quality is controlled by statistical sampling and if necessary rejection of the entire batch. Conversion to the continuous processing mode offers the potential for quality improvement as well as potential cost savings. However, continuous processing requires on-line measurements and active process control and thus the development of Process Analytical Technology (PAT) tools is crucial [1].

In the continuous dry granulation tableting line there are a number of variables that relate to the final product quality. For instance, composition and tablet hardness are among the CQAs in this process. In the dry granulation line, the roller compactor serves as a granulation method. In this unit, the powders are fed into a hopper and then conveyed by the feed screw to two counter rotating rolls. The powder is gripped by the rolls and pulled into the nip area of the compacting rolls before being compacted into a ribbon. Then the ribbon is milled into granules by a hammer mill [2]; generating a granular feed for the tablet press, which has favorable flow and compaction properties.

The milling step is an important unit operation because the granules particle size distribution (PSD) will affect compaction properties and subsequently the tablet hardness. Therefore, the granules PSD is also a CQA. In the roller compaction process, an increase in roll pressure will translate to an increase in ribbon density. The ribbon density has a relation to granule PSD [3]. As a consequence, the ribbon density can be considered an intermediate CQA. The relationship between ribbon density and granule PSD is difficult to establish from first principles, but a correlation can be established by conducting a series of experiments. In this design of experiments, the roll pressure is varied, the ribbon is milled at different speeds and the PSD measured through either off-line means such as sieve analysis or else on-line using laser based methods.

In continuous operations, it is desirable to measure all of the CQAs online, so as to permit the implementation of appropriate process control logic. In the tableting line installed at Purdue University, the roller compacted (RC) ribbon properties such as density and composition can be measured online using either a microwave (MW) and/or a Near-Infrared (NIR) sensor [4]. The PAT tool for online PSD measurement that we have adopted in our work is a high speed imaging camera (Innopharma Eyecon Camera), with which our team has had some previously experience in wet granulation applications [5]. This equipment counts and images particles with a sampling time of 2s and has a scanning rate of $10 m.s^{-1}$.

The milling unit operation is commonly modeled using a population balance model [6]. The key model component of the population balance model (PBM) is the breakage kernel. This component represents the process of creation of smaller particles, as larger particles are impacted by the mill hammers or pins which causes the fracture. A wide range of breakage kernels have been proposed, most of purely empirical or semi-empirical nature. Each of these models requires the estimation of model specific parameters which will depend on feed material properties as well as mill operating variables [7].

This project has three objectives. The first one is to calibrate the Innopharma Eyecon camera. The second objective is to establish the parameters of the breakage kernel and thus the PBM model and the third objective is to develop the empirical relation between the ribbon density and granules PSD both measured using the on-line PAT tools. The first objective is achieved by generating samples of granules prepared under different conditions then relating the Innopharma Eyecon Camera output to the sieve analysis results. Since the sieve analysis measures fractions mass while the camera measures particle size, there is a conversion required which depends on particle density. The data for meeting the needs of the second and third objectives is determined through a series of experiments in which the mill speed, the composition of the powder and the density of the ribbon in the roller compactor and PSD are measured on-line. This data is gathered in runs when the process reaches steady state.

The results of this three-part study enable the capability for on-line monitoring of the combined compactor-mill subsystem. The PBM model and ribbon density-PSD relationship are important components of a data reconciliation model for the entire line. In this paper we report the results of these studies and demonstrate their utility in data reconciliation.

References

[1] S. L. Lee, T. F. O. Connor, X. Yang, C. N. Cruz, L. X. Yu, and J. Woodcock, "Modernizing Pharmaceutical Manufacturing : from Batch to Continuous Production," no. 1, pp. 1–25, 2015.

[2] R. W. Miller and P. J. Sheskey, "Roller Compaction Technology for Pharmaceutical Industry." Marcel Dekker Inc., 2003.

[3] S. Yu, "Roll Compaction Of Pharmaceutical Excipients," University of Birmingham, 2012.

[4] A. Gupta, J. Austin, S. Davis, M. Harris, and G. Reklaitis, "A novel microwave sensor for real-time online monitoring of roll compacts of pharmaceutical powders online - A comparative case study with NIR," *J. Pharm. Sci.*, vol. 104, no. 5, pp. 1787–1794, 2015.

[5] A. S. El Hagrasy, P. Cruise, I. Jones, and J. D. Litster, "In-line size monitoring of a twin screw granulation process using high-speed imaging," *J. Pharm. Innov.*, vol. 8, no. 2, pp. 90–98, 2013.

[6] E. B. Litster, J. D., *The Science and Engineering of Granulation Processes*, vol. 53, no. 9. 2013.

[7] J. K. Pettersen and K. L. Sandvik, "Estimating the breakage and selection functions for a continuous mill," *Int. J. Miner. Process.*, vol. 35, no. 3–4, pp. 149–158, Aug. 1992.

468735 Reducing Cut Size of Uniflow Cyclones with Optimized Swirl Vane Design

Abstract:

Uniflow cyclones are constructed as cylindrical separators with gas and particles passing through them in only one direction. Advantageous compared to reverse flow cyclones is their compact design and their easy implementation into piping systems [1]. These characteristics minimize manufacturing and installation costs. Approved calculation models for uniflow cyclones, which are valid for a wide range of applications, are missing up to now. MCI intends to develop design fundamentals for this cyclone type on the basis of systematic experimental and numerical studies. Results from these investigations provide useful insights into the operation of such devices and means to identify design criteria. The influence of the swirl vane inlet (SVI) to the separation performance is mainly associated with following effects. The direct relation between small vane angles leading in higher separation efficiencies due to higher swirl intensities [2,3]. A second effect is the flow velocity within the vane channels itself which is accounted by geometry of the SVI. Vane optimizations in this work show high separation efficiencies linked with significant lower pressure drop based on reduced flow velocities within the swirl vane channels [4]. This reduction of energy demand due to reduced flow velocities leads in optimized vane angles to force cyclones cut size to a lower level. All investigations have been carried out at a test stand with a cyclone diameter of 0.1m. The results of the performance tests showed good compliance with additional optical investigations within the separation chamber carried out with high speed camera equipment. Moreover, a critical sedimentation path within the separation chamber could be determined, which is correlating to relevant vane design parameters.

[1] Hoffmann, A.C., Stein, L.E., Gas Cyclones and Swirl Tubes - Principles, Design and Operation. Berlin, Heidelberg: Springer, 1. au. ed., 2002.
[2] Kraxner, M., Empirische Ermittlung von Auslegungskriterien für Gleichstromzyklone in Multizyklonblöcken, phd thesis, TU Munich, 2013.

[3] Kraxner, M., Muschelknautz, U., Wechner, S., Ackermann, S., Greif, V., Bolda, J., Influence of the Inlet Vane Geometry On the Uniflow Cyclones Performance (639f). AIChE 2012 Annual Meeting, 2012.
[4] Kraxner, M., Skarke, B., Kofler, T., Pillei, M., Pressure Drop in Uniflow Cyclones: Investigation on an Empirical Calculation Model. CFB11, 2014.

DEVELOPMENT OF FRAMEWORK FOR UNDESIRED AGGLOMERATION IN AN AGITATED FILTER DRYING PROCESS

Hong Lee Lim, Monash University, Clayton, Victoria, Australia
Karen Hapgood, Monash University, Clayton, Victoria, Australia
Brian Haig, SunPharmaceuticals, Port Fairy, Victoria, Australia
**karen.hapgood@monash.edu*

1.0 Introduction

In the manufacturing of active pharmaceutical ingredients (API), the final products are often isolated as solids through crystallization after a series of chemical transformation steps. The solid phase is then separated by filtration and further purified via a series of cake washing step. The drying of the API is often performed following the final filtration step to remove the solvents from the solid phase to meet the residual solvent specification. In the pharmaceutical industry, the agitated filter dryer (AFD) is often used for the filtration, washing and drying of API to avoid the need to transfer the filter cake between filtration and dryer units. The AFD also provides a well contained environment which minimizes the risks of personnel exposure to toxic API while also prevents product contamination. However, the AFD is widely known to produce undesired agglomerates during the agitated drying of wet powders. Formation of hard, unbreakable lumps as well as product degradation due to residual moisture content are two of the most important quality issues caused by this phenomenon. The entrapped moisture in the core of the agglomerates may also lead to substantial increase in drying time due to mass transfer limitation.

In recent years, there is an increased awareness in understanding the inter-dependent relationship between the structure, properties, processing and performance of the API to achieve the desired drug performance [1]. While the performance of the API is determined by its physical properties and chemical structure, the desired properties is essentially governed by how well the processing step is executed. As a result, this has spurred great interests from various researchers to investigate the impact of the filter dryer operation on the undesired agglomeration of powder. A number of investigation works has since been carried out which leads to the proposal of various agglomeration prevention strategies [2-6]. However, it is understood that the suggested methods can be specific depending on the powder-solvent combination. In addition, much less is known on the impact of some phenomena which are commonly neglected such as condensation of vapour in the equipment and detachment of wall deposits.

A lack of generalization of the suggested mitigation method and poor understanding of the different agglomeration mechanisms that may arise for different powders types has led to the need to develop a comprehensive framework which incorporates all the root causes and mechanisms of undesired agglomerations. The agglomeration framework proposed in this work was developed based on bibliographic research comprising of both experimental and theoretical results from literature (ranging from 2003 to 2016), providing the current state of understanding on the agglomeration issues faced during an agitated drying process.

2.0 Agglomeration Framework

The agitated filter drying process is a multi-step operation (e.g. cake wash, filtration, blowdown, agitated drying etc.) which comprised of a wide range of parameter that has to be carefully controlled. Agglomeration for certain powder types can be avoided if these process parameters are managed and maintained consistently at the correct level, without being offset by any disturbances. Appropriate control of temperature, agitation speed, blowdown period and fill volume are some of the most important parameters that have been studied extensively by various researchers.

2.1 Temperature

Figure 1 depicts the potential relationship between temperature control and agglomeration of powder. Many studies have reached a similar conclusion where agglomeration associated with high drying temperature is caused by the increased dissolution-recrystallization (solid bridging) effect and also the melting of crystals in some case. However, it is known that drying high temperature will also lead to drips induced nucleation and growth of agglomerates due to the condensation of excessive vapour on the cold surfaces (e.g. sight glass and spray bars) in the dryer as the drying rate was enhanced [2]. In addition, overheating due to temperature overshoot at the initial heating stage and friction caused by excessive agitation at low clearance may also contribute to the solid bridging and the formation of lumps even when the drying is conducted at a temperature low enough to avoid dissolution and melting. The initial heating of dryer can be done by ramping the temperature at a predefined rate to reach the setpoint value. By carefully controlling the preheating stage, both the temperature overshoot and generation of excessive vapour at the start of drying (where moisture content of powder bed is at its highest) can be avoided. Condensation can also be avoided by applying a "bleed" where stream of dry gas (e.g. nitrogen or air) is flown across the vessel headspace to remove the saturated solvent vapour from the dryer.

On the other hand, drying at low temperature especially for heat sensitive materials are also susceptible to agglomeration and is often neglected in most cases. Low drying temperature suggests that a prolonged period of agitated drying is required to reach the desired moisture content specification. This in turn led to excessive agitation of powder bed at critical moisture content region (where the tendency of wet powder to agglomerate is at its highest), leading to severe agglomeration. When the drying rate is limited by the heat sensitivity of the material, a change in wash solvent with lower boiling can be considered to speed up the drying process.

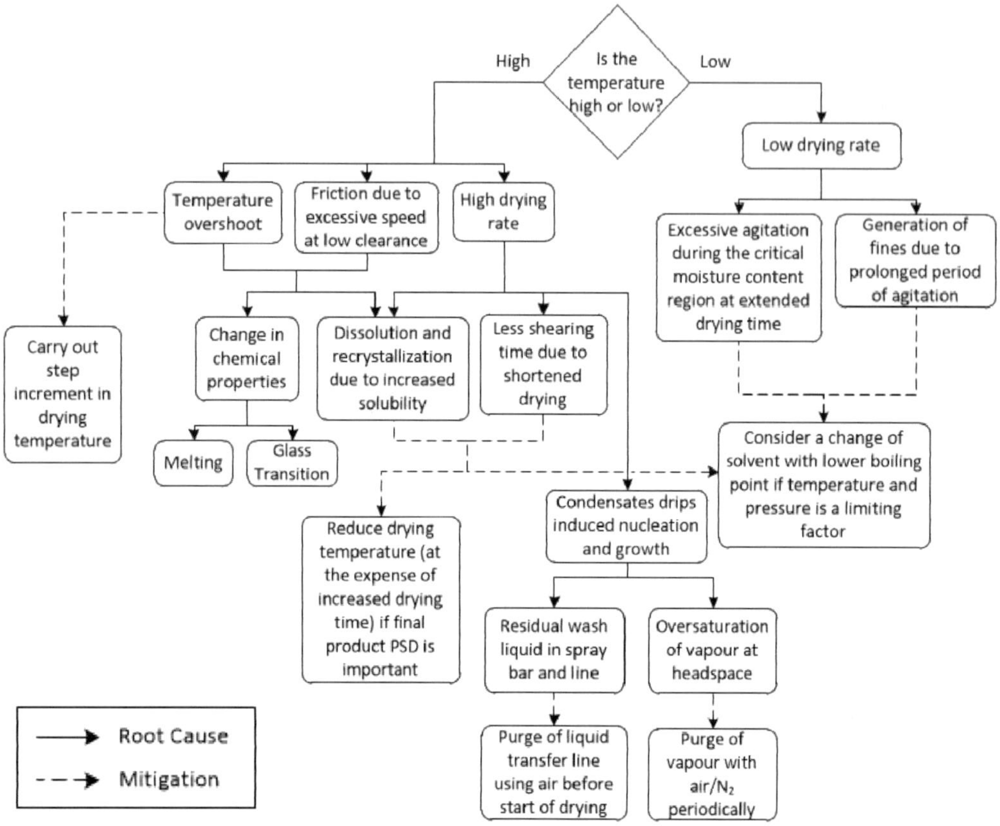

Figure 1: Framework for temperature related agglomeration.

2.2 Agitation Speed

An appropriate choice of agitation speed is also essential in preventing the formation of agglomerates as depicted in Figure 2. High agitation speed often leads to severe compression as well as generation of fines (for fragile crystals) especially when the bed moisture content is high. Excessive generation of fines may contribute to the hardening or cementing of agglomerates where fine particles filled the gaps between the larger primary particles and is further consolidated under high agitation speed. In contrast, low agitation speed tends to produce insufficient shear force for agglomerates breakage purposes while also result in poor mixing. Therefore, consideration must be given to the physical properties of the powder such as compressibility, shape and hardness when selecting the correct impeller speed. It is also worth noting that the control of agitation may not be as simple as selecting a target speed which runs constantly throughout the drying process. Specific drying protocol or regime may be required, depending on the type of powder used. This includes periodic axial raising or lowering of agitator, increase or reduction of agitation speed as well as intermittent stirring to minimize the disruption of bed without hindering its heat transfer capability. It is also important to avoid agitation when the moisture content of the powder bed is high (often at the start of the drying process). The use of intermittent stirring will also prevent the generation of excessive fines to maintain the balance between agglomeration and attrition.

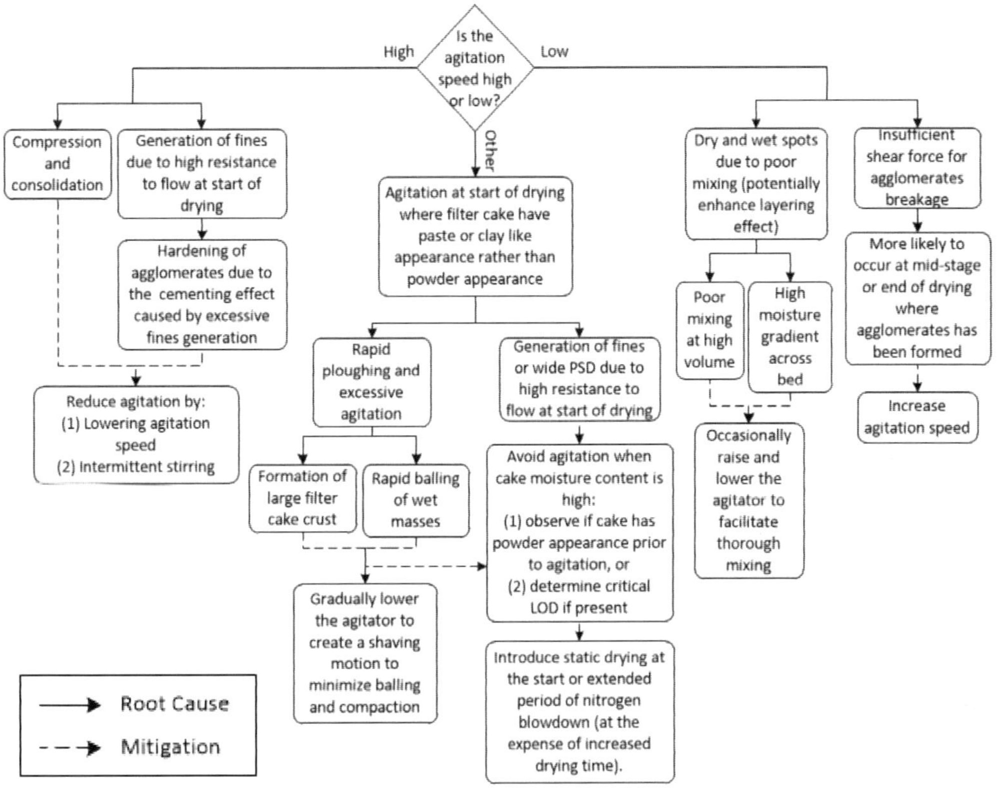

Figure 2: Framework for agitation speed related agglomeration.

2.3 Blowdown Period

The extended blowdown step often serves as an effective way to further reduce the filter cake moisture content prior to the agitated drying step. However, the selection of blowdown pressure can be highly dependent on the properties of the filter cake (see Figure 3). Highly compressible cakes subjected to excessive blowdown period or pressure may increase the cake's resistance to fluid flow (reduction in cake voidage), leading ineffective moisture removal. This in turn produces filter cake with higher initial LOD which is one of the main root causes of agglomeration in an agitated drying process. Excessive cake pasting to removal the filter cake cracking as well as high blowdown pressure may also lead to cake compaction, resulting in the formation of lumps as well. Crystals that are not susceptible to breakage and compression may be subjected to a more severe blowdown condition. If high moisture content persists, an investigation has to be carried out on the crystal structure where its high liquid retention capability may be linked to microscopic properties such as excessive hydrophilic functional group, high surface area and pore network distribution.

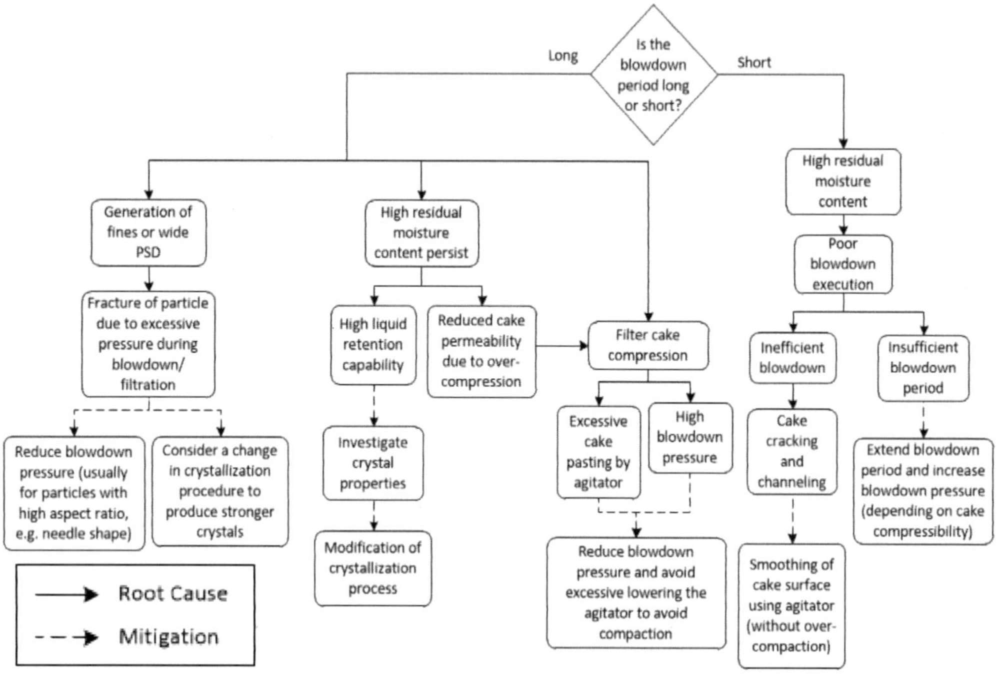

Figure 3: Framework for blowdown related agglomeration.

2.4 Fill Volume

Another aspect of the agitated drying operation which is often overlooked by powder manufacturers is the fill volume of the dryer (see Figure 4). The amount of powder if often dictated the upstream processes such as extraction and other chemical transformation process where certain batches may have a lower yield due to the variability of the raw materials. Agitation of powder bed with low fill volume may produce higher amount of lumps due to the greater collision frequency as well as its higher energy input per unit mass. A lack of hydrostatic pressure to restrict the movement of agglomerates (especially for spherical agglomerates) will also cause the agglomerates to escape the shear impact as it rolls over the impeller. On the other hand, high fill volume may reduce the contact between impeller and particles, giving rise to insufficient power input to the material as well as irregular movement of mass. The "Brazilian nut effect" will also cause the agglomerates to avoid impact of agitation as the agitator lies primarily at the bottom of the bed. It is therefore important to raise and lower the agitator periodically for optimal mixing and drying while ensuring that the agglomerates are exposed to impeller shearing for breakage. In addition, the agitator may also be positioned at various height for a set period of time to ensure powders at different sections of the dryer are subjected to sufficient agitation.

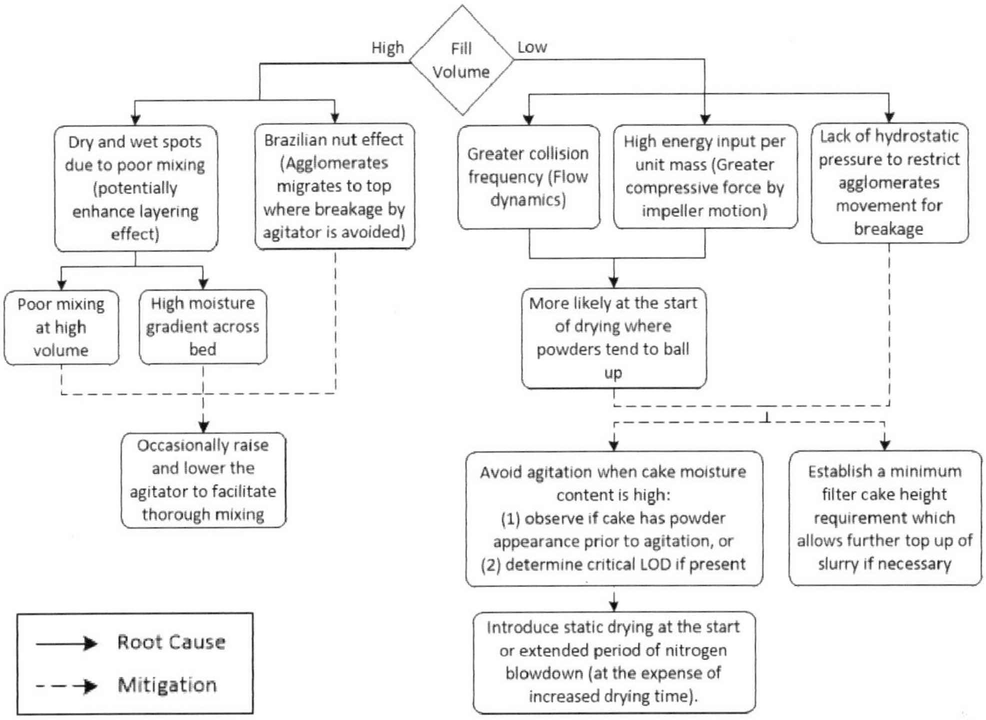

Figure 4: Framework for fill volume related agglomeration.

2.5 Filter Cake Washing and Solvent Composition

The elimination of agglomeration using different solvent combinations has gained significant attention in recent years (see Figure 5). Solvents with high particle solubility profile will promote solid bridging between particles due to the recrystallization of dissolved solids. In addition, residual solvents with high surface tension such as water may encourage the formation of liquid bridging due to their extensive hydrogen bonding capability. It was reported that a secondary solvent type can be added to reduce both the overall surface tension and solubility which was proven to be effective in minimizing the formation of agglomerates for certain powder-solvent combination. The use of multi-component solvent may also promote the removal of multiple impurities during the cake washing step. However, the choice of solvent combination has to be carefully selected based on their compatibilities in boiling point. Wash liquids comprised of solvents with varying boiling point will lead to enrichment of the less volatile solvent. Solvent properties may be reverted to its undesirable state as the concentration of the more volatile solvent decreases during drying, especially if the less volatile solvent has a higher solubility or surface tension. To prevent solvent enrichment, an additional wash can be carried out to isolate as single solvent after the main cake washing procedure has been completed.

While the contribution of dissolution for crystals with low solubility may be negligible towards solid bridge formation, it is speculated that the number of wash cycles and amount of wash solvent used may change the physical properties of the particles that will subsequently affect its downstream processing behaviour. Crystals that undergo several wash cycles may exhibit a more irregular faces with eroded edges suggesting that washing may not only remove the target impurities, but also dissolves the produced crystals. The potential impact of eroded crystal surface on agglomeration is not fully understood but the presence irregular surface due to erosion may increase the specific surface area of the crystals, allowing the crystals to retain a greater amount of liquid on their surfaces while also increases its particle interlocking tendency (or reduced flowability).

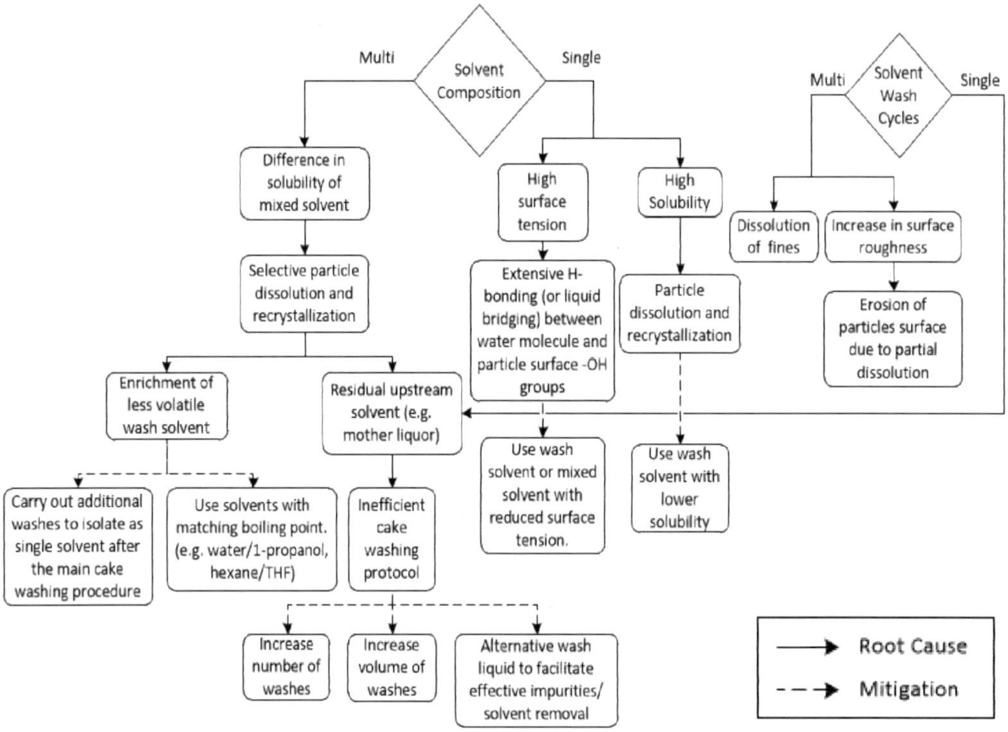

Figure 5: Framework for filter cake washing related agglomeration.

3.0 Future Direction

The framework has outlined the potential agglomeration root causes that may arise for different powder types. While this may serve as a useful tool to overcome any existing agglomeration issues faced by the manufacturers, a guide to identify the early symptoms of the various agglomerations during the initial product development stage is still lacking. Further work is required to incorporate different characterization methods or small scale testings into the framework which will serve as a comprehensive guide to identify the potential agglomeration root causes during the early development stage. Early detection of potential agglomeration prior to up-scaling allows the agglomeration issues to be addressed in the lab scale where optimization of the process condition can be carried out to ensure a smooth and successful transition to large scale production. This also prevents the over-reliance on operators to manually adjust the agitator to break any visible lumps formed during the process which can cause significant delay in production cycle.

4.0 Conclusions

This work reports the preliminary stage of ongoing studies in developing a compiled mechanism framework that will serve as a guideline in troubleshooting the agglomeration during an agitated filter drying process. While this work may not cover entirely the potential causes of powder agglomeration in the pharmaceutical industry, it is understood that this framework will provide a better understanding in controlling the occurrence of agglomeration contributed by some of the common root causes that has been identified to date. This work is the subject of further ongoing research by the authors and will be further revised as more data become publicly available.

Acknowledgement

The authors would like to acknowledge GlaxoSmithKline and Monash University for providing financial support and resources for this study. Note that since September 2015, the Port Fairy facility where this study was conducted is now owned and operated by SunPharmaceuticals.

References

1. Sun, Calvin C. (2009), "Materials science tetrahedron—A useful tool for pharmaceutical research and development," in *Journal of Pharmaceutical sciences*, 98(5), pp. 1671-1687.
2. Lim, H.L., Karen P. Hapgood, and Brian Haig (2016), "Understanding and preventing agglomeration in a filter drying process," in *Powder Technology*, 300, pp. 146-156.
3. am Ende, D., Birch, Melissa Birch, Steven J. Brenek, and Mark T. Maloney (2013), "Development and application of laboratory tools to predict particle properties upon scale-up in agitated filter-dryers," in *Organic Process Research & Development*, 17(10), pp. 1345-1358.
4. Birch, M. and Ivan Marziano (2013), "Understanding and Avoidance of Agglomeration During Drying Processes: A Case Study," in *Organic Process Research & Development*, 17(10), pp. 1359-1366.
5. Kim, Soojin, Bruce Lotz, Mark Lindrud, Kevin Girard, Terence Moore, Karthi Nagarajan, Mario Alvarez et al. (2005), "Control of the particle properties of a drug substance by crystallization engineering and the effect on drug product formulation," in *Organic process research & development*, 9(6), pp. 894-901.
6. Kougoulos, E., C.E. Chadwick, and M.D. Ticehurst (2011). "Impact of agitated drying on the powder properties of an active pharmaceutical ingredient," in *Powder technology*, 210(3), pp. 308-314.

472581 Modeling of the Pressure Drop Across Polydisperse Packed Beds in Depth Filtration

Abstract:

Depth filtration is the process of separation that involves flow of solid-liquid (or liquid-liquid) mixture through a fixed (packed) bed of porous (or particulate) materials. The flow through such a device is commonly modeled using the so-called Kozeny-Carman equation in order to predict the pressure drop during the filtration process. In this work we assess the predictive power of this approach for systems operated at low Reynolds numbers and comparably high pressure drops (on the order of several bar). We find substantial agreement between model and experiment only for systems that result in well-ordered particle packings (i.e., those that have a tight distribution of void sizes). Dramatic disagreement is observed for particle beds that exhibit wide void size distributions. The cake structure is primarily influenced by the size ratio of the particles that compose the cake; specifically, particles with a size ratio where Rsmall/Rlarge is larger than 0.5 do not typically form an ordered pore structure. We propose a modified approach, based on a bimodal void distribution, by introducing two factors: the fraction of expanded voids (kappa) and the ratio of void sizes (beta). Discrete Element Method (DEM) simulations of the packing of poly-disperse spheres are used to analyze the bed structure for different size ratios of binary mixtures. Based on the simulation results, void size distributions of the simulated beds can be extracted by means of a radical Delaunay tessellation. The void structure is quantified in terms of probability density functions of pore and constriction sizes. By fitting the simulated void size distributions to a bimodal (two normal) distribution, the factors kappa and beta can be calculated based on different mean void sizes and probability density. The predicted flow dynamics from the modified equation with factors extracted from the simulation results are found to be much more similar to the experimental flow rates than those calculated using the unmodified Kozeny-Carman equation.

465585 Measurement of the Axial Dispersion of Cohesive Powders in Rotary Kilns

Abstract:

With applications in a wide range of solids manufacturing processes including blending, drying, and disposing of waste, the rotary kiln has been established as an essential device in chemical and metallurgical industries. In the preparation of catalysts in particular, a better scientific understanding of rotary kilns will improve continuous calcination processes for increased throughput with improved product quality. While widely used in large-scale catalyst manufacturing, fundamental understanding of rotary calcination is absent from scientific literature. Thus, the goal of this research is to improve fundamental understanding of rotary calcination processes to aid in scale-up from laboratory and pilot plant scales to the manufacturing scale. For successful calcination to occur, the residence time of the particles must exceed the time required for heating and subsequent treatment. The optimal residence time therefore depends on both of these competing time scales, which are functions of feed material properties, kiln geometry and kiln operating conditions. For uniform treatment of the feed, the particles must also exhibit low axial dispersion. In this work, the residence time distribution and axial dispersion coefficient for two cohesive fluid catalytic cracking powders were measured in pilot plant kilns using a tracer study developed by Danckwerts. Results were successfully matched to the Taylor fit of the axial dispersion model and the Sullivan prediction for mean residence time. It was found that an increase in feed rate, kiln incline and rotary speed decreased mean residence time and overall dispersion. The axial dispersion coefficient was found to vary with kiln conditions. Such results have not been previously reported for cohesive powders.

464496 Residence Time Distribution of Fine to Coarse Particles in Rotary Kilns

Abstract:

Rotary kilns are gas-solid reactors commonly used in industry to achieve a wide array of material processing operations. Rotary kilns are used for applications such as reduction of oxide ore, pyrolysis of hazardous waste, calcining of petroleum coke, conversion of uranium fluoride into uranium dioxide for the manufacture of nuclear fuel, and so on. When operated at atmospheric pressure, they consist of a cylindrical shell that can be inclined, into which the solid charge is fed continuously at one end and discharged at the other. They can be equipped with lifting flights or lifters, and/or an exit dam at the kiln outlet end. They usually require very little labor to operate.

Though the operational cost of these units is usually high, their design is often conservative due to the lack of fundamental physical understanding of both solid flow and heat transfer. The objective of this presentation is to provide a new model to predict the mean residence time, axial dispersion coefficient and hold-up of fine to coarse solid particles within the rotary kiln based on a dimensional analysis. Flow of material through a rotary kiln depends on many factors: length and diameter of the kiln, design and number of lifters distributed around the circumference, rotational speed and slope of the kiln, exit dam height at the kiln outlet end, flow rate and physical properties of the material.

Experiments were carried out on two pilot-scale rotary kilns at room temperature, whether equipped or not with different kinds of lifters or fitted with a dam at the outlet end. These experiments aimed at determining the effects of most of the factors listed above on the Residence Time Distribution of solid particles. The tracer impulse-response technique was used to establish residence time distribution curves. Four granular solids having different properties were used: fine and medium sand, broken rice and beech chips. The other operational parameters were also varied, maintaining the cascading (tumbling) motion.

We have already presented* the results on the smaller kiln with two types of particles. In the present study, we will show that the correlation established with these earlier results had a good predictive ability over the larger kiln with coarser particles. New results with fine particles will also be presented. These new experiments will be used for the consolidation and validation of the previous model.

* Alex Stéphane Bongo Njeng, Stéphane Vitu, Marc Clausse, Jean-Louis Dirion and Marie Debacq « Modeling of Mean Residence Time of Solid Particles in Rotary Kilns » 14AIChE annual meeting, Atlanta, November 2014 [paper 374307]

466810 Evaluation of Small-Scale Powder Flow Characterization Tests

Abstract:

In pharmaceutical development, powder flow characteristics are notoriously difficult to predict. Although tests exist that aim to determine the flowability of powders, many have not been definitively correlated with a powder's performance in downstream processes. Part of the challenge is that powder flow must occur over a broad range of consolidating stresses, including low stress flow regimes such as the filling of a die cavity in a tablet press and high stress flow regimes such as the deformation associated with tablet compression. The objective of this work was to characterize several placebo formulations utilizing a variety of standardized and non-standardized experiments and contrasting those results with downstream tablet compression performance. Angle of repose, bulk and tap density, Hausner ratio, Carr's compressibility index, a visual powder flow assessment, a mass flow rate test, flodex, and powder shear testing were evaluated on several different blends of placebo powders to determine a predictive test for downstream process performance. The placebo powders consisted of a standard base unit formula, mixed with various levels of four different "pseudo-Active Pharmaceutical Ingredients (API's)" (i.e. excipients that mimic poorly flowing Active Pharmaceutical Ingredients). All formulations were evaluated by the individual flow tests and then compressed on a rotary tablet press. Formulations were deemed failures if the relative standard deviation of tablet weight was > 2%, material stuck to punches (low lubrication), or exhibited ejection forces high enough to trigger automatic shutdown of tablet press. The 2% RSD limit is based on internally developed best practices for transferring tablet compression processes from development into manufacturing. The results of the tablet compression process were then compared to the results of the flowability tests to detect any correlations. The Hausner ratio and powder shear test results discriminated marginally between successful and failed batches and were highly correlated with each other.

456386 Assessment and Predictive Modeling of Pharmaceutical Powder Flow Behavior in Small-Scale Hoppers

Abstract:

The propensity of powder to bridge in hoppers and bins is important for a wide variety of pharmaceutical processes, yet it remains difficult to predict large-scale performance from small-scale powder measurements. Despite the availability of various powder flow testing techniques, the prediction of powder bridging has remained elusive. To address this issue, powder flow performance was characterized for several pharmaceutical powders out of small-scale conical hoppers under various conditions. Specifically, the impacts of powder properties, hopper design, hopper fill level and vibration prior to discharge were characterized. The hopper flow regimes were categorized as mass flow, funnel flow, rat-holing or bridging. A Projection to Latent Spaces (PLS, a.k.a partial least squares regression) model was developed to predict flow behavior from easily measured powder properties such as the Hausner ratio and measured hopper dimensions. The model correctly predicted flow initiation for new materials for ~90% of tests. The variables of Hausner ratio, vibration prior to discharge, and hopper orifice diameter were seen to be most predictive of flow. Surprisingly, powder shear test data such as flow function and unconfined yield strength along with other flow properties were not predictive of hopper performance for the conditions studied. A useful metric, *tan(Ï•)/D*, was developed utilizing the hopper semi-apex angle (*Ï•*) and orifice diameter (*D*), and was considered as a first-pass scale-up model.

461681 Application of PAT-Based Feedback Control for the Crystallization of Pharmaceuticals in Porous Media

Abstract:

The development of new drug delivery methods to tailor the desired drug release have been of major interest such as substrates, polymer films and microspheres. The drug release can be affected by various attributes of the polymer properties such as molecular weight, size and size distribution, porosity, morphology and make-up. However, the purpose of this work is to evaluate the benefits of controlling the bulk crystallization in the presence of alginate porous particles. The main objective is to increase the drug loading (DL) in the alginate beads and minimizing the heterogeneous nucleation occurring in the bulk system. Two types of controlled cooling profiles and the model-free Direct Nucleation Control (DNC) strategy were evaluated. A mean DL from 49 to 69 % was achieved through the various scenarios. The application of DNC increased or maintained the mean DL achieved compared to the cooling only scenarios, while a decrease of about 75 % was observed in the batch-to-batch variability. SEM images showed that crystals of different crystal size distribution (CSD) were obtained for the various cooling scenarios, which can be related to the different operating trajectories in the phase diagram achieved during the feedback control approaches. The possibility of manipulating the CSD of particles within the polymer beads can be used to tailor dissolution profile and bioavailability. The work demonstrates the benefits of using feedback control approaches to increase DL, decrease batch-to-batch variability and control dissolution behavior of drug-loaded polymer microspheres.

470315 Optimization of the Bottom-up Precipitation of Nano-Particles of an Ultra-Fast Crystallizer through the Qualitative Characterization of API-Polymer Interactions

Abstract:

Formulations of poorly soluble drugs, seeking to improve bio-availability by API size reduction into the nano-range, typically rely on top-down comminution techniques, e.g. media-milling. While effective, media-milling is costly, time and energy-intensive as well as presenting some isolation challenges. Alternative methods, aiming to access the nano-range through control of the precipitation process, usually rely on very high-intensity mixing in specialty devices like impinging jets and multi-inlet vortex mixers.

For a special class of compounds, crystal nucleation and growth are so fast that even very rapid solvent-antisolvent mixing is insufficient to limit the final crystal size to the nano range.

This work explores the use of polymeric additives to quench the crystal growth of a very fast crystallizer in an attempt at bottom-up nano-crystal precipitation. Lack of first principles understanding of the exact process, through which polymers suppress crystal growth typically leads to high-throughput screening methods for the identification of the optimal additive. However, the large number of potential polymeric candidates, coupled with the parameters controlling the complex high-shear precipitation process lead to a design space too large to explore through brute-force techniques.

Instead, chemometric analysis of the API-polymer interactions was correlated to the performance during precipitation, decoupling polymer optimization from the precipitation process. Identification of specific functional group interactions between the polymer and API also sheds some light on the nature of the crystal growth suppression effect. The combination of high-shear solvent-antisolvent mixing in a rotor-stator wet-mill and polymer additive effect enabled the scalable bottom-up production of nano-particles.

469039 Core-Shell Nanocomposites for Fast Release of Hydrophobic Active Pharmaceutical Ingredients: Impact of Carrier Size

Abstract:

The bioavailability of a large percentage (estimated to be between 40% and 70%) of newly discovered active pharmaceutical ingredients (APIs) is limited by their poor water solubility–hydrophobicity and slow dissolution rate. Over the last decade, major research efforts have been focused on the development of API nanocrystals in suspension form and subsequently converting it into dry core-shell nanocomposite microparticles (NCMPs) to overcome the aforementioned challenges. Preparing NCMPs via fluidized bed (FB) coating of the API nanosuspensions onto inert excipient particles has been used in the manufacturing of nanoformulation based solid dosages. Recent work shows that if the suspensions are properly stabilized, core-shell NCMPs having a water-soluble core can enhance the dissolution rate of poorly water soluble APIs signficantly[1,2]. However, the impact of the carrier particle size has not been investigated. In this study, formation of core-shell nanocomposites of Fenofibrate and Itraconazole, model poorly water soluble APIs, via FB coating of their well-stabilized high drug loaded nanosuspensions is investigated. Specifically, the extent of dissolution enhancement, when fine carrier particles (sub-50 mm) as opposed to the traditional large carrier particles (>300 µm) are used, is examined. This allows testing the hypothesis that greatly increased carrier surface area and corresponding thinner shell at the same API loading for finer carriers can significantly increase the dissolution rate. Fine sub-50 µm lactose (GranuLac® 200) carrier particles were made fluidizable via dry coating with nano-silica, enabling decreased cohesion, fluidization and nanosuspension coating. For both APIs, 30% loaded suspensions were wet-stirred media milled along with a polymer and a surfactant as stabilizers. The stabilizer amounts were screened to achieve stable suspensions with small particle size. The suspensions were FB coated onto hydrophilic nano-silica (M5P) dry coated sub-50 micron lactose (GranuLac® 200) carrier particles or larger carrier particles of median size >300 µm (PrismaLac®40). The resulting finer composite powders (sub-100 mm) based on GranuLac®200 were freely flowing, had high bulk density, and exhibited much faster, immediate release of the poorly water-soluble drugs, in particular for Itraconazole. This is attributed to a much higher specific surface area of the carrier leading to corresponding thinner coating layer for fine carriers as opposed to those for large carrier particles.

Keywords: Active Pharmaceutical Ingredients/drugs, Poorly Water Soluble, Wet-Stirred Media Milling, Fluidized Bed Coating, Fine and large Carrier Particles, Fast Release/Dissolving Nanocomposite Powders

References:

1. C. Knieke, **M. Azad**, D. To, E. Bilgili, R. Dave, Sub-100 micron Fast Dissolving Nanocomposite Drug Powders, **Powder Technol.** 271 (2015) 49–60.
2. A. Bhakay, **M. Azad**, E. Bilgili, R. Dave, Redispersible Fast Dissolving Nanocomposite Microparticles of Poorly Water-Soluble Drugs, **Int. J. Pharm. Sci.** 461(2014) 367–379.

469735 *'designer'* Hydrogel Particles for Controlled Crystallization and Release of Hydrophobic Active Pharmaceutical Ingredients

Abstract:

Nearly 40% or more of drugs currently in the pipeline of the pharmaceutical companies are poorly water soluble, which introduces a significant hurdle for formulating such major class of drugs with an effective dosage. Utilization of innovative biocompatible materials in formulations of hydrophobic drugs to enhance their solubility-limited oral bioavailability and absorption is a rapidly growing area in pharmaceutical materials research. In this work, we develop a novel bottom-up approach for producing and formulating *nanocrystals* of poorly water-soluble active pharmaceutical ingredients (APIs) using core-shell composite hydrogel beads. Organic phase nanoemulsion droplets stabilized by polyvinyl alcohol (PVA) and containing a model hydrophobic API (fenofibrate) are embedded in the alginate hydrogel matrix and subsequently act as crystallization reactors. Controlled evaporation of this composite material produces core-shell structured alginate-PVA hydrogels with drug nanocrystals (500 - 650 nm) embedded within the porous confinement environment of the core. Adjustable loading of API nanocrystals up to 83% by weight is achieved with improved dissolution behavior. We also develop a quantitative model and experimentally validate that the drug release patterns of the fenofibrate nanocrystals can be modulated by controlling the thickness of the PVA shell and drug loading. In summary, our composite hydrogel enabled encapsulation system allows for controlled crystallization, adjustable drug loading, and customizable dissolution profiles ranging from fast to delayed release by utilizing the unique microstructures of coreâ€"shell hydrogels, thus provides a versatile oral delivery system.

Dissolution of Carbamazepine Crystallised directly onto Excipients

Vivek Verma, Clare M Crowley, Peter Davern, Sarah Hudson & B. Kieran Hodnett
Synthesis and Solid State Pharmaceutical Centre, Department of Chemical and Environmental Sciences, Materials and Surface Science Institute, University of Limerick, Ireland
Email ID: vivek.verma@ul.ie

For BCS Class II drugs, such as carbamazepine (CBMZ), *in vivo* bioavailability is significantly affected by dissolution rate [1]. Conventional approaches to improving the latter have relied on mechanical crystal size reduction techniques, such as attrition milling, but this can lead to degradation of active pharmaceutical ingredients (APIs), non-uniform crystal size distributions and the incorporation of impurities [2]. To circumvent these issues various 'bottom-up' approaches (*e.g.*, spray drying, antisolvent precipitation, sonoprecipitation, *etc.*) to crystal size reduction have been investigated [3].

A potential alternative approach to reducing crystal size is the heterogeneous nucleation of APIs in the presence of dispersed excipient particles. Chadwick *et al* [4] showed an increase in the nucleation rate of acetaminophen (AAP) in the presence of α-Lactose Monohydrate (α-LMH) and β-D-Mannitol (β-D-Man). Increasing nucleation rates should concomitantly reduce the crystal size distribution of the APIs and may, therefore, obviate the need to mill API batches. In this study CBMZ was selected as the model API and the influence of the process conditions (*i.e.,*time and supersaturation) and dispersed, insoluble excipients on its crystallisation from metastable CBMZ-methanol solutions and its subsequent dissolution was examined. Unlike the AAP system studied by Chadwick et al, the presence of excipients in a range of supersaturation and crystallisation time had no impact on the final crystal size distribution (CSD) of CBMZ.

PXRD and Raman spectroscopy of the isolated solids fractions confirmed the formation of CBMZ FIII from metastable CBMZ-methanol solutions (supersaturations (S) of 1.22, 1.34 and 1.55) in the absence and presence of excipients (α/β-Lactose (α/β-Lac), β-D-Mannitol (β-D-Man), microcrystalline cellulose (MCC) and carboxymethyl cellulose (CMC)). The crystal size distribution (CSD) of the FIII crystals, assessed by image analysis of the SEM micrographs, showed similar CSD in the presence and absence of dispersed excipient (Table 1 and Figure 1). However, the dissolution rate of CBMZ FIII was enhanced 2-fold upon crystallisation in the presence of dispersed excipients as compared to CBMZ FIII recrystallised in the absence of excipients under the same conditions (Figure 2). The dissolution rate of CBMZ FIII crystallised in the presence of dispersed excipients is higher than the physical mix of CBMZ FIII and excipients, mostly evident with α/β-Lac and β-D-Man (Table 2) reflecting the superiority of the process. SEM-Raman indicates the presence of low concentration of CBMZ on the surface of excipient facilitating faster dissolution rate at initial times. Figure 2 also presents the comparitive dissolution of CBMZ FIII crystallised in the presence of dispersed β-D-Man and the physicsal mix of CBMZ FIII and β-D-Man indicating the improved dissolution rate throughout the dissolution profile.

Table 1: Mean crystal size distribution (CSD) (μm) of CBMZ FIII crystals formed in the presence and absence of excipients

S	Excipient				No Excipient
	α/β-Lac	β-D-Man	MCC	CMC	
1.22	16±11	13±14	17±11	13±7	22±10
1.34	23±14	13±7	14±7	32±22	23±10
1.55	17±9	18±10	19±10	20±10	27±12

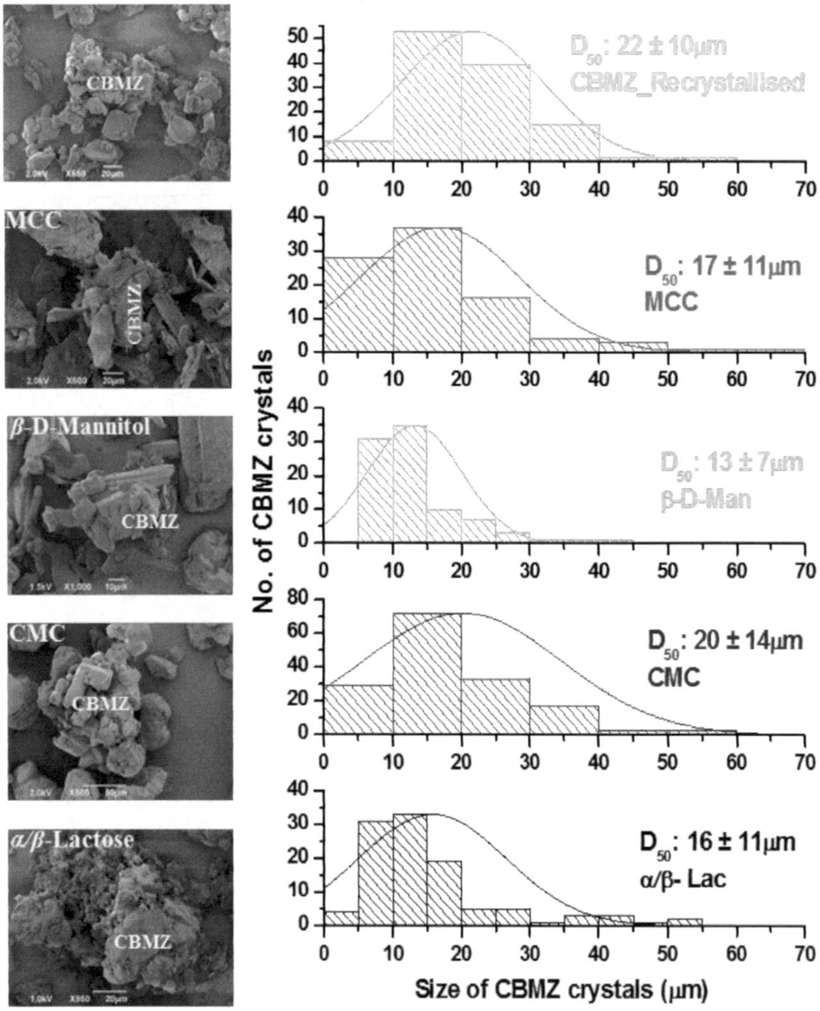

Figure 1: SEM micrographs and CSDs of CBMZ FIII crystallised in the absence and presence of excipients from metastable CBMZ-MeOH solution (t=3 hr aged samples; S = 1.22)

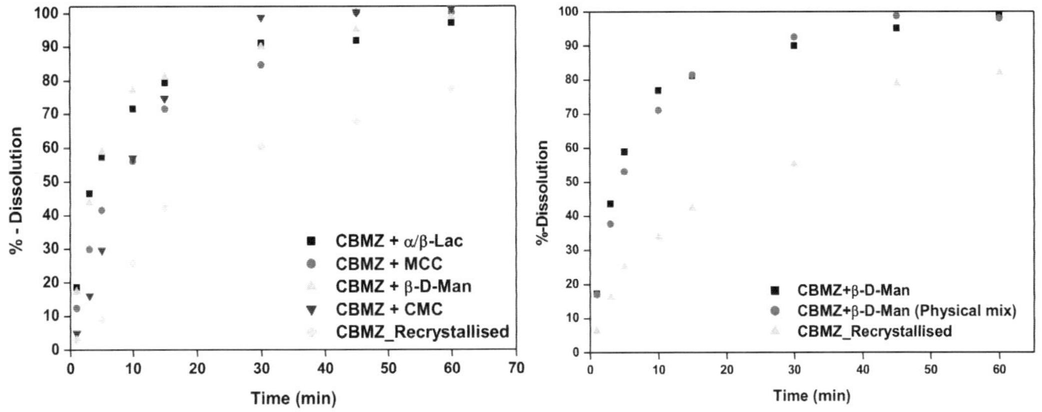

Figure 2: (Left) Comparison of the dissolution rates of CBMZ FIII crystallised in the absence and presence of excipients from metastable CBMZ-MeOH. (Right) Comparision of the dissolution rates of CBMZ crystallised in presence of β-D-Man and the physical mix of CBMZ FIII and β-D-Man; (S=1.22, t=3hr)) (dissolution media: PBS, pH = 7.4; sink conditions ($[CBMZ]_{max}$= 40 mg/L))

Table 2: Dissolution rate of CBMZ FIII over 5 min in the presence and absence of excipients

Excipients	S	Dissolution rate over 5 min
None	Recry	3.47
None	Recry	3.47
None	Recry	3.47
None	Recry	3.47
CMC	1.22	5.75
CMC	Phymix	6.01
CMC	1.55	6.12
CMC	1.34	6.45
MCC	1.34	8.36
MCC	1.22	8.85
α/β-Lac	Phymix	9.84
MCC	Phymix	10.79
β-D-Man	Phymix	11.31
β-D-Man	1.34	11.97
MCC	1.55	12.2
α/β-Lac	1.34	12.3
β-D-Man	1.22	12.65
α/β-Lac	1.22	12.72
α/β-Lac	1.55	13.62
β-D-Man	1.55	15.74

Acknowledgements

This work was funded by Science Foundation Ireland under Grant 12/RC/2275. This work was also conducted under the framework of the Irish Government's Programme for Research in Third Level Institutions Cycle 5, with the assistance of the European Regional Development fund.

References

1. Lindenberg M., S. Kopp and J.B. Dressman, European Journal of Pharmaceutics and Biopharmaceutics, 2004, **58**(2): p. 265-278.
2. Horn D. and J. Rieger, Angewandte Chemie International Edition, 2001, **40**(23): p. 4330-4361.
3. Poornachary S. L. et al., Crystal Growth and Design, 2016, **16**(2):p. 749-758.
4. Chadwick, K.; Chen, J.; Myerson, A. S.; Trout, B. L., Toward the Rational Design of Crystalline Surfaces for Heteroepitaxy: Role of Molecular Functionality. *Crystal Growth & Design* 2012, **12** (3), 1159-1166.

*** Formatting Issue - Best Available Paper/Graphic ***

467271 Modeling Spray Dried Dispersion: From Droplet to Particle

Abstract:

Modeling Spray Dried Dispersion: From Droplet to Particle

Jaime L. Curtis-Fisk[1], Priti Jain[2], Jodi Mecca[1], Bart Rijksen, [3] Shrikant Khot[1], William W. Porter III[1]
[1]The Dow Chemical Company, Midland, MI USA

[2]Dow Chemical International Private Limited, Mumbai, India

[3] Dow Benelux BV, Terneuzen, Netherlands

Purpose:

Poorly soluble APIs are often formulated as a solid amorphous dispersion in combination with an excipient polymer to enhance their solubility. A common route to producing the dispersion is through spray drying. This process involves spraying a solution of polymer, active pharmaceutical ingredient (API) and solvent into a heated drying gas where the liquid spray droplets dry into solid particles of the amorphous dispersion. Hydroxypropyl methylcellulose acetate succinate (HPMCAS), such as AFFINISOL™ HPMCAS, is one polymer shown to significantly enhance API solubility when in an amorphous dispersions. The purpose of this work is to develop a droplet drying model connecting the fundamental properties of HPMCAS polymer solutions to the spray drying particle formation kinetics and the resulting particles' morphology and compositional uniformity. With this modeling capability, it may be possible to predict and gain insight into the observed properties of amorphous spray dried dispersions.

Approach and Methods:

A particle formation model was developed considering the balance between solvent evaporation and solute diffusion to predict particle morphology and composition. The model, as illustrated in Figure 1, considers the diffusion of API and excipient towards the core of the droplet and surface recession of the solvent as the two main counteracting mechanisms in the drying of droplets. Methods for predicting particle formation are known, but do not fully address the complexities of producing spray dried dispersions from a solution containing API and polymer.[i] This model framework builds upon these references and addresses the non-ideality of the liquid phase due to the use of polymeric excipients, including the interaction between solutes and the change in evaporation rate due to non-ideal vapour liquid equilibrium. This added complexity will provide a more accurate route to predicting the impact of formulation and spray drying conditions on particle morphology.

To provide inputs to the model, fundamental properties including viscosity and diffusivity of HPMCAS-API solutions were characterized in a wide range of solvents and solvent blends. Additionally methods to characterize HPMCAS-API interactions were used to understand potential effects on mass transport within the solution droplet.

*** Formatting Issue - Best Available Paper/Graphic ***

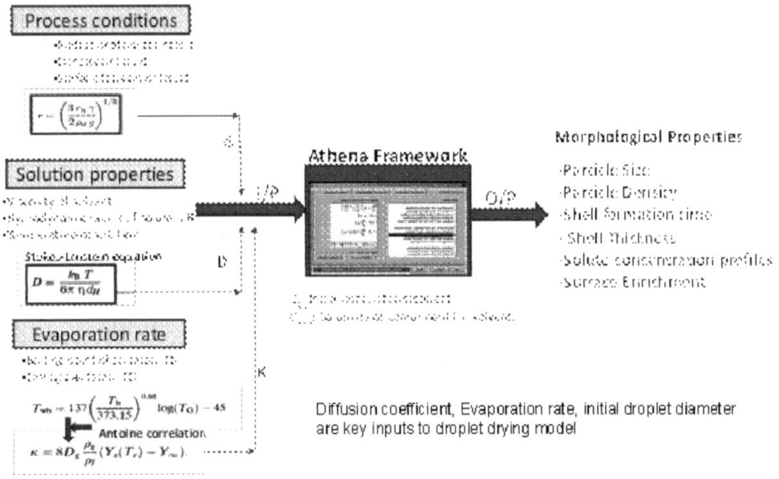

***Figure 1.* The model framework combines process conditions, solution properties, and evaporation rate in an Athena framework to generate predictions of particle morphological properties**

Results:

This particle formation model was shown to compare well against experimental results in the literature. Figure 2 illustrates a comparison of the model predictions of spray dried trehalose against results from Vehring et al.[ii] The samples evaluated cover a low Peclet number range of 0.44 to 1, with an initial droplet diameter of 20 μm resulting in a dry particle diameter of 1.7 to 1.8 μm. The model accurately predicts the measured dry particle density, which is near the density of amorphous trehalose indicating solid particles.

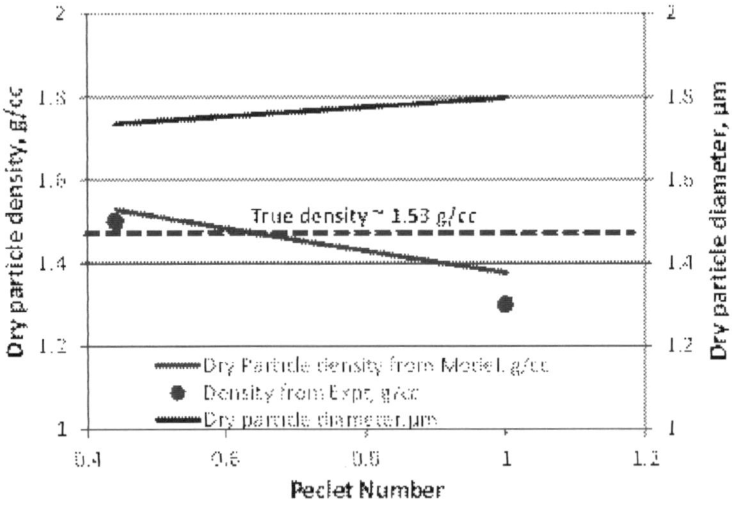

Figure 2. Comparison of experimental results from Vehring et al and model predictions of trehalose particles formed through spray drying.[ii]

It was also found to predict well the experimental results from Vehring et al.[ii] for spray drying of glycoprotein solution as shown in Figure 3. As the Peclet number increases, the particle diameter increases while the density decreases. Both the experimental and modeled particle densities (0.1 to 0.5 g/cc) are significantly lower than the density of solid glycoprotein. This indicates hollow particles with void space, a result confirmed through microscopy of the generated particles.

***** Formatting Issue - Best Available Paper/Graphic *****

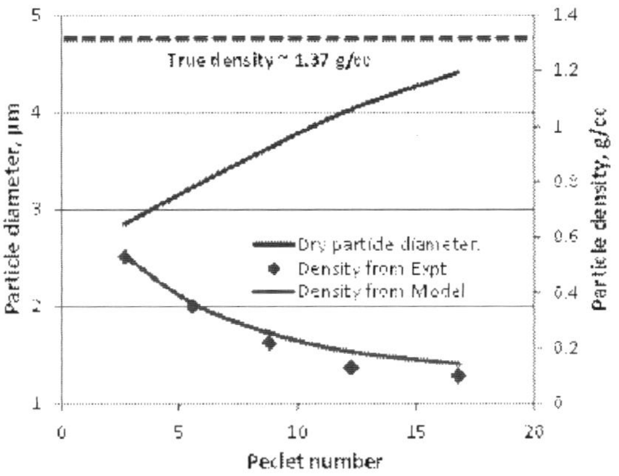

Figure 3. Comparison of experimental results from Vehring et al and model predictions of glycoprotein particles formed through spray drying.[ii]

To apply the droplet drying model to spray drying API-HPMCAS dispersions, fundamental solution properties were characterized. The viscosities of such solutions were found to vary strongly based on the specific HPMCAS polymer and spray drying solvent. As shown in Figure 4, at similar concentrations a wide range of viscosities were observed indicating very different polymer configurations in solution. Similar measurements were conducted to quantify the diffusivity of polymer and API solutes in solution and the surface tension of these solutions. Having quantified these properties, predictions around the droplet drying process and resulting particle properties will be shared. It will be shown that depending on the solvents selected for spray drying, diffusion controlled or surface recession controlled drying kinetics can occur resulting in very different particle morphologies and compositional uniformity.

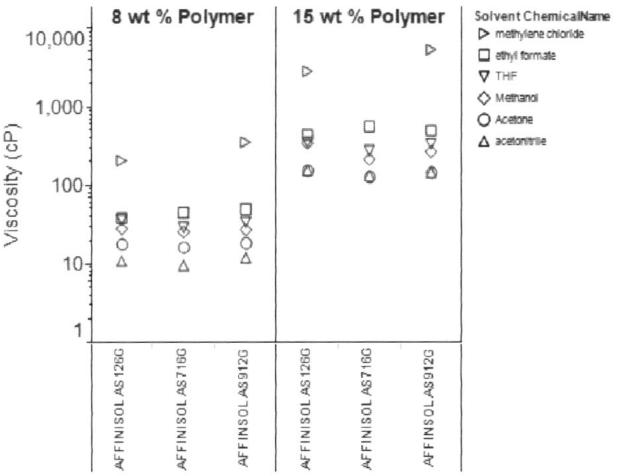

Figure 4: Viscosity of Polymer/Solvent Solutions that serve as inputs for particle formation modeling

Conclusions:

Experimental analysis of formulation solution properties and polymer chemistry provide a solid framework for building modeling capabilities to predict the formation of solid particles from liquid droplets in the spray dried dispersion process. The two examples presented highlight the dramatically different potential outcomes from a spray

***** Formatting Issue - Best Available Paper/Graphic *****

drying process, from solid particles of high density to low density hollow particles. The ability to predict the nature of the particles formed based on formulation and process properties provides the formulator with a valuable tool to optimize this process to hone in on a targeted outcome prior to conducting time consuming experiments. Particle morphology can play a critical role in downstream processing and ultimately the application performance. Combining polymer chemistry and formulation expertise with modeling capabilities provides formulators with a streamlined route to product development.

---

[i] Chen, Xiao Dong; Sidhu, Harvinder; Nelson, Mark; "Theoretical probing of the phenomenon of the formation of the outermost surface layer of a multi-component particle, and the surface chemical composition after the rapid removal of water in spray drying," Chem. Eng. Sci. 66 (2011) 6375-6384.

[ii] Vehring, Reinhard; Foss, Wiillard R.; Lechuga-Ballesteros, David; "Particle formation in spray drying," Aerosol Science 38 (2007) 728-746.

458912 Modeling Dissolution of Polymer Strip Films Loaded with Poorly Water-Soluble Drug Nanoparticles

Abstract:

Thanks to a wide variety of excipients, APIs, and their interactions, reliable prediction of the performance of any solid dosage form is a significant challenge. This is especially true for swellable polymeric strip films containing poorly water-soluble drug nanoparticles, as several different mechanisms may potentially control the rate of drug release. These limiting mechanisms include diffusion of the dissolved drug out of the swollen polymer gel layer, erosion of the polymer matrix, and dissolution of the poorly water-soluble drug particles. The objective of this work is to develop a predictive model for drug release from nanoparticle-laden polymer strip films based on various formulation parameters and inherent drug/excipient properties. This model utilizes first principles to estimate evolution of film thickness, dissolved drug concentration, drug particle size, and water concentration throughout the film over time during the dissolution process. Diffusion of dissolved drug and solvent, drug particle dissolution, polymer erosion, and film swelling/shrinkage are all taken into consideration. Results from this model will be used to better understand the fundamental driving forces behind drug release from poorly water-soluble drug particle-laden polymer strip films, allowing for more intelligent formulation design of targeted applications.

468415 A Novel Method for Grindability Assessment of Pharmaceutical Powders

Abstract:

In the early stages of development of a new Active Pharmaceutical Ingredient (API), insufficient material quantity is usually available for addressing process engineering issues, such as milling. Mechanical properties that are accountable for size reduction are hardness, toughness and stiffness (for semi-brittle failure), and these can be characterised using a small sample quantity by nanoindentation. However, the use of nano-indentation may not be appropriate for strain-rate sensitive solids, and hence impact testing, backed up by breakage analysis, can give better representative information on grindability. Unfortunately, there is no commercially-available single particle impact testing for fine particulate solids. In contrast, dry powder dispersers, such as the Scirocco powder disperser of Malvern Mastersizer 2000, are widely available. In this device, particles are dispersed by an air jet in a Venturi eductor configuration, impacted at a bend and presented to the laser light beam for size analysis by laser light diffraction. So this device has the potential to be used for testing the breakage propensity (grindability) of particles. However, particles of different sizes accelerate to different velocities, due to short acceleration length, and break to different extents; hence the results will be 'fuzzy', unless the relation between particle size and impact velocity and nozzle pressure is established.

Three dimensional multiphase CFD simulations of the Scirocco disperser have been carried out to analyse the air flow field, following which particle trajectories and impact velocities have been calculated. These calculations are then used in the estimation of the breakage propensity parameter. The extent of breakage and change in the surface area are then related to the breakage propensity parameter, for which a good correlation is shown for a range of particle sizes of several organic particles. The obtained information shows the ease of grindability of different materials relative to each other.

471687 Feedforward Control of Continuous Pharmaceutical Manufacturing Process

Abstract:

Continuous manufacturing has been in use in many industries for a long time, however the interest in pharmaceutical continuous manufacturing is relatively new and most pharmaceutical companies are now moving from batch to continuous. With continuous manufacturing we have shorter processing times, increased efficiency and increased product quality assurance in real time. The product quality is of utmost importance in pharmaceutical manufacturing and in order to control that, we need to make sure that effect of input disturbances are proactively mitigated before they affect end product quality. Therefore it is desired that a combined feedforward/feedback control system be implemented over a traditional feedback control system to dissipate and control the disturbances from entering the process. The feedforward controller measures and takes corrective actions for the disturbances before they affect the process while the feedback controller ensures the consistency of the output [1].

In this work, feedforward control loops have been designed for continuous pharmaceutical manufacturing process via wet granulation (WG), roller compaction (RC) and direct compaction (DC) and their performance have been evaluated through a flexible integrated process flowsheet model. For WG line, a combined feedforward/feedback control system has been developed to control the drug concentration at twin screw granulator (TSG) outlet. This process involves a batch pre-blending step where API and intragranular components are mixed, and subsequently fed into a continuous wet granulation (TSG) processing step, together with second stream containing excipients. The commingling of these two streams in the twin screw granulator leads to an overall low API concentration. This concentration of the API is very small and hence, difficult to be measured by a Near Infrared (NIR) sensor. Thus, the role of the residence time distribution (RTD) model in predicting the concentration at the outlet of the TSG as well as to determine the disturbance dissipation along the process becomes significant. Along with this, RTD can also be used for the raw material traceability and scale-up requirements. A major source of drug concentration variability is the potential lack of homogeneity of the stream entering the granulator. Pre-blends discharged by a tumbling blender are known to fluctuate fairly significantly in composition and due to limited back mixing in the twin screw granulator, these fluctuations are propagated to the granulator exit, affecting the content uniformity of API in the granules. In turn, such variability can affect product content uniformity (CU). Since the composition of the stream discharged by the feeder can be monitored instantaneously using PAT method, it is therefore possible to adjust the ratio of the API-bearing stream and the excipient-only stream, and their flow rates, to compensate for composition fluctuations entering, and therefore exiting, the granulator. This has been achieved through a feedforward control mechanism combined with a feedback control loop. For RC line, a feed forward control loop has been developed to mitigate the effect of lubricant concentration. This loop has been integrated with feedback control loop for ribbon density which is cascade with hydraulic pressure feedback (slave) control loop, and a feedback control loop for throughput. For DC line, three feedforward control loops have been developed to proactively mitigate the effects of variations of powder bulk density, lubricant concentration and drug concentration on tablet weight & hardness, tablet hardness, and tablet potency respectively. However, maximum two feed forward control loops can be activated at a time. These feedforward loops have been integrated with feedback control loops [2].

The aim of this presentation is two-fold. First to highlight the feedforward control loops involved in continuous pharmaceutical manufacturing via WG, RC and DC and second to demonstrate the performance of combined feedforward/feedback control system over only feedback control system.

References:

[1]. Singh, R., Muzzio, F., Ierapetritou, M., Ramachandran, R. **(2015)**. A combined feed-forward/feed-back control system for a QbD based continuous tablet manufacturing process. PROCESSES Journal, 3, 339-356.

[2]. Singh, R., Sahay, A., Karry, K. M., Muzzio, F., Ierapetritou, M., Ramachandran, R. (2014). Implementation of a hybrid MPC-PID control strategy using PAT tools into a direct compaction continuous pharmaceutical tablet manufacturing pilot-plant. International Journal of Pharmaceutics, 473, 38â€“54.

471742 Simultaneous Measurement of Granule Disintegration and API Dissolution Profiles As a Tool for Maintaining Product Quality during Process Scale-up

Abstract:

The release rate of an active pharmaceutial ingridient (API) is a crucial atribute of a solid dosage form. In vitro dissolution tests with clearly defined conditions are used to assess quality of the solids. A dissolution profile provides information about the amount of dissolved API from the solid dosage form in time. Comparison of the different profiles could serve as a tool to assess a success of a transfer or scale-up of a given process. Granulation proces is a key step during the manufacture of solid dosage forms, therefore the granules are tested by in vitro dissolution experiments. However, single dissolution profile of API from the granules does not provide deep insight into dissolution process. While classical in vitro dissolution tests are not always discriminatory, their combination with the simultaneous monitoring of other variables, such as the mechanical propertier of disintegrating granules before and during dissolution, may be more sensitive to in consisitncies in the product quality â€“ either between batches or during scale-up or transfer.

The aim of this work is to investigate the dissolution mechanism of the granules consisted of 6-component formulation with a high load of API (58 % wt.) and produced at qualitatively different conditions of the granulation process (various main impeller speed, expressed by Froude number ranging from 0.26 to 2.04) and on different scales (granulator volume: 4, 25 and 400 L). A novel metodology which combines granule disintegration online monitoring with a measurement of API dissolution profile was developed. An evolution of a granule size distribution is analyzed online using a laser particle size analyzer (Horiba Partica LA950) by static light scattering. A 300ml circuit with flow cell and input cone for load and to take the samples for offline measurement on UV spectrophotometer in order to obtain dissolution profile coupled with information about granule disintegration. The inner structure of the granules was investigated using x-ray micro-CT tomography and their mechanical properties were analyzed by CT3 Texture Analyzer. Three possible mechanisms of API release from granules were proposed: leaching, surface erosion and granule breakage. These mechanisms were identified in different stages of the dissolution process and different contribution of each mechanism was observed for granules produced at different Froude number granulation set-up. In a more detailed part of study, where granules sieved into coarse (over 400 μm) and fine (under 400 μm) fractions were investigated, breakage of the coarse granules into considerably stable sub-granules was observed, where the prevailing mechanism of release was leaching of API. This demonstrates that a sub-population of granules exists in each batch regardelss of the process conditions, which shows very robust and repeatable properties. The challenge for future QbD research will be to identify mechanisms that lead to the formation of this sub-population.

471422 Particle Characterization Technique to Enable Rapid Pharmaceutical Formulation and Process Optimization

Abstract:

Pharmaceutical processing consists of several unit operations that transform active ingredients and excipients to improve the processing and handling characteristics for downstream processing into the desired solid dosage form. These same processes can impart physical changes to an active an ingredient that can potentially affect the in vivo performance of the formulated product.

The current in vitro analytical methodology used by industry to predict in vivo absorption is often times poorly correlated to true product performance. The standard USP/Ph Eur in vitro dissolution methods do not allow for a complete understanding of dissolution kinetics and underlying particle-engineering principles that may be at the root cause of a potential processing issue.

Understanding the performance of the drug product is of utmost importance to establish the fundamental mechanisms of drug release. The drug release is governed by the formulation composition and to a larger extent the processing conditions for a given formulation. If a drug substance has low solubility, for example, then the particle size will be a governing factor of drug absorption.

The focused beam reflectance measurement (FBRM) probe is an integral tool for formulation and process development at Nucleo Life Sciences. The FBRM is implemented in formulation optimization, qualification of API particle size, evaluation of formulation performance, and scale-up qualification. The measurements are simple and quick to analyze and provide a wealth of information on the particulate system.

Evaluating formulations using the FBRM particle size characterization provides valuable insight to formulation

development. The ability to monitor tablet and granule disintegration behavior allows for an understanding of the underlying mechanisms of product performance. All of these dynamic events can be monitored in situ. Our goal is to assist companies with implementing the data generated by FBRM into regulatory submissions to support the selection of the in vitro dissolution method development or even use the FBRM methodology as a routine in process release test.

468287 HPMC Cloud Point: Exploring Hydroxypropylmethyl Cellulose Behavior in Pharmaceutical Formulations

Abstract:

HPMC cloud point: Exploring hydroxypropylmethyl cellulose behavior in pharmaceutical formulations

Maryam Mohammadpour[1,2,*], Brian Kerins[1], Mary Ellen Crowely[1], Gavin W. Walker[2], Abina M. Crean[1,2]

School of Pharmacy, University College Cork, Ireland[1]

Synthesis & Solid State Pharmaceuticals Centre (SSPC), Bernal Institute, University of Limerick, Ireland[2]

*Tel: + 353 (21) 4901683, email: marym.mohammadpour@ul.ie

INTRODUCTION

Hydroxypropylmethyl cellulose (HPMC) is widely used in pharmaceutical applications as a sustained release agent (Viriden, Wittgren et al. 2011), tablet coating, nanoparticulate stabilisers and nucleation inhibitor (Ghosh, Bose et al. 2011). In these applications its gelation behaviour is key to its performance and therefore needs to be understood within each system. Aqueous solutions of HPMC display two characteristic temperatures: a thermal gelation temperature (TGT) and a lower critical solution temperature (LCST). The gelation behaviour of HPMC is influenced by the HPMC degree of substitution, formulation additives (such as salts, surfactants, drug molecules) and temperature (Zhang, Wang et al. 2015). [i]

HPMC begins to precipitate out of solution at the LCST due to the loss of water of hydration at increased temperatures (Viridén, Wittgren et al. 2011). This phase separation into a polymer rich and polymer depleted phase, caused by association of hydrophobic groups leading to aggregation, allows for the cloud point to be determined as the polymer rich phase is capable of scattering light. Though there seems to be no direct relationship between TGT and cloud point, it seems that both properties are affected by electrolytes in the same manner i.e. if an electrolyte increases the TGT, it will also increase the cloud point (Mitchell, Ford et al. 1990).

The objective of this paper is to demonstrate the ability of cloud point measurements to identify the impact of formulation additives on the behaviour of HPMC within formulations during dissolution. The additives investigated include; a surfactant, docusate sodium (DOSS) and a model drug carbamazepine. The goal of this work is to use cloud point measurements to understand formulation interactions with HPMC which will increase understanding of variability in formulation behaviour during dissolution.

*** Formatting Issue - Best Available Paper/Graphic ***

Materials and Methods

The HPMC grade investigated was HPMC substitution type 2910, label viscosity 3 cp. In comparison a 2nd HPMC derivative was also investigated, HPMCAS Grade AS-MG viscosity type 3 cp. HPMCAS is a cellulose polymer which does not exhibit a cloud point. The goal of the study was to investigate both polymers in aqueous systems with and without docusate sodium and carbamazepine to assess the additive effect on cloud point behaviour.

Aqueous polymeric solutions were prepared by adding HPMC to an aqueous solution of additive while stirring with a magnetic stirrer at 500 rpm. The solutions were made up to the required volume with ice cold deionised water and left mixing for 20 minutes. The solution was degassed by a sonicator for one hour and then moved to a refrigerator to hydrate overnight.

A Chirascan circular dichroism spectrometer performed cloud point assays. Percentage transmission was examined at 800nm over a range of 25 – 91 °C at a rate of 1.0°C/min. Sample was present in a 0.5mm quartz cuvette. An air background calibration was carried out prior to testing each day.

Results

To ensure the accuracy of the produced results a centre point (6% HPMC, 0.5% DOSS) was ran in triplicate over 3 separate days. The average CP was calculated to be 73.25 ± 0.725 °C. The low standard deviation indicates the Chirascan instrument is capable of generating reproducible CP measurements.

There is a clear trend evident that as the concentration of DOSS increases the the CP increases. This we attribute to the increased energy needed to overcome the solubilisation of the hydrophobic groups of the HPMC by the surfactant (Figure 1). When ionic surfactants and non-ionic polymers are mixed, two competing phenomena take place: 1) Self aggregation of surfactant molecules to form micelles or 2) interaction of individual surfactant molecules with monomer units of the polymer chain (Acevedo, Takhistov et al. 2014). The free surfactant molecules bind to the hydrophobic parts of the polymer through adsorption until a state of saturation is reached. The priority binding of the surfactant produces polar outshells, increasing the energy barrier for the crosslinking of the HPMC chains to occur, by hindering the free access to HPMC chains. The thermal energy requirements for dismantling this shell and for the aggregation of the polymer is determined by the chemical structure and electrostatic interaction between the surfactant and HPMC (Joshi 2011).

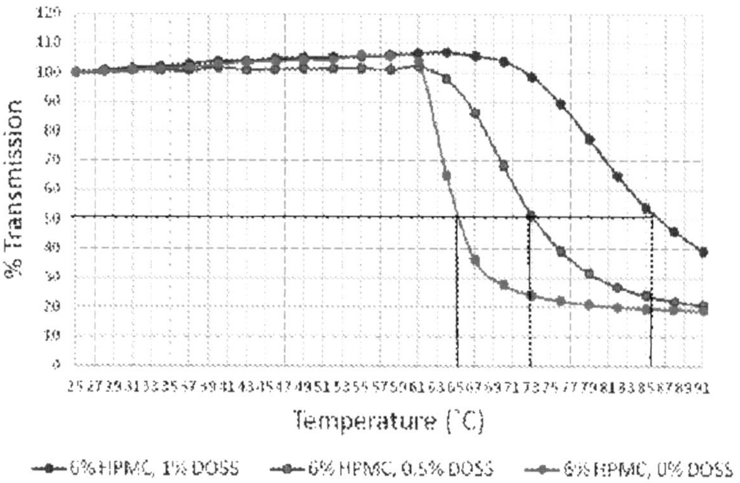

Figure 1: The effect of varying concentration of DOSS on HPMC cloud point (Cloud point was reached when he overall % transmittance of light reaches 50% of that compared to the same sample at 25°C (Silva, Pinto et al. 2008).

Additional results will be presented looking at the effect of carbamazepine level on HPMC cloud point. Also the corresponding behaviour of HPMCAS polymer will be presented. Studies to date have not detected a cloud point for aqueous solutions of HPMCAS alone. To show the relevance of these cloud point studies, dissolution studies was be present from a range of these formulation compositions.

Conclusions

Formulation additives such as docusate sodium were shown to positively influence the cloud point of HPMC. Future work will explore how this alteration of the polymer self-assembly behaviour and cloud point can inform the HPMC-formulation behaviour during dissolution studies.

References

Acevedo, A., et al. (2014). "Thermal gelation of aqueous hydroxypropylmethylcellulose solutions with SDS and hydrophobic drug particles." Carbohydrate polymers **102**: 74-79.

Joshi, S. C. (2011). "Sol-Gel behavior of hydroxypropyl methylcellulose (hpmc) in ionic media including drug release." Materials **4**(10): 1861-1905.

Mitchell, K., et al. (1990). "The influence of additives on the cloud point, disintegration and dissolution of hydroxypropylmethylcellulose gels and matrix tablets." International Journal of Pharmaceutics **66**(1): 233-242.

Silva, S. M., et al. (2008). "Aggregation and gelation in hydroxypropylmethyl cellulose aqueous solutions." Journal of colloid and interface science **327**(2): 333-340.

Viriden, A., et al. (2011). "The consequence of the chemical composition of HPMC in matrix tablets on the release behaviour of model drug substances having different solubility." European Journal of Pharmaceutics and Biopharmaceutics **77**(1): 99-110.

Zhang, L., et al. (2015). "Rheological and gel properties of hydroxypropyl methylcellulose/hydroxypropyl starch blends." Colloid and Polymer Science **293**(1): 229-237.

452573 QbD for Generic Modified Release Drug Products

Abstract:

As stated in the White paper recently published on FDA website - FDA Pharmaceutical Quality Oversight: One Quality Voice, a number of challenges to pharmaceutical quality continue to manifest. Among all the problems, product recall and defect reporting data is listed as the number one problem. The high occurrences of problems can be attributed to inherent defects in product and process design; these data further indicate failures in the implementation of manufacturing process scale-up as well as routine production.

FDA's approach to quality oversight has evolved in recent years, from an emphasis placed on quality control using empirically derived specifications, moving towards an increasing emphasis on in-depth product and process understanding. However, for a while, a disproportionate amount of regulatory attention is devoted to low-risk products and issues, diverting resources needed for the assessment of high-risk products. Therefore, risk based Quality by Design (QbD) approach is necessary. For simply designed immediate release oral products, traditional quality control approach is adequate. However, for more complex modified (oral) release products, risk based QbD approach should be adopted.

More and more drug products have been developed as modified release dosage forms in order to maintain therapeutic activity for an extended time, reduce toxic effects, protect the active substance against degradation due to low pH, etc. As these pharmaceutical products have increased their design complexity, approaches adopted to ensure the equivalence and quality of generic drug products, must likewise evolve. This presentation will focus on the quality assessment of modified release drug products based on their risk level, product design, and process understanding.

461268 Utilization of Quality By Design Principles to Define Formulation Best Practices for a Direct Compression Excipient

Abstract:

Utilization of Quality by Design Principles to Define Formulation Best Practices for a Direct Compression Excipient

Kathryn O. Hewlett (KOHewlett@Dow.com), True L. Rogers (TLRogers@Dow.com), Karen M. Balwinski (KMBalwinski@Dow.com), Jaime L. Curtis-Fisk (JLCurtisFisk@Dow.com), Robert L. Schmitt (SchmitRL@Dow.com), Shrikant N. Khot (SNKhot@Dow.com)

Introduction: Quality by Design (QbD) initiatives enable pharmaceutical manufacturers to proactively design quality and performance into drug products, with the ultimate goal to maximize safety and efficacy for the patient. Hydrophilic matrix tablets represent one of the most prevalently utilized modified-release oral drug delivery systems, which are relatively straightforward to develop and cost-efficient to manufacture. Hydroxypropyl methylcellulose (HPMC) is a common rate-modifying polymer incorporated into hydrophilic matrices. The water-soluble HPMC particles are uniformly distributed throughout the matrix tablet. Upon contact with aqueous gastrointestinal media, the particles hydrate and swell. The particles coalesce to form a swollen polymer network layer. The release of active pharmaceutical ingredient (API) from the tablet is modulated by both diffusion through the swollen network and erosion of the outermost surface.

Matrix tablet performance can be impacted by a number of variables, such as API physicochemical properties (solubility, crystalline morphology), physicochemical properties of the rate-modifying polymer (particle shape and size, viscosity grade), formulation composition (filler, glidant, lubricant selection) and manufacturing methodology (blending, direct compression vs. granulation, tableting conditions). The Dow Chemical Company has recently begun commercial-scale production of a new direct compression grade HPMC, METHOCEL™ DC2, to complement the finer particle size, but less flowable controlled release grade, METHOCEL™ CR. We used QbD principles to characterize properties and performance of matrix tablets containing METHOCEL™ DC2 in order to determine and balance optimal performance regimes for tablet physical properties and modified-release (MR). The results of these investigations will be presented along with formulation guidances and best practices. Particular emphasis will be placed on filler selection, HPMC molecular weight and morphology grade, and polymer concentrations utilized to ensure robust MR performance.

Methods: Gliclazide is a poorly soluble sulfonylurea used to treat non-insulin dependent *diabetes mellitus* and was used as model API. HPMC concentrations of 10, 20, 30, 40, and 50% (w/w) were investigated to cover the performance design space. The new K100LV, K4M, and K100M DC2 grades of HPMC were utilized as rate-modifying polymers, and the corresponding CR grades were used as comparative controls. Fillers investigated were Flow-Lac lactose (Meggle), Avicel PH102 microcrystalline cellulose (FMC), Starch 1500 pregelatinized starch (Colorcon), DiTab dicalcium phosphate (Rhodia), and Manogenn powdered mannitol (SPI Polyols). Tablets were produced via direct compression on a pilot-scale Manesty Beta Press at 50-rpm turret speed. Tablet physical properties, such as weight and tensile strength, were characterized using a Sotax HT 100 - 500NV instrument. Modified-release performance was characterized using the USP II paddle method in either a Varian VK 7010 dissolution system equipped with a Varian Cary 50 UV visible diode array spectrophotometer (Agilent Technologies, Santa Clara, CA USA) or a Distek 2100 single-bath dissolution system (North Brunswick, NJ USA) equipped with an Agilent 8453 diode array spectrophotometer (Agilent). Each tablet tested for MR performance was positioned in a hanging basket to minimize dissolution measurement error.

Results: Figure 1 shows the modified release performance of METHOCEL™ K100LV CR and DC2 as a function of time for formulations with differing HPMC content. Burst release occurred at low HPMC concentrations (10 and 20%) with both morphology grades. Increasing the level of HPMC in the formulation resulted in a shift into a robust MR performance regime. Above this robustness transition, modified release performance from both CR and DC2 matrix tablets was consistent. Further increase of HPMC level minimally impacted API release rate once the robust regime was reached, so there is a diminishing impact with further increase in HPMC concentration once inside the robust regime.

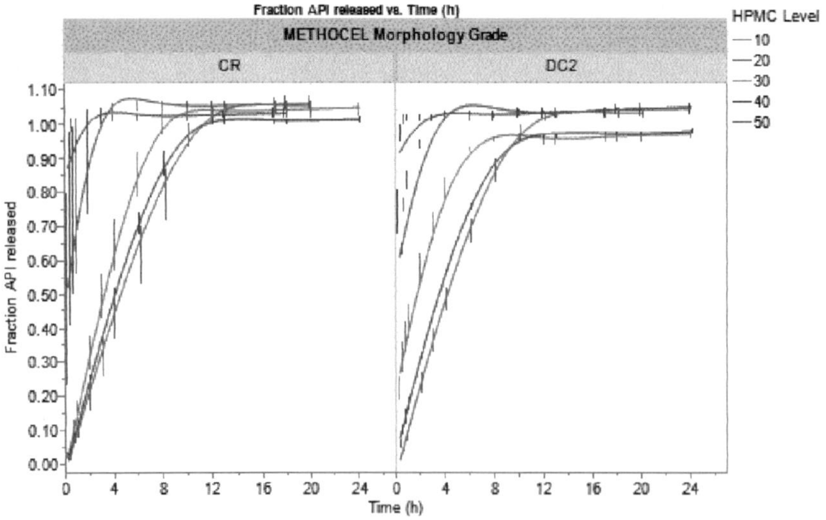

Figure 1. Release of gliclazide over time, as a function of METHOCEL grade and formulation level.

Optical imaging reflected the trends observed from dissolution testing. Figure 2 shows images taken after 4 hours of swelling for tablets comprising 10% (a) and 40% (b) METHOCEL™ K100LV CR. At lower HPMC concentration, the polymer content was insufficient to achieve a continuous swollen network. As a result, the swollen network was unable to maintain structural integrity, causing burst API release as the tablet rapidly eroded. Higher HPMC concentration delivered a continuous swollen network. Dark regions in the figure represent water and areas devoid of swollen network, and the dry or partially hydrated matrix appears as grey or white. The white line in approximately the center of each image represents the location of the initial dry tablet surface. After 4 hours, the swollen network of the 10% HPMC formulation was noncontiguous (Fig. 2a). A large region near the top of the image was devoid of swollen network. The presence of voids in the swollen network is consistent with the observed burst release at lower HPMC concentrations. By contrast, the 40% formulation had a contiguous, swollen network layer surrounding the tablet and delivered robust MR performance (Fig. 2b).

(b)

(a)

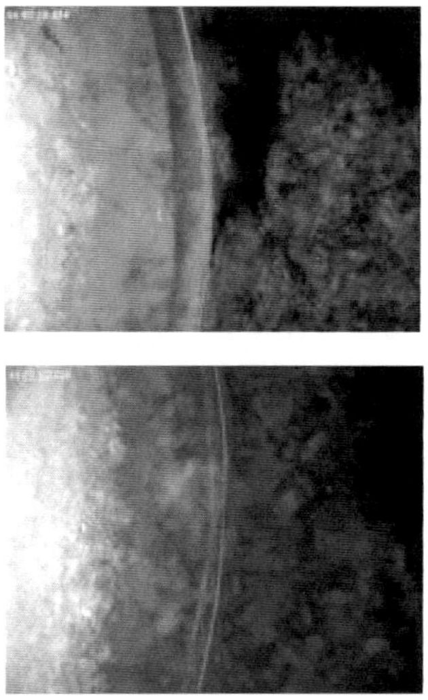

Figure 2. Optical images of (a) 10% and (b) 40% CR tablets after 4 hours of swelling.

Tablet filler selection also impacts MR performance. Figure 3 shows MR profiles of the gliclazide formulation containing 30% METHOCEL™ K4M DC2 with a range of soluble and insoluble fillers. Depending upon filler solubility, the API release rate can be increased or decreased. Once inside the robust regime, filler solubility can be used to adjust MR performance to the desired target.

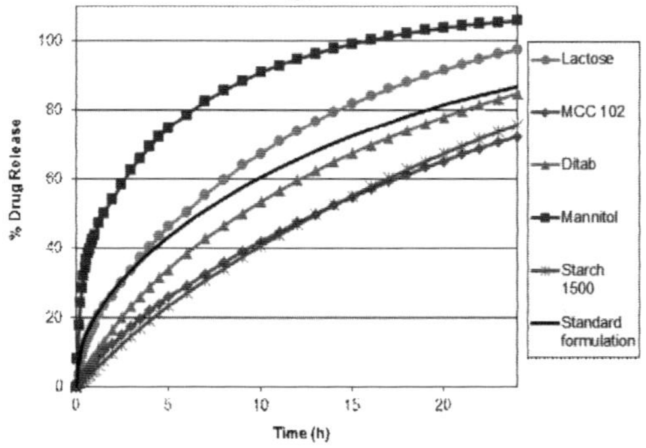

Figure 3.The effect of filler solubility on gliclazide release rate. Tablets were formulated with 30% METHOCEL DC2.

Conclusions: When working with matrix tablets, many factors must be considered to obtain robust MR performance. This work focused on examining the role of filler selection, the molecular weight and morphology grade of HPMC, and polymer concentration necessary to ensure robust MR performance. A new, direct-compression grade HPMC, METHOCEL™ DC2, was compared to METHOCEL™ CR, and both morphology grades enabled robust MR performance at comparable levels. As the HPMC concentration was increased, the formulation transitioned into a robust regime of consistent MR performance. Further increase in HPMC concentration had diminishing impact once inside the robust MR regime. Filler solubility also impacted MR performance and could be used to fine-tune the desired release rate. The systematic framework utilized while conducting this study enabled us to define formulation guidances and best practices for the new direct compression grade of HPMC. There is now a clear understanding of the influences of HPMC grade, polymer level, and filler selection on MR performance.

455234 Microcrystalline Cellulose Does Wood Pulp Source Impact the Direct Compression Performace of This Excipient?

Abstract:

INTRODUCTION

In a quality by design (QbD) environment knowledge and understanding of raw material critical quality attributes (CQAs) are essential [1]. Excipient variability is an important consideration particularly when blended with challenging APIs where manufacturing may occur close to the edge of failure, i.e. close to the acceptance limits for critical quality attributes. In such cases small variations in an excipients attribute could possibly impact the end product quality, even if the variation is within the excipient's certificate of analysis release specifications. Continuous process improvement, continuous manufacturing and regulatory requirements for improved process understanding as covered by ICH Q10 [2] drive a need to understand the impact of excipient variability on finished product performance.

Microcrystalline cellulose (MCC) was chosen as the excipient of study due to its widespread use as a diluent/filler or binder in solid oral dosage form manufacturing. MCC has been available as a pharmaceutical excipient since 1964 [3] and consequently there has been much previous research into its material attributes. Moisture [4], crystallinity [5-7], degree of polymerisation (DP) [3], surface area and porosity [4] have been suggested in the literature as material attributes for consideration when processing MCC.

In order to understand the possible variability in MCC it is important to understand its manufacturing process. Pharmaceutical cellulose is derived from wood pulp. Trees are broken down into wood chips which undergo a chemical process known as the 'Kraft process'[8] which removes the lignin that links cellulose chains together as wood. Wood pulp is the starting material bought by MCC manufacturers. The type of wood source (soft/hard woods) will affect the chemical content (the hemicellulose content and crystallinity) of the wood pulp.

MCC manufacturers treat the wood pulp with acid to breakdown the long cellulose chains into smaller chain lengths (reduce the degree of polymerization). Acid hydrolysis cleaves the long cellulose chains at amorphous sites where hemicellulose sugar chain branches are located. Therefore how the cellulose is hydrolysed is dependent on its chemical content and the number of amorphous regions. The length of the hydrolysis time and final DP of a batch of MCC is influenced by the wood pulp type as different wood pulps have characteristic leveling off degrees of polymerization e.g. 180-210 for hard woods and 210-250 for soft woods [3]. By blending different wood pulps together before hydrolysis the DP can be predicted by manufactures. This is an important control as it can influence the MCC final bulk density[3]. The resulting shorter cellulose chains are thus more crystalline as the amorphous regions have been cleaved, hence the name microcrystalline cellulose. Post hydrolysis steps (spray drying/air stream drying and sieving) offer manufactures the opportunity to manipulate the final lot bulk density, PSD and mean particle size to produce product within specification.

The study presented aims to understand if samples of MCC grade PH102 produced from different wood pulp mixes show differences in physicochemical properties and hence compaction performance during a direct compression process.

<>**Materials and Methods**

Three batches of MCC PH102 produced from a range of pulp sources were supplied by FMC. The pulp types have not been disclosed due to commercial sensitivity. All samples were produced using the commercial scale process and meet with supplier and pharmacopeia specifications.

- Sample 1 - 75% Pulp X: 25% Pulp Y
- Sample 2 - 50% Pulp X: 50% Pulp Z
- Sample 3 - 100% Pulp X

MCC PH102 lots were characterized and compared for differences in particle size distribution (sieve analysis), moisture content (thermogravimetric analysis), crystallinity (FTIR) and surface area (BET N_2 adsorption), degree of polymerization and scanning electron microscopy. Two compaction studies were carried out on an instrumented Piccola™ rotary tablet press. The first compaction study was on samples 'as received'. The second compaction study was performed on a sieved size fraction (106-250 µm) of each lot in order to remove the confounding factor of particle size distribution differences. Compacted material hardness, weight and thickness were measured on a Pharmatron™ Smart Test 50 and compaction profile generated.

RESULTS AND DISCUSSION

The particle size distribution (PSD) profiles of each sample as received did indicate a significant difference in PSD particularly the % fines. The sieved size fraction (106-250 μm) of each sample was then compared in order to eliminate the influence of PSD and % fines as a possible reason for differences in the compressibility of the lots.

Characterization of the 106-250 μm size fractions for each sample identified differences in the surface area, crystallinity and bulk density (Table 1) and the PSD (Figure 1). All lots had a moisture content determined by loss on drying in the range 3.74 - 4.86% w/w. It was assumed that within this range moisture was not an influencing factor [4].

When comparing compression profiles of samples (Figure 2), Samples 1 and 2 which were manufactured from pulp mixes were similar. However Sample 3, manufactured form 100% of pulp X, showed a significantly higher compressibility.

The study results show that particle size is a major factor that influences MCC compressibility. Following correction for particle size, Sample 3 produced from a single pulp source showed superior compression properties. The particle properties resulting in this increase in compaction are currently inconclusive and being explored. Further studies will focus on degree of polymerization and particle mechanical parameters during compression.

CONCLUSIONS

Previous studies [9, 10] have discussed wood pulp as a source of variability between batches of MCC. Mixing wood pulps minimizes this variability. In this study a MCC sample produced from 100% of one particular wood pulp during manufacture of MCC produced material which performed differently during direct compression compared to samples from pulp mixes when corrected for difference in particle size between samples.

The impact of this variability in MCC will be formulation specific. It is envisaged for blends with a high % API loading which is difficult to compact, this variability may be significant to drug product manufacturers. This also highlights the importance of using a number of different lots of excipients during drug development to capture excipient variation where it is potentially a risk factor.

ACKNOWLEDGMENTS

This research is funded by the Synthesis and Solid State Pharmaceutical Centre (SSPC) and Science Foundation Ireland (SFI) under grant number 12/RC/2275. PH102 was donated by FMC Corporation.

REFERENCES

1. Chen, Z., D. Lovett, and J. Morris, *Process analytical technologies and real time process control a review of some spectroscopic issues and challenges.* Journal of Process Control, 2011. **21**(10): p. 1467-1482.

2. ICH, *Pharmaceutical Quality System Q10*, in *ICH Q10*, I.C.o.H.o.T.R.f.R.o.P.f.H. Use, Editor 2008, International Conference on Harmonisation of Technical Requirements for Registration of Pharmaceuticals for Human Use.

3. Thoorens, G., et al., *Microcrystalline cellulose, a direct compression binder in a quality by design*

environment-A review. Int J Pharm, 2014. **473**(1-2): p. 64-72.

4. Sun, C.C., *Mechanism of moisture induced variations in true density and compaction properties of microcrystalline cellulose.* Int. J. Pharm., 2008. **346**(1-2): p. 93-101.

5. Park, S., et al., *Cellulose crystallinity index: measurement techniques and their impact on interpreting cellulase performance.* Biotechnol Biofuels, 2010. **3**: p. 10.

6. Mihranyan, A., et al., *Moisture sorption by cellulose powders of varying crystallinity.* Int. J. Pharm., 2004. **269**(2): p. 433-442.

7. Rowe, R.C., A.G. McKillop, and D. Bray, *The effect of batch and source variation on the crystallinity of microcrystalline cellulose.* Int. J. Pharm., 1994. **101**(1-2): p. 169-72.

8. Chakar, F.S. and A.J. Ragauskas, *Review of current and future softwood kraft lignin process chemistry.* Industrial Crops and Products, 2004. **20**(2): p. 131-141.

9. Landin, M., et al., *Effect of country of origin on the properties of microcrystalline cellulose.* Int. J. Pharm., 1993. **91**(2-3): p. 123-31.

10. Landin, M., et al., *Effect of batch variation and source of pulp on the properties of microcrystalline cellulose.* Int. J. Pharm., 1993. **91**(2-3): p. 133-41.

Table 1. Characterization results of the 106-250 μm size fraction for each of the three samples before compression.

	SAMPLE 1	**SAMPLE 2**	**SAMPLE 3**
Moisture %w/w	4.41 ± 0.68	4.86 ± 0.64	3.74 ± 0.36
Bulk density g/cm^3	0.32	0.32	0.34
IR Crystallinity (***Nelson et al 1964,)***	0.86	0.90	0.94
Surface Area m^2/g	1.232	1.148	1.289
D50 μm	174 ± 0.00	212 ± 3.46	186 ± 1.53

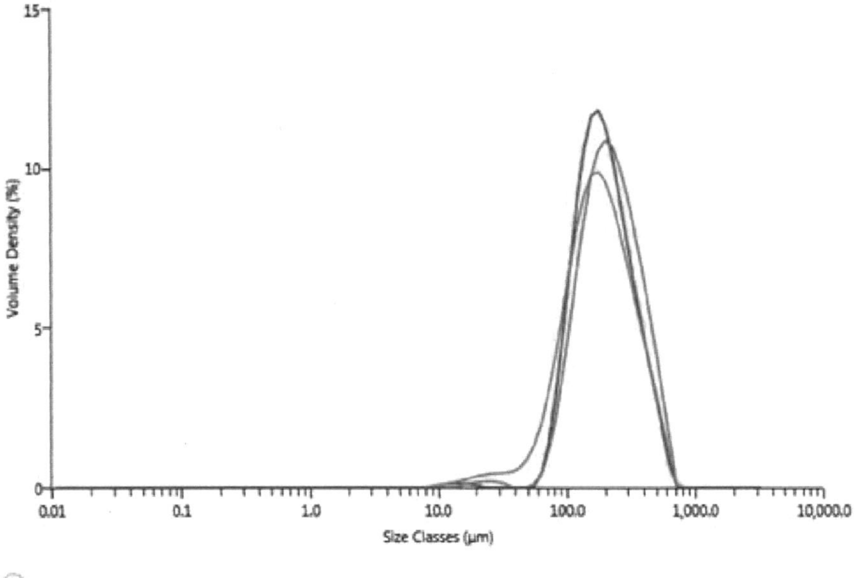

Figure 1. Malvern PSD of 106-250 μm size fraction comparison for the three samples. Sample 1 (green), sample 2 (blue) and sample 3 (red).

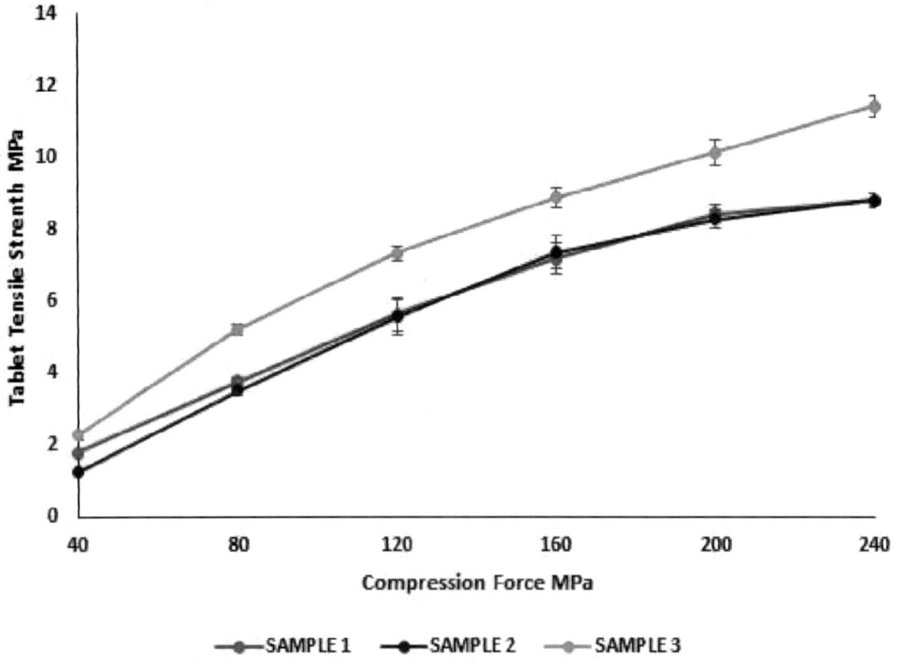

Figure 2. Compression profiles for the three samples. Tensile strength for N=20 tablets

457458 Understanding the Role of Hpmcas in Phenytoin Solid Dispersion Formulation Using Coarse Grained Molecular Dynamics Simulations

Abstract:

Many orally administrated drug candidates currently being developed are highly lipophilic, which necessitates the use of an excipient to stabilize and increase their solubility in the gastrointestinal tract. Hydroxypropyl methylcellulose acetate succinate (HPMCAS) has been identified to be a very effective excipient; however, to date very little understanding has been established on how HPMCAS interacts with drug molecules at a molecular level, and how different structural variables such as molecular weight and substitution pattern affect the performance of HPMCAS as an excipient. In this study, we employed coarse-grained (CG) molecular dynamics (MD) simulations to model HPMCAS and a drug candidate, phenytoin, in an implicit water environment. We parameterized our CG model using the radial distribution functions (RDFs) from atomistic simulations of short HPMCAS oligomers and applied the model to simulate long chains. We studied the effect of individual substitution type (e.g. methyl, hydroxypropyl, acetate etc.) by simulating homogenous polymer chains with drug molecules. We found the conformation and the structural details of the polymer-drug complex are affected by monomer substitution type, the amount of polymer loading, and polymer length. Specifically, bulky hydroxypropyl group results in a more loosely packed polymer matrix, thus allowing drug molecules to diffuse through the matrix more easily. The use of high polymer concentration (>80 wt%) can significantly reduce the size of the drug aggregates in the complex. We determined the drug release rate for different polymer-drug complex by running simulations for over 50Î¼s. We also modeled the heterogeneous copolymer and investigated the effect of deprotonated succinate substitution on the release behavior. We found that the polymer-drug complex swells when succinate groups switch from protonated to deprotonated state, thus facilitates the release of the drug molecules. Our simulation box size is comparable to the size of polymer-drug complex identified from the experiment. This study provides insights to help optimizing the composition of HPMCAS (e.g. ratio of acetate to succinate group) for any given drug.

464851 Assessing the Efficacy of Thermo-Responsive Poly(N-isopropylacrylamide) in Phenytoin Solid Dispersion Formulation Using Atomistic Molecular Dynamics Simulations

Abstract:

Abstract for AICHE Annual Meeting 2016

Submission Deadline: May 9th, 2016

Session: 26004 Amorphous Solid Dispersion for Drug Product

Category: Pharmaceutical Discovery, Development, and Manufacturing Forum

Assessing the Efficacy of Thermo-Responsive Poly(N-isopropylacrylamide) in Phenytoin Solid Dispersion Formulation Using Atomistic Molecular Dynamics Simulations

Soroush Moghadam[1] and Ronald G. Larson[2]

[1]Department of Mechanical Engineering, University of Michigan, Ann Arbor, MI, 48109

[2]Department of Chemical Engineering, University of Michigan, Ann Arbor, MI, 48109

For an active pharmaceutical ingredient (API) to be absorbed from the gastro-intestinal tract into the bloodstream, it must be sufficiently soluble in the GI fluid. However, more than 40% of new chemical entities in the pharmaceutical industry are poorly soluble, leading to drug aggregation and crystallization, which necessitates the use of an excipient to stabilize and increase the solubility. Solid dispersions can maintain the API in an amorphous form and result in a supersaturated API solution upon its dissolution in the GI tract. As the hydrophilic component, thermo-sensitive poly(N-isopropylacrylamide), namely pNIPAAm, has been suggested to provide desired controlled release properties due to its lower critical solution temperature (LCST~32°C), biocompatibility and also large design space for its functional groups. Although many simulation studies analyzed the LCST behavior of pNIPAAm chains in water, to date there has not been a computational analysis of how pNIPAAm interacts with drug molecules at molecular levels as a function of temperature and how we can tune these properties to reach desired stability of polymer-drug complex. In this study, we performed all-atom molecular dynamic (AA-MD) simulations to model pNIPAAm and a drug candidate, phenytoin, in an explicit water environment. AMBER forcefield and solvated ensemble averaging of partial charges have been implemented. After validating the forcefield parameters using the well-known lower critical solution behavior of pNIPAAm, we simulated the polymer-drug complex in water and its behavior at temperatures below (295K) and above the LCST (310K). The effect of copolymerization with hydrophilic elements (e.g. Dimethylacylamide (DMA)) on the excipient behavior has been studied. We found that existence of hydrophilic comonomers in a pNIPAAm chain dramatically affects its temperature transition dependence. However, It affects affinity of the excipient functional groups to interact with drug molecules as well. Using radial distribution functions, we found that there is an optimum comonomer molar fraction of around 20-30% DMA at which interaction with phenytoin drug molecules is strongest. Furthermore, maintaining the weight percents of the components, using shorter polymer chains close to pNIPAAm reported persistence lengths can significantly influence the drug-polymer interaction upon heating. Although the length and time scales of our simulations are not high enough to be conclusive, this study provides insights to help optimizing the composition of pNIPAAm for any given drug.

KEYWORDS: Polymer Drug Aggregates, Molecular Dynamics, pNIPAAm

474782 Solid State Form Understanding Applied to the Design of a Spray Dried Dispersion Process: A Quality By Design Approach to Optimization of Feed Solution Loading

Abstract:

Spray drying is a common unit operation across many industries due to the ability to control bulk powder properties and isolate and dry product from a liquid feed in a single operation. For the pharmaceutical industry, spray drying offers the control of physical properties of crystalline or amorphous active pharmaceutical ingredients (APIs), as a neat or formulated drug product that can lead to improved bioavailability of the drug candidate. Drug candidate molecules are often formulated as a solid dispersion, incorporating both API and a polymer excipient for enhanced dissolution performance and improved physical stability. In order to reliably achieve solid form control to the amorphous state during the spray drying operation, the API must be fully solubilized in the spray dryer feed solution, eliminating crystalline API material. The crystalline form of the API may or may not be the most stable form in the selected spray drying solvent. To ensure robust solid state form control of the spray dried dispersion (SDD), not only should the form landscape of the API in the spray drying solvent be well understood the process control capabilities during feed solution preparation must be considered as well. Numerical modeling of the solubility of relevant API forms, in conjunction with feed solution loading and process control capability should be used to determine an optimal feed solution loading. The operating point can be selected through optimization based upon on selected risk-based criteria and control capability. Applying this methodology to Compound A, the throughput of the spray drying process was improved two-fold, while ensuring a robust control strategy to deliver the desired API in an amorphous form as a spray dried dispersion.

468548 Ternary Dispersions of Itraconazole: A Comparison of Hot Melt Extrusion and Spray Drying

Abstract:

Ternary dispersions of itraconazole: a comparison of hot melt extrusion and spray drying

Mark T. Davis[1]*, Catherine B. Kelly[1], & Gavin M. Walker[1, 2]

[1]Synthesis and Solid State Pharmaceutical Centre (SSPC), Bernal Institute,

University of Limerick, Limerick, Ireland

[2]Pharmaceutical Manufacturing Technology Centre, PMTC, University of Limerick, Limerick, Ireland

**Tel.: +353-(0)-61- 213096, E-mail: mark.davis@ul.ie*

ABSTRACT

BCS Class II drug itraconazole (ITR) represents a difficult formulation challenge due to its very low aqueous solubility and poor dissolution kinetics. It has been extensively studied in an attempt to achieve higher bioavailability and therefore better dissolution. The amorphous state represents a disordered structure relative to the crystalline form which has been shown to have an advantage of higher dissolution rates and better oral absorption. Much work has been carried out on the performance of single polymers in the formulation of the amorphous form as an amorphous solid dispersion however evidence exists that ternary amorphous solid dispersions (ASDs) containing two rationally chosen biocompatible polymers may exhibit better amorphous stability and dissolution profiles than their binary equivalents.

In this study, ITR was formulated into a number of amorphous solid dispersions using hot melt extrusion (HME) and spray drying (SD). The systems comprised ternary formulations of itraconazole with Soluplus® and/or HPMCP. Initial studies within our group have indicated that although HPMCP is an enteric polymer it interacts more favourably with ITR than Soluplus does and so should offer amorphous stabilisation benefits. Soluplus, however, is an excellent aqueous phase solubiliser for ITR so contributes to the dissolution properties of the ASD.

A drug loading of 40% was chosen to produce tablets of 500 mg containing 200 mg of active pharmaceutical ingredient (API). Equivalent physical mixtures were prepared to measure the solubility benefit of the amorphous state. All formulations were shown to be completely free of either liquid-crystalline or crystalline ITR using both wide-angle XRD and hyper differential scanning calorimetry (DSC). Further analysis of physical characteristics was carried out using SEM, laser scattering and density measurements.

Dissolution testing was carried out on both the tablets and the physical mixture using USP dissolution apparatus 2 (paddle at 37 ^{o}C) in biorelevant media. 5ml aliquots were removed at regular intervals and, following production and validation of an absorbance - concentration calibration curve for ITR, dissolved drug was quantified using UV-Vis spectrophotometry. Differences in dissolution rate and extent were rationalised in terms of the preparation methods, hydrogen bonding and the material properties.

469561 Tunable Drug Loaded Stable and Fast Release Amorphous Composites Production Using Mesoporous Neusilin ®UFL2

Abstract:

Literature has revealed that there may exist possibility of greatly reduced mobility of drug molecules due to confinement, which could occur through several different manners. Impregnation of poorly water drug solutions in to mesoporous materials is one such attractive approach for nano-confinement that leads not only to creation of amorphous state but also its stability due to what may be the poor molecular mobility. Such as approach is examined here to increase the aqueous solubility and thus the bioavailability of poorly soluble drugs since it allows formation of drug amorphous state which generally exhibit higher apparent solubilities than their crystalline counterparts. Unfortunately, mesoporous materials such as silica are expensive and proprietary, and not easily available in bulk quantities. In this work, Neusilin® UFL2 as commercially available mesoporous amorphous materials was used to prepare amorphous drug composites of the model poorly soluble drugs Fenofibrate and Itraconazole via impregnation. Drug loading could be tuned as low as 1.0% (w/w) to as high as 30%. PXRD, DSC, and Raman spectroscopy were used to assess the amorphousness/crystallinity of the drug in the impregnated composites after formation as well as after storage. It was shown that the impregnated drug composites are stable for three months even stored at 40°C and 75% relative humidity. Re-crystallization of the amorphous drug was completely suppressed in the composites due to the confinement effect provided by the mesopores of Neusilin® UFL2. The amorphous drug composites result in higher apparent solubility and faster release/dissolution rate of the model drugs compared to crystalline counterparts. UV imaging technique, a real-time and sample sparing method, was used to qualitatively differentiate the amorphous and crystalline drug dissolution besides the conventional USP II dissolution. In conclusion, preparation of amorphous drug composites using Neusilin® UFL2 appears to be promising as a stable amorphous drug delivery vehicle providing a rapid drug release and subsequently bioavailability enhancement.

Keywords: Mesoporous Neusilin®, Impregnation, Amorphous Drug Composites, Stability, Dissolution Enhancement

458072 Hot Melt Extrusion of Ternary Solid Dispersions of Itraconazole: Amorphous *Vs* Liquid Crystal Intermediate

Abstract:

Abstract

It has been demonstrated that ternary solid dispersion compositions can be more efficient than binary compositions in terms of their excellent physical stability and ability to inhibit the recrystallization of API's in bio-relevant media. This study is mainly focused on investigating the preparation of ternary solid dispersions of itraconazole, a BSC Class (II) model drug, containing a water-soluble polymer and/or a water-insoluble polymer using hot melt extrusion. The formation, physio-chemical characterization, stability, and phase transformation in the context of drug formulation and development of amorphous/liquid crystals of itraconazole was investigated in detail.

Five combinations of ternary solid dispersions comprising itraconazole, hydroxy propyl methyl cellulose phthalate (HPMCP-HP50) and/or soluplus were prepared by hot melt extrusion (HME) using a 16 mm pharmalab co-rotating twin screw extruder and L/D ratio of 40: 1 (Thermo Fisher Scientific, Karlsruhe, Germany). Information regarding the miscibility of the drug with the excipient, physical state, phase transformation, possible interactions and proper storage conditions were obtained by employing small angle powder X-ray diffractometry (SAXS), differential scanning calorimetry (DSC), Raman spectroscopy and infrared spectroscopic (FT-IR) measurements for specific polymer-polymer and drug-polymer interactions.

The SAXS results showed that itraconazole in the solid dispersion was in the amorphous state and indicated the absence of liquid crystalline itraconazole. The Raman and FT-IR analysis indicated the formation of hydrogen bonding between itraconazole and HPMCP-HP50/soluplus molecules and that itraconazoleâ€'HPMCP-HP50 interactions were stronger than itraconazoleâ€'soluplus interactions. From the DSC scans melt extrudates of itraconazole and HPMCP-HP50/soluplus exhibited one glass transition (Tg) and an absence of an endotherm around the melting point of itraconazole confirming that the API is present only in its amorphous state and that each formulation consists of a single homogenous amorphous phase of Itraconazole and HPMCP-HP50 and/or soluplus.

470461 Investigation of the Dfferences in Material Properties Between Amorphous Dispersions Created By Conventional Methods and High-Shear Co-Precipitation

Abstract:

High-shear co-precipitation is a technique for the manufacturing of API-polymer amorphous solid dispersions, and is an alternative to the typical spray-drying and hot-melt extrusion processes. The technique relies on the rapid contacting of a solution of polymer and API with a common anti-solvent, resulting in a stable suspension where the precipitated amorphous solid dispersion is below its glass transition temperature.

In this work we present the characterization and analysis of a co-precipitated solid dispersion exhibiting significant differences in properties compared to equivalent dispersions, prepared via spray-drying and hot-melt extrusion. These differences extend across the full spectrum of "fundamental" characteristics, including morphology, true density, glass transition temperature and enthalpy, as well as indirect, "performance-based" properties, such as compactability and dissolution. A large collection of "orthogonal" techniques is brought to bear in the analysis of the material and the elucidation of the physical processes underlying these differences in properties.

469250 Melt Rheology of Cellulose Ethers Designed for Hot Melt Extrusion

Abstract:

Melt Rheology of Cellulose Ethers Designed for Hot Melt Extrusion

Authors: Tirtha Chatterjee, Kevin O'Donnell, Yongfu Li, Uma Shrestha, Robert L. Sammler, Shrikant Khot

Formulation of new drug entities is increasingly challenging due to poor solubility of the drug and limited polymers that can produce amorphous solid dispersions (ASDs). One of the leading technologies to manufacture ASDs is hot melt extrusion (HME) in which the solid or melted drug is mixed with a molten polymeric carrier. In order to be processed, the polymeric carrier must have adequate melt rheology at temperatures acceptable for pharmaceutical applications. The AFFINISOL™ cellulosic polymer family has been engineered to have a melt viscosity that enables their use as excipients in HME.

In order to design a successful HME process the formulator must understand the excipient melt rheology as a function of operating conditions. In the present study, the melt rheology of AFFINISOL™ HPMC HME (hydroxypropyl methylcellulose) and HPMCAS HME (hydroxypropyl methylcellulose acetate succinate) is presented. Traditionally the melt rheology is characterized using a rotational rheometer, however, this was found to be inadequate for this family of polymers. To overcome this challenge an alternative method was adapted allowing rheological analysis directly in line with a twin screw extruder. This enables analysis under conditions identical to those a formulator would experience. Melt viscosity of the cellulose ethers was found to be dependent upon temperature and shear rate with all polymers exhibiting a power law shear thinning behavior. The rheology of select formulations containing model active pharmaceutical ingredients will also be presented.

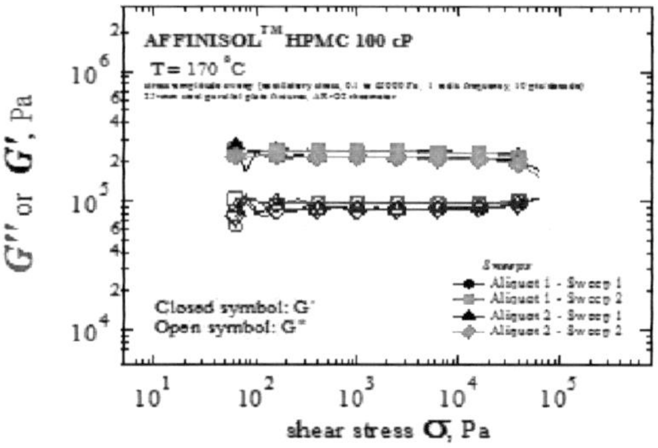

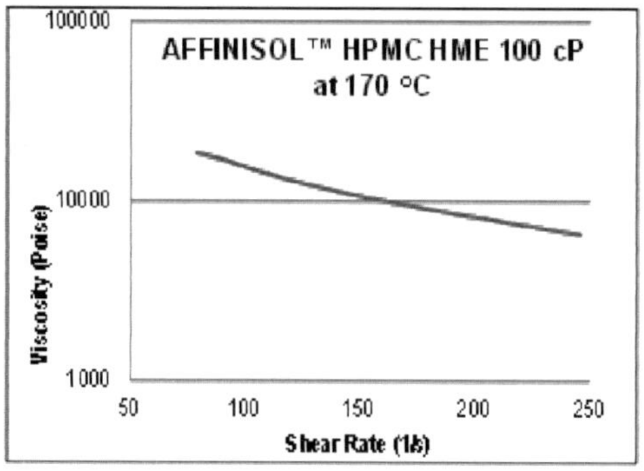

Figure 1: (Left) Rotational rheology of AFFINISOL™ HPMC HME at 170 °C exhibiting solid like behavior despite ability to extrude at temperatures well below this and (Right) alternative in-line method demonstrating melt viscosity as a function of shear rate at 170 °C.

471573 The Water Sorption Behaviour of Amorphous Solids: Experimental and Theoretical Insights Using a Model Protein BSA

Abstract:

This paper considers the water sorption behaviour of complex amorphous solids using advanced experimental investigations as well as a number of thermodynamics models for the formation of solid- solute solutions. The water sorption behaviour of amorphous materials is especially complex kinetically as the loss or gain of water vapour by the solid is controlled by two nominally independent kinetic phenomena. The first is the diffusion kinetics of moisture sorption and the second factor is the relaxation kinetics of the glassy amorphous state.

Experimental evidence for a glassyâˆ'rubbery hydration mechanism of bovine serum albumin (BSA) is reported where BSA is used as an amorphous model protein substrate. Experimental results corroborate that the hydration mechanism in amorphous protein materials are more complex than anticipated. We have experimental evidence indicating that BSA can be experimentally trapped into distinct conformational states during water desorption. In considering the dimension of time during hydration or dehydration of a protein, a significant problem in determining and interpreting a water sorption isotherm is posed by the â€˜sample historyâ€™, a time effect, of a lyophilised amorphous biological products. Its influence, both prior to the sorption experiment as well as produced during the actual measurement, has not been previously clarified.

A water vapour sorption experiment is typically carried out in one of the three modes: (a) interval sorption,where the solvent activity is subsequently raised in a stepwise fashion through one continuous measurement; (b) integral sorption, where the solvent activity is raised from zero up to one specific value and the mass uptake is recorded (typically with a new specimen for each measurement); or (c) differential sorption, where an activity ramp is applied in one continuous measurement. It should be understood that each of these three sorption modes creates a different experiment-induced time dependent behaviour. This paper will provide recommendations in best practise for

determined water sorption properties of amorphous solids.

The final section of this paper will introduce modern equilibrium and non-equilibrium thermodynamics models for analysing water sorption isotherm behaviour for amorphous solids.

469149 A Kinetic Approach to Polymorph Prediction:Â Polymorph Specific Alpha and Beta Glycine Nucleation Rates and Solubility Curves from Molecular Dynamics

Abstract:

Predicting polymorphism, the ability of a molecule to self-assemble into multiple solid-state crystalline forms, continues to elude both experimentalists and molecular modelers alike. Landscape energy minimization techniques, although elegant in their own right, fail to include standard design crystallization variables such as solvent, temperature, or pressure. Furthermore, the effect of crystal size, a critical variable for molecules that display size dependent polymorphism, remains unaddressed in these optimization algorithms. Molecular dynamics (MD), which encapsulates the aforementioned kinetic design variables naturally, has suffered from time scale challenges associated with activated rate processes, such as nucleation. However, with the advent of modern supercomputers, and HPC hardware such as xeon phi coprocessors, these types of simulations are just now becoming within reach of computational scientists. To date, only a handful of attempts to determine the effects of the nucleation pathway on polymorphism of organic molecules have been performed using MD .

In this work, a first principles molecular dynamics approach to the study of homogeneous nucleation rates and solubility of polymorphs is presented for both alpha and beta glycine in water as a model system. To this end, a rare-event molecular dynamics simulation methodology is used in combination with a newly derived stochastic model for one-dimensional nucleation barrier crossings. This works seeks to assess the degree to which polymorph specific nucleation kinetics and solubility, both at the nanoscale and bulk scale (>1000nm), can be used for polymorph prediction. Bulk solubility curves over a 45 K temperature range are predicted directly from molecular dynamics. Both the trend-line estimated and MD predicted alpha glycine solubility values are within 5% to 11% agreement with experimental values. Beta glycine solubility estimation is achieved purely by MD prediction, as the solubility of this polymorph is not measurable in water due to rapid recrystallization kinetics. From the solubility calculation, alpha glycine is shown to be less soluble than beta glycine over the entire 45K temperature window at bulk length sales. This is in agreement with experimental evidence that shows the presence of alpha glycine at equilibrium in aqueous solvent for micron-sized particles. However, it is also shown that beta glycine is less soluble than alpha glycine at the nanoscale. Furthermore, consistent with Ostwaldâ€™s rule of stages, the nucleation kinetics show a strong preference for the formation of beta glycine at ~2nm length scales, due to beta glycine critical clusters being ~40 kcal/mol more stable than alpha glycine. This yields the prediction of beta formation at these length scales as well as the size dependent polymorphism of glycine.

465879 Silica Nanoparticle-Seeded Microfluidic Emulsion Crystalisation - Crystalline Drug Composite Microparticles

Abstract:

We have recently demonstrated the fabrication of co-formulated drug-(polymeric) excipient microparticles with exquisite control over particle size, shape and polymorphism, using microfluidic emulsion crystallization and model hydrophobic drugs such as ROY and Carbamazepine [1]. In some cases, the co-formulation of polymeric excipients and drugs, depending on their mutual interactions, may lead to problems of drug-polymer miscibility and result in amorphous solid dispersions. [2-3] These amorphous solid dispersions may present storage problems, as they can crystallize over time. Here, we demonstrate the use of colloidal silica particles as excipients in microfluidic emulsion crystallization of a model hydrophilic drug. Silica nanoparticles, conventionally incorporated in solid dosage forms at the later stages of formulation, function as glidants or fillers. In contrast, in our work, the silica particles act as heterogeneous seeds for nucleation of the drug, thereby drastically reducing crystallization times, while also acting as an excipient that improve flowability and enable excellent crystallinity. Another notable advantage of using colloidal silica is its general applicability to a wide range of drugs, since the silica particles can easily be functionalized [4]. We envision significant potential for drug-silica composites prepared by microfluidic emulsion crystallisation, which allow rapid crystallization of a variety of drugs and their direct formulation into spherical microparticles without compromising the crystallinity of the drug product.

Word count:213

CONTACT * S. A. Khan, Tel: +65-6516-5133; saifkhan@nus.edu.sg

[1] Leon, R. A. L.; Badruddoza, A. Z. M.; Zheng, L.; Yeap, E. W. Q.; Toldy, A. I.; Wong, K. Y.; Hatton, T. A.; Khan, S. A., Highly Selective, Kinetically Driven Polymorphic Selection in Microfluidic Emulsion-Based Crystallization and Formulation. *Crystal Growth & Design* **2015,** *15*(1), 212-218.

[2]Huang, Y.; Dai, W. G., Fundamental aspects of solid dispersion technology for poorly soluble drugs. *Acta Pharm Sin B* 2014, *4*(1), 18-25.

[3] Marsac, P. J.; Shamblin, S. L.; Taylor, L. S., Theoretical and practical approaches for prediction of drug-polymer miscibility and solubility. *Pharm Res* 2006, *23*(10), 2417-26.

[4] Md. Badruddoza, A. Z.; Toldy, A. I.; Hatton, T. A.; Khan, S. A., Functionalized Silica Nanoparticles as Additives for Polymorphic Control in Emulsion-Based Crystallization of Glycine. *Crystal Growth & Design* **2013,** *13* (6), 2455-2461.

461527 The Effects of Crystal Size and Habit on the Dehydration Induced Polymorphism; The Case of Carbamazepine Dihydrate

Abstract:

The process conditions influence the kinetics and thermodynamics during drying of hydrated materials. The water removal from hydrated crystals induces mechanical stress on the crystal lattice leading to structural deformations. Concurrently, the nucleation of the anhydrous form alters material's properties[1]. Thus, material properties influence dehydration, but also they are influenced by it. This work examines the role of interfacial phenomena of hydrated materials, using carbamazepine dihydrate as model compound.

Three methods were used for the crystallization of carbamazepine dihydrate:

1. 0.075 g*mL^{-1} of anhydrous carbamazepine (as received) were dissolved under heating in a methanol-water (60:40 %v/v) mixture. Then the mixture was left to cool down for 12 hours under stirring (500 rpm). The solution was filtered out and the crystals were collected[2].
2. 0.05 g*mL^{-1} of anhydrous carbamazepine (as received) were dissolved under heating in a methanol-water (50:50 %v/v) mixture. Then the mixture was left to cool down for 12 hours under stirring (500 rpm). The solution was filtered out and the crystals were collected.
3. 0.025 g*mL^{-1} of anhydrous carbamazepine (as received) were dissolved under heating in a methanol-water (40:60 %v/v) mixture. Then the mixture was left to cool down for 12 hours under stirring (500 rpm). The solution was filtered out and the crystals were collected.

X-Ray powder diffraction (XRPD) patterns agreed that all methods result to carbamazepine dihydrate crystals, however optical examination of the crystals, using Scanning Electron Microscope (SEM), revealed differences in their crystal size and habits. Crystals from methods one are smaller in size and they have a smaller aspect ratio, leading to a more compact shape and better flowability and cohesion behaviour, compared with the crystals obtained from the other two methods. It is important, for later on, to mention that the dominant face in the crystals prepared via methods two and three facet (0 2 0) is by far the dominant one.

The three types of carbamazepine dihydrate were exposed in dehydration conditions of 30, 40, 60 and 80°C for 24 hours and the anhydrous material was examined with XRPD and SEM. The material from methods two and three dehydrated to the Triclinic polymorph, the less stable carbamazepine polymorph[3]. The crystals obtained from method one dehydrated concomitantly to the P-Monoclinic and Triclinic polymorphs. Increasing temperature leads to increasing amount of the Triclinic polymorph, as shown from the SEM images. The domination of the Triclinic polymorph, in the high aspect ratio crystals is associated with the formation of well-ordered macroscopic cracks perpendicular to the (0 2 0) facet of the crystal, followed by the growth of Triclinic carbamazepine whiskers. The high aspect ratio of the crystals obtained from methods two and three appears to be quite attractive for crack nucleation and propagation. Triclinic carbamazepine whiskers grow in needle-shaped structures with high surface area to volume ratio to enhance the dissipation of heat from the crystals. In the, smaller and more compact, particles obtained from method one smaller cracks, formed randomly on the crystal facets, seem to be sufficient to dissipate the heat from the system without the need of the whiskers. The smaller size of the particles, probably, contributes to better heat dissipation. From the above, it can be concluded that energy dissipation drives dehydration induced polymorphism in this system.

Material from method one was placed in a glass vial with six milling balls (5 mm diameter)[4-6]left to roll for 12 hours. XRPD examination of the material obtained did not reveal dehydration of the crystals, although the SEM images showed that defects were formed on the crystals. The grinded material was subjected to the aforementioned drying conditions, however the trend in the dehydration-induced polymorphism did not change dramatically, the dried material was again a mixture of the two polymorphs, P-Monoclinic and Triclinic. This highlights the importance of powder's bulk properties (size and habit) over surface properties in the determination of drying induced polymorphism. The effects of stirring were also examined in the context of method one. It was shown that low Reynolds number favors the formation of high aspect ratio dihydrates which give a Triclinic polymorph upon dehydration. Using the same liquid composition as in method one the effect of supersaturation on crystal habit were examined. It was shown that low supersaturation, also, leads to the formation of high aspect ratio dihydrates.

FD-IGC measurements were performed on samples from method 1 dried at the four different temperatures. The shift from lower to higher surface energies is correlated with the increasing amount of Triclinic material observed with SEM. Computational models and FD-IGC measurements on pure P-Monoclinic and Triclinic can be combined to give approximate values of the polymorphic content of each sample.

This work, establishes pathways to obtain pure Triclinic carbamazepine and P-Monoclinic carbamazepine via the dehydration carbamazepine dehydrate. It critically revisits previous studies suggesting that the dehydration follows the Ostwald[2] rule of stages. A mechanism for the dehydration-induced polymorphism, suggesting a concomitant nucleation scheme is proposed. According to this mechanism crystal size and habit determine the dehydration induced polymorphism. The mechanism is driven by the energy dissipation mechanism of the system. This drying mechanism can be mechanistically understood using notions similar to those employed by Machon to describe the influence of defect concentration of nanoparticle's phase transitions using the one dimensional Ginzburg-Landau equation[7,8]. It is, also, demonstrated that FD-IGC measurements can be used to distinguish the polymorphic content of the dried material.

Disclosure:

Imperial College London and AbbVie jointly participated in study design, research, data collection, analysis and interpretation of data, writing, reviewing, and approving the publication. Eftychios Hadjittofis is a graduate student at Imperial College London; Jerry Y. Y. Heng is a professor at Imperial College London. They all have no additional conflicts of interest to report. Geoff G. Z. Zhang is an employee of AbbVie and may own AbbVie stock.

References:

1. Galwey AK. The topochemistry of the reaction interface in decomposition processes in solids. *Reactivity of Solids*. 1990;8(3–4):211-230.

2. Khoo JY, Shah UV, Schaepertoens M, Williams DR, Heng JYY. Process-induced phase transformation of carbamazepine dihydrate to its polymorphic anhydrates. *Powder Technol*. 2013;236:114-121.

3. Grzesiak AL, Lang M, Kibum K, Matzger AJ. Comparison of the four anhydrous polymorphs of carbamazepine and the crystal structure of form I. *Journal of Pharmaceutical Sciences*. 2003;92(11):2260-2271.

4. Wang Z, Shah UV, Olusanmi D, et al. Measuring the sticking of mefenamic acid powders on stainless steel surface. *Int J Pharm*. 2015;496(2):407-413.

5. Shah UV, Olusanmi D, Narang AS, et al. Effect of crystal habits on the surface energy and cohesion of crystalline powders. *Int J Pharm*. 2014;472(1–2):140-147.

6. Shah UV, Olusanmi D, Narang AS, Hussain MA, Hinder SJ, Heng JYY. Decoupling the contribution of surface energy and surface area on the cohesion of pharmaceutical powders. *Pharmaceutical Research*. 2014;32:248-259.

7. Machon D, Melinion P. Size-dependent pressure-induced amorphization: A thermodynamic panorama. *Physical Chemistry Chemical Physics*. 2014;17(2):903.

8. Machon D, Piot L, Hapiuk D, et al. Thermodynamics of nanoparticles: Experimental protocol based on a comprehensive ginzburg-landau interpretation. *Nano Letters*. 2014;14(1):269-276.

462188 Raman Mapping for Analyzing Multi-Component Solid Dosage Formulations

Abstract:

It has always been challenging to use spectroscopic methods to analyze salt disproportionation in a multi-component tablet matrix due to the spectral interference generated by the various excipients. Although combining Raman spectroscopy and chemometrics can be a powerful approach to study the extent of salt disproportionation, it was found in the present study that bulk measurements and chemometric modeling have obvious limitations when the targeted component is present at low levels in the tablet. Hence, a two-step Raman mapping approach was developed herein to investigate salt disproportionation in tablets with a low drug loading (5% w/w). The first step is to locate the area of interest where the drug particles reside throughout the tablet surface by using a statistically optimized sampling method termed deliberate sub-sampling. The second step, referred to herein as close-step mapping, utilize a step by step mapping of the targeted area to find more details of salt disproportionation in the tablet regions where the drug is concentrated. By using this two-step Raman mapping approach, we successfully detected the existence of minor species embedded in multi-component low drug loading tablet matrices, where bulk measurements from routine techniques usually lack of sensitivity. This approach will help formulation scientists detect and understand salt disproportionation and in situ drug-excipients compatibility issues in low dose solid dosage formulations.

470812 Optimizing Process Yields of a Legacy 25,000L Mammalian Cell Culture Process Using Integrated Upstream and Downstream Process Yield Models

Abstract:

Process monitoring is defined as a continuous effort to evaluate manufacturing parameters across multiple lots to ensure commercial manufacturing processes are operating in-control within specifications. As it becomes increasingly important to reduce the cost of goods for cell culture processes, scientists and engineers require digital data-acquisition and aggregation systems to develop a robust process monitoring approach for the entire process. These automated data acquisition tools assist in rapidly contextualizing data over the course of a process lifecycle and can identify correlations across an entire process (cell culture and purification). In particular, advent of digital plant combined with development of informatics tools to generate, acquire, analyze and visualize large data sets for process monitoring has led to increased process understanding, reduced cycle time, and implementation of process improvements based on analysis performed with the tool.

We will present a case study where predictive models of production bioreactor and chromatography steps were used in tandem to identify sources of variability in overall process yield for a 25,000L fed-batch CHO process. From this analysis, bioreactor culture duration was identified to negatively impact glycosylation levels of the recombinant fusion protein produced from the culture. By implementing calibrated, data-driven modifications to the upstream manufacturing schedule, we were able to realize tighter control of upstream protein quality, optimal downstream process performance, and improved overall process yield.

470647 Intelligent Control Strategies Towards the Intensification of Monoclonal Antibody Production Via Continuous Operation

Abstract:

Current industrial trends encourage the development of sustainable, environmental friendly processes with minimal energy and material consumption. Meanwhile, the increasing market demand as well as the tight regulations in product quality, necessitate efficient operation procedures that guarantee products of high purity. In this direction, process intensification via continuous operation paves the way for the development of novel, eco-friendly processes, characterized by higher productivity compared to batch.
This shift could benefit the market of high value biologics, such as monoclonal antibodies (mAbs), as it would lead to shorter production times, decreased costs, as well as significantly less energy consumption compared to batch [1, 2]. In particular, mAb production comprises two main steps: the culturing of the cells (upstream) and the purification of the targeted product (downstream). Both processes are highly complex and their performance depends on various parameters. In particular, the efficiency of the upstream depends highly on cell growth and the longevity of the culture, while the product quality can be jeopardized in case the culture is not terminated timely. Similarly, downstream processing, whose main step is the chromatographic separation, relies highly on the setup configuration, as well as on the composition of the upstream mixture. Therefore, in order to be able achieve a sustainable and efficient continuous operation, it is of vital importance to design intelligent computational tools that form a solid basis for the: (i) execution of cost-free comparisons of various operating strategies, (ii) design of optimal operation profiles and (iii) development of advanced, intelligent control systems that can maintain the process under optimal operation, rejecting disturbances.

This work focuses on the development of advanced control strategies for: (i) a cell culture system in a bioreactor and (ii) a semi-continuous purification process. More specifically, we consider fed-batch culturing of GS-NS0 cells producing a chimeric IgG4 antibody (based on the feeding of five key nutrients) and the semi-continuous Multicolumn Countercurrent Solvent Gradient Purification (MCSGP) [3-5] process for the purification of the upstream mixture. For the controller development we follow the generic framework & software platform PAROC that considers: (i) development of a high-fidelity process model, (ii) approximation of the complex, process model, (iii) design of the multi-parametric controller, (iv) 'closed-loop', in-silico validation of the controller against the process model [6]. PAROC allows 'in-silico' testing of the advanced controllers against the high-fidelity process model and evaluate their performance before operating them online. The results from this study indicate that the in-silico upstream/downstream integration can be achieved and the designed controllers succeed in maintaining both systems under optimal operation. In particular, the upstream system is characterized by a significantly increased mAb titer under the operation of the controller. Moreover, the controller indicates the time point that the culture needs to be terminated in order to minimize the risk of product degradation. Similarly, the downstream control scheme manages to maintain a stable operation throughout the process cycle, tracking efficiently the predefined setpoints. In addition, the controller is immunized against the uncertainty that results from the composition of the upstream mixture by considering the latter as measured disturbance. Finally, the suggested controller input indicates decreased use of utilities, while maintaining process efficiency.

Acknowledgements

The authors would like to thank Mr R. Oberdieck and Miss A. Quiroga-Campano for their contribution in the assessment of the controller validation. The authors would also like to thank Mr. Fabian Steinebach & Prof. M. Morbidelli from ETH Zurich, as well as Dr. Thomas Mueller-Spaeth from ChomaCon AG for their valuable input on the understanding of the MCSGP process. Financial support from the European Commission (OPTICO/G.A. No.280813) & Texas A&M University are also gratefully acknowledged.

References

1. Konstantinov, K.B. and C.L. Cooney, *White Paper on Continuous Bioprocessing May 20-21, 2014 Continuous Manufacturing Symposium.* Journal of Pharmaceutical Sciences, 2015. **104**(3): p. 813-820.

2. Xenopoulos, A., *A new, integrated, continuous purification process template for monoclonal antibodies: Process modeling and cost of goods studies.* J Biotechnol, 2015. **213**(0): p. 42-53.

3. Aumann, L. and M. Morbidelli, *A continuous multicolumn countercurrent solvent gradient purification (MCSGP) process.* Biotechnology and Bioengineering, 2007. **98**(5): p. 1043-1055.

4. Krättli, M., F. Steinebach, and M. Morbidelli, *Online control of the twin-column countercurrent solvent gradient process for biochromatography.* Journal of Chromatography A, 2013. **1293**: p. 51-59.

5. Papathanasiou, M.M., et al., *Advanced control strategies for the multicolumn countercurrent solvent gradient purification process.*AIChE Journal, 2016.

6. Pistikopoulos, E.N., et al., *PAROC—An integrated framework and software platform for the optimisation and advanced model-based control of process systems.* Chemical Engineering Science, 2015. **136**: p. 115-138.

470938 Development of a Scalable Process for Production of a Broadly Protective Influenza Antigen

Abstract:

While most influenza infections cause only mild illness, the World Health Organization (WHO) estimates that, on average, 3 to 5 million people worldwide develop respiratory or heart complications related to seasonal influenza infections each year. The WHO also estimates 250,000-500,000 influenza associated deaths yearly, with a large percentage occurring in developing world children with inadequate access to vaccines. Such mortality would be much worse for a virulent pandemic influenza strain. The current standard of care for influenza immunization is a trivalent whole-inactivated virus vaccine consisting of two influenza A strains (H1N1 and H3N2) and one influenza B strain. Due to antigenic drift, the influenza virus is constantly changing and experts must predict the dominant strains for the upcoming flu season roughly 7-9 months in advance of distribution. As a result, the actual circulating viruses may differ from those included in the vaccine, causing an increase in the number of influenza cases and the number of influenza-related deaths as was the case during the 2014-15 flu season. Once the three strains are selected, the majority of the vaccine is produced in eggs and treated for inactivation. This egg-based method takes roughly 6 months. As a result of the lengthy production times, vaccines cannot be updated to more accurately reflect the circulating strains in the case of a poor prediction.

A broadly protective influenza vaccine that would elicit strong protective responses for most or all possible circulating strains would dramatically reduce influenza-related hospitalizations and deaths by conferring life-long immunity. Our proposed vaccine would address many deficiencies in the current vaccine strategy by targeting a conserved region of the influenza hemagglutinin protein to protect against both seasonal and pandemic influenza strains. We will use a stabilized virus-like particle scaffold to present the antigen in a multivalent manner which more closely mimics the presentation of the antigen on the natural virus. By providing broadly protective immunity worldwide against seasonal and pandemic influenza, we would decrease the cost of medical care for influenza related illnesses, and dramatically reduce the number of influenza related deaths.

We have developed a stabilized hemagglutinin stem antigen which has the potential to provide broad protection against seasonal and pandemic strains of influenza. Here we discuss the development of a scalable refolding and formulation process for the production of our stabilized hemagglutinin stem antigen. By evaluating the kinetic and molecular forces required for the proper folding and trimerization of the hemagglutinin stem antigen, we have transitioned a number of lengthy, industrially unattractive refolding steps to a more reliable, scalable process. We are also working to further stabilize our hemagglutinin stem antigen for improved conformational stability with the ultimate goal of producing a cold-chain independent vaccine which would allow for improved distribution worldwide. We plan to evaluate both lyophilization and spray drying techniques for producing solid drug substance compatible with long term room temperature storage.

467571 Application of Process Monitoring Data Management Systems to Make Multiple Incremental Improvements to Cell Cultures Processes at the Manufacturing Scale

Abstract:

Process monitoring data management systems have been important tools to capture the large number of parameters and conditions that influence both the product quality and productivity of cell culture processes. Recent improvements in process analytical technologies and refinements in process control strategies have improved the biopharmaceutical industryâ€™s ability to gather large, multidimensional data sets that can be used for statistical process monitoring. In this work, we discuss the implementation of a data management platform and its application to build understanding around mammalian cell culture processes at the manufacturing scale. The database comprises of the end to end collection of process information from vial thaw to drug substance release for several commercial biologics produced using mammalian cell culture.

We explore the use of this database to make several incremental improvements to the cell culture process through a series of case studies including responses to health authority requests, technology transfers between multiple facilities, and evaluations of upstream parameters to understand their impact on final drug substance output. These case studies highlight the gains in process knowledge that can be achieved using these data management tools and illustrate some of the challenges in developing these capabilities. While data management tools facilitated the rapid retrieval of available data, there were several obstacles associated with data structure, data acquisition, and the subsequent data analysis. The current work demonstrates the successful application of data management tools to improve the performance and robustness of existing cell culture processes at the manufacturing scale.

471374 Scalable Production of Glioblastoma Cancer Stem Cells in 3 Dimension Thermoreversible Hydrogels for Drug Discovery

Abstract:

There is growing interest in developing drugs that specifically target glioblastoma cancer stem cells. Current cell culture methods, however, cannot cost-effectively produce the large numbers of glioblastoma cancer stem cells required for drug discovery and development. In this paper we report a new method that encapsulates patient-derived primary glioblastoma cancer stem cells and grows them in 3 dimension thermoreversible hydrogels. Our method allows long-term culture (~ 50 days, 10 passages tested, accumulative ~>10^{10}-fold expansion) with both high growth rate (~ 20-fold expansion / 7 days) and high volumetric yield (~ 2.0 x10^7 cells/ml) without the loss of stemness. The scalable method can be used to produce sufficient, affordable glioblastoma cancer stem cells for drug discovery.

470399 Levers to Modulate Product Quality Attributes of Product-a in CHO Cell Culture

Abstract:

Levers to modulate two product quality attributes were investigated as part of process development of Product-A produced in Chinese Hamster Ovary (CHO) cell culture. The addition of a chelating agent to the cell culture fluid was identified as a lever to minimize product fragment formation which indicates that metallo-proteases could potentially be contributing to fragment formation. A study to understand the nature and formation of product size variants identified that aggregate formation was predominantly intracellular and occurred prior to secretion and the fragments were formed in the cell culture fluid after secretion. In addition, media additives and process conditions were evaluated for potential levers to modulate product quality attribute-2. Process parameter-1 was determined to be a robust lever for controlling product quality attribute-2 across various production scales.

468324 Data-Driven Optimization Using an Evolutionary Design of Dynamic Experiments for Biopharmaceutical Processes

Abstract:

Biopharmaceutical processes are substantially complex and are not understood well enough to enable the development of an accurate knowledge-driven model for process optimization purposes. In such a case, a data-driven modelling approach provides a very attractive alternative, which, if designed properly, can quickly optimize the process with a small set of experiments. The recently introduced Design of Dynamic Experiments (DoDE) [1], a generalization of the traditional Design of Experiments (DoE) [2, 3], has demonstrated that it is an effective data-driven modeling and optimization approach [1, 4, 5].

Here we apply the DoDE method in an evolutionary manner to optimize biopharmaceutical processes while satisfying budgetary and developmental time constraints on the number of experiments in two representative biopharmaceutical processes; Penicillin fermentation [6] and Hybridoma cell culture [7]. We maximize the amount of product produced by determining the optimal feeding policy and batch duration.

To minimize the number of experiments we initially consider only a few dynamic sub-factors and the simplest Response Surface Model (RSM). After the initial results are analysed and an approximate RSM is estimated, additional experiments are added sequentially to further improve the data-driven model and the process performance. We demonstrate that with a reasonable amount of evolutionary experiments one can come very close to the model-based optimum. The model-based optimization [8], however, requires a very accurate knowledge-driven model that is rarely at hand.

References

1. Georgakis, C., *Design of Dynamic Experiments: A Data-Driven Methodology for the Optimization of Time-Varying Processes.* Industrial & Engineering Chemistry Research, 2013. **52**(35): p. 12369-12382.

2. Box, G.E.P. and N.R. Draper, *Response Surfaces, Mixtures, and Ridge Analysis*. 2007, Hoboken, NJ: Wiley.

3. Montgomery, D.C., *Design and Analysis of Experiments*. 8 ed. 2013, New York: Wiley. 729.

4. Fiordalis, A. and C. Georgakis, *Data-driven, Using Design of Dynamic Experiments, versus Model-driven Optimization of Batch Crystallization Processes.* Journal Of Process Control, 2013. **23**(2): p. 179-188.

5. Makrydaki, F., C. Georgakis, and K. Saranteas. *Dynamic Optimization of a Batch Pharmaceutical Reaction Using the Design of Dynamic Experiments (DoDE): The Case of an Asymmetric Catalytic Hydrogenation Reaction*. in *Proceedings of the 9th International Symposium on Dynamics and Control of Process Systems*. 2010. Oude Valk College, Belgium.

6. Bajpai, R.K. and M. Reuss, *A Mechanistic Model for Penicillin Production.* Journal Of Chemical Technology And Biotechnology, 1980. **30**(6): p. 332-344.

7. De Tremblay, M., et al., *Optimization of Fed-Batch Culture of Hybridoma Cells Using Dynamic Programming Single and Multifeed Cases.* Bioprocess Engineering, 1992. **7**(5): p. 229-234.

8. Biegler, L.T., *An Overview of Simultaneous Strategies for Dynamic Optimization.* Chemical Engineering And Processing, 2007. **46**(11): p. 1043-1053.

465077 In silico Formulation Development of Oral Solid Dosage Forms Using a Material Property Database

Abstract:

Motivated by the desire to lessen the empirical nature of oral solid dosage formulation and to pursue a quality by design framework, a material property database (MPD) and a material science interface (MSI) were recently developed at Abbvie. The MPD contains critical material properties of common excipients along with the lot-to-lot variability and humidity dependence of these properties. The centralization of these material properties in the MPD, collected in a standardized and reliable manner, also facilitate the use of predictive modeling via the MSI. The MSI is a user-friendly graphical interface which allows access to the MPD and associated modeling tooling used to aid in formulation development. Currently, the MSI contains models to predict blend uniformity, powder flowability, and compactability of pharmaceutical formulations intended for a direct compression process by making use of material properties contained within the MPD. Examples utilizing the MSI models will be given for the formulation development of a model active pharmaceutical ingredient (acetaminophen). The examples given herein show the utility of the MPD and MSI for providing in silico exploration of formulation design space, better guidance and scientific rational for experimentation, and improved understanding in the relationship between material properties, manufacturability, and product performance.

Disclosures:

All authors are employees of AbbVie. The design, study conduct, and financial support for this research was provided by AbbVie. AbbVie participated in the interpretation of data, review, and approval of the publication.

466851 Prediction of Tablet Dissolution from API and Excipient Polymer Properties

Abstract:

Prediction of Tablet Dissolution from API and Excipient Polymer Properties

Jaime L. Curtis-Fisk[1] (JLCurtisfisk@dow.com), True L. Rogers[1] (TLRogers@Dow.com), Karen M. Balwinski[1] (KMBalwinski@Dow.com), Kathryn O. Hewlett[1] (KOHewlett@Dow.com), Robert L. Schmitt[2] (SchmitRL@Dow.com), Shrikant N. Khot[1] (SNKhot@Dow.com)

[1] The Dow Chemical Company, Midland, MI USA

[2] The Dow Chemical Company, Collegeville, PA USA

Purpose: Modified-release (MR) matrix tablets are typically produced using hydroxypropyl methylcellulose (HPMC) as rate-modifying polymer. HPMC effectively sustains release of a broad range of active pharmaceutical ingredients (APIs), but specific dissolution behavior varies depending on the API properties, excipient properties, and formulation. Recently, a new line of HPMC, METHOCEL™ DC2 grades, have been introduced to facilitate direct compression formulations given its enhanced flowability. The ability to predict performance based on properties of components could streamline formulation development and provide a more direct route to troubleshooting challenging APIs. Combined such a capability with a direct compression manufacturing technique would mean faster and less expensive development and manufacturing. The intent of this study was to identify the API and polymer properties that most directly impacted dissolution performance, connect the observed performance to fundamental API and polymer chemistry, and develop predictability that connects this fundamental understanding to efficient formulation design.

Methods: The impact of API and polymer properties on dissolution was investigated by evaluating dissolution performance of APIs covering a broad range of solubility, shown in Figure 1. Tablets contained a standard formulation of 50% API, 30% METHOCEL DC2 HPMC, 19% impalpable lactose, and 1% magnesium stearate. A standard formulation was utilized for all APIs to make a direct comparison of the structure/activity relationship between API properties and dissolution performance. Tablets were prepared by hand-feeding powder formulations into the tablet press to remove variability in die fill due to powder flow differences between API formulation blends.

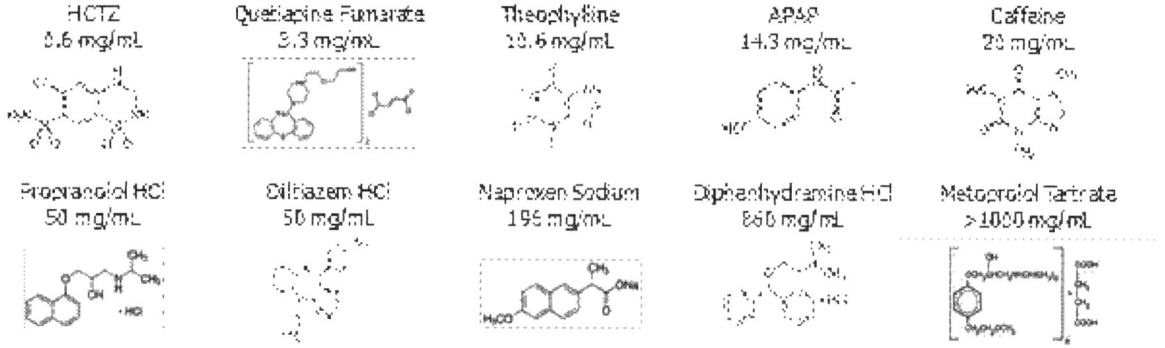

Figure 1. APIs with a broad range of solubility were evaluated in order to connect drug properties to dissolution performance.

Results: API dissolution performance was evaluated for three viscosity grades of METHOCEL DC2 (K100LV, K4M, K100M). Figure 2 displays the results of tablet formulations containing METHOCEL K4M DC2; percent API release at 2, 4 and 12 hours is plotted as a function of API solubility. The amount of API released at a given time point is greater as API solubility increases. The sensitivity to API solubility is most dramatic for intermediate solubility, with very low or high solubility API reaching a plateau where small changes in solubility to do not significantly impact release.

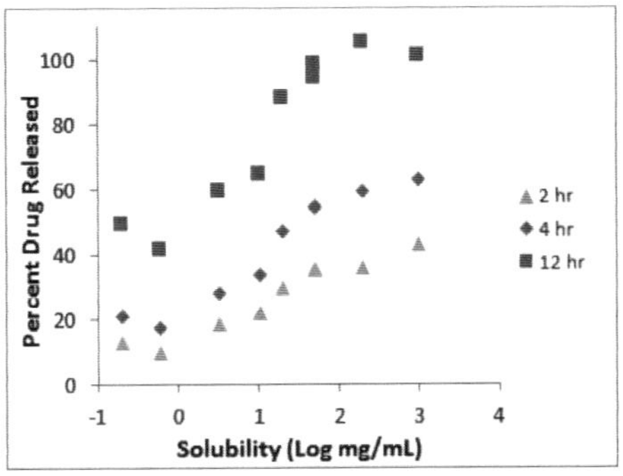

Figure 2. Percent API release as a function of API solubility, demonstrating a similar correlation at 2, 4, and 12 h dissolution time points.

The correlation of API solubility with drug release rate was also evaluated using t_{50} as a measure of release rate. This different perspective on the same datasets provides new insight into dissolution mechanisms of the tablet and the implications to formulation optimization. Figure 3 displays t_{50} values as a function of API solubility for three viscosity grades of METHOCEL DC2. All three viscosity grades demonstrate similar behavior; at low solubility there is a strong correlation between solubility and t_{50}, but at higher solubility the observed t_{50} value is constant. This behavior can be explained considering the competing mechanisms that occur during tablet dissolution. For poorly soluble APIs, the rate limiting step is API exposure to the greater sink of dissolution media as the outermost layer of swollen matrix erodes, followed by API dissolution into the media. When API solubility is greater, the limiting factor becomes diffusion of dissolved API through the swollen polymer layer and subsequent release into the greater sink of dissolution media. While K100M results in greater t_{50} values for all APIs evaluated, followed by K4M and lastly K100LV, the differentiation between grades was greatest for poorly soluble APIs. The difference in performance indicates that for low solubility APIs, polymer viscosity is a critical variable in formulation design.

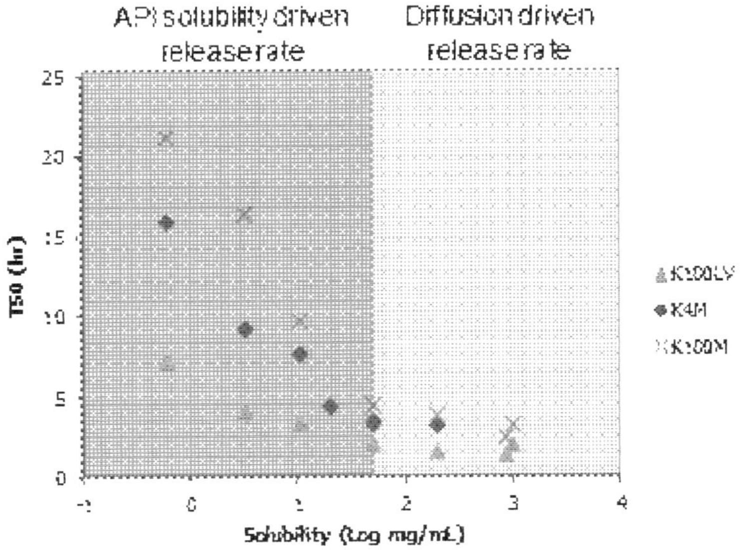

Figure 3. Dissolution rate described by t_{50} ***as a function of API solubility for three viscosity grades of METHOCEL DC2, demonstrating the correlation to API solubility.***

Conclusions: A key learning from this study is that API release from matrix tablets prepared with the METHOCEL DC2 product line is strongly impacted by API solubility and polymer viscosity. APIs of low solubility are released through a dissolution mechanism limited by API solubility and erosion of the outermost surface of the swollen polymer layer, but at higher solubility the release is driven by diffusion of dissolved API through the swollen polymer network. While higher viscosity polymers demonstrated greater extended release, the impact of this effect is more pronounced for lower solubility APIs and became negligible for higher solubility APIs. These relationships provide formulation guidance on selecting the appropriate viscosity grade of METHOCEL DC2 based upon API solubility.

463230 Connecting Processed Materials Properties to Process Development Strategy

Abstract:

A quality-by-design (QbD) process development case study would be presented using an example of a high drug load wet granulation formulation. This case study will identify the dependence of drug product performance quality attributes on material properties of intermediate granules. The translation of this mechanistic product and process understanding to process development, process analytical technology (PAT), and process control will be highlighted. The correlation of granule size versus porosity with tablet dissolution will be discussed. The implications of this correlation on the use of focused beam reflectance measurement (FBRM) versus drag flow force (DFF) sensor for process monitoring and control will be discussed. This talk will highlight the importance of material understanding to process development.

458431 Decoupling of a Low-Dose Dosator Capsule Filling Process in Dynamic and Static Mode Tests to Understand the Extend of the Effect of Powder and Process Parameters on Capsule Quality Attributes

Abstract:

Introduction

The precise filling of capsules with small powder quantities is essential for a broad range of industrial operations and a number of low-dose capsule filling machines are currently available. Filling principles can be categorized in either volumetric (e.g., the dosator nozzles, vacuum drum filler, vacuum dosator and tamp filler) or gravimetric (e.g., micro-dosing) methods. Capsule filling using the nozzle dosator technique has been widely investigated and is an important technology applied in the pharmaceutical industry today especially for the filling of capsules for inhalation, as the doses need a controlled degree of compaction to ensure an efficient dose delivery [1] [2] [3]. It is known that a large number of powder and processing parameters affect the quality of filled capsules [4] [5] [6] [7] [8]. However, the precise dosing of small powder quantities still poses significant problems. Thus, the objective of this study is to decouple the dosing event, which consists of multiple steps in order to identify and understand in-depth the complex relationship between material attributes and process parameters and their influence on critical quality attributes (fill weight and fill weight variability). Supported by *in silico* simulation, we will link those material properties to process performance and develop a mechanistic understanding in order to optimize the low-dose dosator capsule filling process.

Materials and Methods

Three different grades of alpha lactose monohydrate (Lactohale100, 200, 220 – DFE Pharma) specially designed for inhalation application were used as received. The following material attributes were characterized in triplicate: particle size (Qicpic OASIS/L dry dispersing system Sympatec, Germany), bulk (BD) and tapped density (TD) (Pharmatest PT-TD200), true density (AccuPyc II 1340, Micromeritics, Norcross, USA). The compressibility (CPL), air permeability (PD), flow function coefficient (FFc), cohesion (C), angle of internal friction (AIF), basic flowability energy (BFE), wall friction angle (WFA) were measured with the FT4 powder rheometer (Freeman Technology, Malvern, United Kingdom). Results are displayed in table 1.

Table 1: Inhalation carrier characterization

	Lactohale 100	**Lactohale 200**	**Lactohale 220**
Particle size and friction properties			
Volumetric mean diameter [µm]	160.02 ±0.62	83.63 ±n.a.	16.38 ±n.a.
Angle of internal friction (AIF) [°]	18.43 ±0.49	21.24 ±0.53	25.73 ±2.66
Wall friction angle (WFA 3 kPa 0.2 Ra) [°]	7.70 ±0.02	11.46 ±0.82	19.97 ±3.82
Bulk powder properties			
Bulk density (BD) [g/ml]	0.6972 ±0.0036	0.6223 ±0.0031	0.3996 ±0.0067
Tapped density (TD) [g/ml]	0.8275 ±0.013	0.9962 ±0.0017	0.7847 ±0.0066
True density [g/ml]	1.5385 ±0.0028	1.5426 ±0.0024	1.5466 ±0.0042
Compressibility (CPL) at 8 kPa (ratio ρ_{comp}/ρ_{BD}) [%]	1.05 ±0.00	12.66 ±0.29	36.95 ±0.16
Air permeability (PD) at 8kPa [mbar]	1.05 ±0.02	7.00 ±0.10	12.04 ±0.55

Flow properties			
Basic flowability energy (BFE) [mJ]	910.67 ±43.00	1722.85 ±27.00	667.32 ±10.80
Flow function coefficient (FFc)	6.58 ±0.01	4.04 ±0.01	1.65 ±0.01
Cohesion (C) [kPa]	0.24 ±0.02	0.39 ±0.01	1.05 ±0.13

Dynamic capsule filling: The powders were filled into hard gelatin capsules of size 3 (supplied by Capsugel) with a lab scale capsule filling machine (Labby, MG2, Bologna) adopted with a special low dose equipment. Two volumes of dosing chamber (2.5, 5 mm), two different dosator sizes (1.9, 3.4 mm), two different powder bed heights (5, 10 mm) and two different filling speeds (500, 3000 cph) were used for the experiments. The study was performed under humidity controlled conditions (45-55 % relative humidity). To create an initial powder bed without densification, the layer was adjusted manually. After setting all machine parameters, capsules were collected at time point 0, after 25, 50, 75, 100 rotations for a total run time of 30 minutes in order to sample from the same position inside the rotating container. Powder feeding was adjusted according to the amount of powder collected during capsule filling. Subsequently, the weight of filled and empty capsules was measured with a Denver (SI-234A) scale.

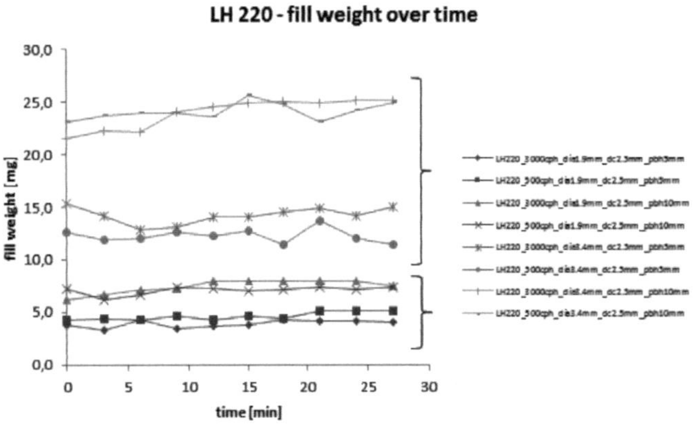

Figure 2: Capsule fill weight of Lactohale220 at different combinations of process settings

Static capsule filling: The static mode capsule filling experiments were conducted under the same combination of process settings and environmental conditions as the dynamic mode tests. The difference here is that the powder is picked up from a static powder bed not a rotating drum in order to study the effect of powder layer homogeneity on capsule quality attributes and to get a mechanistic understanding about the powder compaction behavior inside the dosing chamber.

Simulation and Modeling

In addition to the experiments, DEM simulations of the static capsule filling experiments are performed. Therefore, a model of the dosator is implemented in the DEM software LIGGGHTS® and different particle contact properties are used to describe the filling behavior of different powders. The simulations show the evolution of the powder mass inside the nozzle during filling and therefore help to understand the dosator process in more detail.

Conclusion and Outlook

In this research, various material attributes of different grades of lactose were determined and their performance during capsule filling using a lab scale low dose dosator nozzle system was investigated in dynamic mode. Obtained results will be correlated with experiments conducted with the static stand-alone device and the simulation work to understand the influence of particular process parameters on the filling performance, which will furthermore lead to an improved filling process.

References

[1] B. E. Jones, "The filling of powders into two-piece har+d capsules," *Int. J. Pharm.*, vol. 227, no. 1, pp. 5–26, 2001.

[2] M. Llusa, E. Faulhammer, S. Biserni, V. Calzolari, S. Lawrence, M. Bresciani, and J. Khinast, "The effect of capsule-filling machine vibrations on average fill weight," *Int. J. Pharm.*, vol. 454, no. 1, pp. 381–387, 2013.

[3] M. Llusa, E. Faulhammer, S. Biserni, V. Calzolari, S. Lawrence, M. Bresciani, and J. Khinast, "The effects of powder compressibility, speed of capsule filling and pre-compression on plug densification," *Int. J. Pharm.*, vol. 471, no. 1, pp. 182–188, 2014.

[4] S. B. Tan and J. M. Newton, "Powder flowability as an indication of capsule filling performance," *Int. J. Pharm.*, vol. 61, no. 1, pp. 145–155, 1990.

[5] R. Patel and F. Podczeck, "Investigation of the effect of type and source of microcrystalline cellulose on capsule filling," *Int. J. Pharm.*, vol. 128, no. 1–2, pp. 123–127, 1996.

[6] F. Podczeck and J. M. Newton, "Powder and capsule filling properties of lubricated granulated cellulose powder," *Eur. J. Pharm. Biopharm.*, vol. 50, no. 3, pp. 373–377, 2000.

[7] E. Faulhammer, M. Llusa, C. Radeke, O. Scheibelhofer, S. Lawrence, S. Biserni, V. Calzolari, and J. G. Khinast, "The effects of material attributes on capsule fill weight and weight variability in dosator nozzle machines," *Int. J. Pharm.*, vol. 471, no. 1–2, pp. 332–338, 2014.

[8] E. Faulhammer, M. Fink, M. Llusa, S. M. Lawrence, S. Biserni, V. Calzolari, and J. G. Khinast, "Low-dose capsule filling of inhalation products: Critical material attributes and process parameters," *Int. J. Pharm.*, vol. 473, no. 1–2, pp. 617–626, 2014.

467102 Quantitative Investigation on Segregation Phenomenon Using Discrete Element Method in Large Scale Pharmaceutical Compression Process

Abstract:

Segregation is a common issue in pharmaceutical manufacture of solid dose drug product. Especially during manufacture scale-up, pharmaceutical powder with segregation tendency has potential risk to fail specification of blend or content uniformity in large scale process. It may result in significant waste of drug substance or cause possible quality issues for the drug product. Moreover, experimental investigation on large scale segregation is usually inefficient to determine root cause or provide quantitative improvement plan while the consumption of drug substance was still significant. In this work, quantitative Discrete Element Method (DEM) simulation method was developed and applied on the prediction of segregation tendency of powder blend in large scale pharmaceutical unit operations.

In the large scale drug product manufacture of Project X, segregation was consistently observed in the last quarter of the tablet compression process. For the two major blend components in this project, material properties such as particle size distribution, true density, and flowability were measured and imported into simulation. The previous established correlation between shear flow of experiment and simulation was applied to guarantee that the flow behavior of simulated DEM particles was representative of real pharmaceutical powder [1]. Several slice geometries were developed to represent the dimensions of the current blender- hopper combination and alternative candidates for geometry improvement. The blender discharge and hopper feeding process in the bulk tablet compression of project X was then simulated using the commercial DEM software Star CCM+® v11.02 by CD-Adapco® and with processors of dual CPUs, Intel® Xeon® E52640 @ 2.50 GHz.

Based on the simulation results, difference on true density of the blending components was determined as the root cause of segregation in the compression process of Project X. Geometries show different segregation tendency with the maximum CU range 103% - 120% due to different powder flow and segregation mechanics in these geometries. It was observed that segregation mainly took place on the surface of inclined material bed during the discharging and feeding processes and was significantly influenced by the relative location of outlet. The method was then validated by performing feeding experiment of Project X final blends with 3D printed blender-hopper geometries, which match well with the segregation behavior of corresponding simulations. Suggestions were therefore made on the setting of hopper geometry for further manufacture activity. The case study shows robust applicability of the proposed simulation method on pharmaceutical powder flow and segregation. This method has potential benefit on significant saving of cost and drug substance, as well as risk reduction of batch failure especially in scale-up of solid dose manufacture.

1. Gao Y., Chalasani, S., Mittal, B., Investigation on Key Parameters of Common Pharmaceutical Blending Unit Operations Using Discrete Element Method, AIChE Annual Meeting, Salt Lake City UT, Nov 2015, paper 412e.

454354 Two-Phase Flow (solids: interstitial air) Modeling of a Press Feed System

Abstract:

Production rate limitations can develop when feeding fine powders to a press, granulator or mill. These powder feed rate limitations arise when the powder interacts with entrained air, resulting in adverse â€œtwo-phaseâ€ (powder-air) flow effects. Two-phase flow problems can result in out-of-specification tablets, requiring operating at below target production rates and can occur in a continuous or batch process. A critical step in developing a robust process, including conducting Quality by Design (QbD) studies, is modeling two-phase flow behavior in the press feed system.

Two-phase flow risks can be accounted for during process design by:

- Measuring key flow properties to use as inputs in the two-phase flow model;
- Computational modeling the two-phase flow effects based upon first principle models;
- Evaluating potential corrective actions through two-phase modeling.

The two-phase modeling software uses key powder flow properties, including:

- Permeability
- Compressibility
- Specific gravity of a particle
- Wall friction angles (phi-prime, fâ€™)
- Effective angle of internal friction (delta, d) measured via a cohesive strength test.

The two-phase model as includes the press feed system geometry and process parameters as inputs. The model outputs

include parameters that dictate two-phase flow behavior, such as vertical solids stress, interstitial gas pressure, bulk density and normalized gas pressure gradient. These modeling outputs are used to predict whether flow problems will occur and analyze potential corrective actions, including modifying the geometry of the press feed system, increasing the permeability of the powder and adding venting.

462120 Use and Misuse of the Diametrical Compression Test

Abstract:

The diametrical compression (DC) test has been used extensively to derive the tensile strength of pharmaceutical tablets from the measured breaking force. DC-derived tensile strengths provide a good approach to measuring lot-to-lot variability in the mechanical properties of tablets. For these quality control type applications, method precision is required, but accuracy is not. Additionally, DC has been used to calibrate parameters of the Druker Prager Cap (DPC) model, a yield criterion expressing the failure of a powder compact under arbitrary 3D loading conditions. For this application, the DC method must not only provide suitable precision, but also accuracy. In this work we explore the accuracy of the DC method in the measurement of tensile strength and conclude that the true tensile strength of a powder compact is approximately double the DC-derived value. While historical literature assumes that tablets fracture under tension along the centerline of the tablet, analysis of the stress state suggests that tablets are likely to fracture under shear. The impact of this ~50% error should be considered when accuracy of the tensile strength result is required.

AUTHOR INDEX

Abbasi, Akram 225
Abbott, Erin K. 329
Abebe, Admassu 136
Abercrombie, Johnathan 106
Abramov, Yuriy 49
Acevedo, David 281
Adack, Paige 97
Adjiman, Claire S. 40
Agarwal, Siddhartha 177
Agastin, Sivaprakash 101, 326
Aigner, Isabella 73, 114, 135
Ainslie, Kristy M. 188
Akpinar, Fulya 42
Akseli, Ilgaz 2
Albadarin, Ahmad 72, 76
Albadarin, Ahmad B. 115, 316
Al-Haiek, Alexandre 165
Ali, Muzammil 294
Almeida, Lucilla 174
Aloia, Marie 4
Anderson, Daniel G. 195
Androulakis, Ioannis P. 69, 82
Ang, Ou W 330
Anita, Lalloo 116
Aon, Juan 220
Arai, Hiroaki 134
Armani, Mina 4
Armenante, Piero M. 162, 264
Ashley, Jonathan D. 187
Ataíde, Filipe A.P. 35, 159, 207
Axelson, June 329
Aydogan, Nihal 230, 231
Aytar, Selin 190
Azad, Mohammad 283, 315
Badawy, Sherif 132
Badruddoza, Abu Zayed Md 284
Bajpai, Rakesh 48
Balderrama, Michael 259
Balwinski, Karen 303, 334
Barks, Edward 255
Barnum, Roger 343
Barrett, Mark 111, 222
Barton, Paul I. 139
Barua, Dipak 247
Barua, Sutapa 247
Bastian, Rainer 261
Baumann, John 234
Baxter, Thomas 343
Bear, Marie-Maud 64
Behkam, Bahareh 245
Beitle, Bob 258
Bermingham, Sean K. 30
Bernthal, Harold 199
Bertels, Johny 68
Bharadwaj, Rahul 174
Bhatnagar, Sumit 212
Biegaj, Karolina 3
Biegler, Lorenz 36
Bielser, Jean-Marc 263
Bilgicer, Basar 187
Bilgili, Ecevit 1, 4, 129, 283, 293
Bird, Andrew 161
Biserni, Stefano 338
Bizios, Rena 158
Blackshields, Caroline A. 86
Blakely, Kyle 313
Bock, Daniel 103
Bommarius, Andreas S. 157
Bonakdar, Tina 294
Bordawekar, Shailendra 320
Borghard, William G. 277
Bose, Arijit 225
Bostijn, Nils 57, 150
Bothun, Geoffrey D. 225
Bowers, John 218, 224
Braatz, Richard D. 181
Brady, Sonia K 127
Brandenbusch, Christoph 44
Brau, Christopher 217
Bredael, Gerard 264
Brocker, Caitlin 247
Brodean, Eric 98
Broly, Hervé 263
Brueggemeier, Shawn 190
Bruggraber, Sylvaine 256
Brune, Ellen M. 258
Brunskill, Andrew 317
Brunsteiner, Michael 81, 90
Buenrostro, Denise 191
Bullard, Joseph 28, 122, 173
Bunchatheeravate, Pongpumin 122, 173
Buonocore, Gregory 173
Burcham, Christopher L. 31, 112, 313
Burns, Mark A. 154
Butté, Alessandro 263
Byrne, Mark 249
Calabrese, Richard V. 163
Caldwell, Anna 191
Callegari, Gerardo 123
Calzolari, Vittorio 338
Capece, Maxx 333
Cappuyns, Philippe 71
Caracci, Stephen 64
Carr, Dearbhla 97
Carson, John 343
Cha, Bumjoon 117
Chadwick, Keith 281
Champion, Julie A. 189, 242
Chang, Jane 167
Chang, Kevin 219

AUTHOR INDEX

Chatterjee, Tirtha 318
Chaudhary, Garima 103
Chaudhry, Mina 126
Chauhan, Niharika 179
Chehreghanianzabi, Yasaman 209
Chen, Jonathan J. 77
Chen, Wei 160
Chen, Weifeng 36
Chen, Xiaodong 205
Cheng, Zhijian 214
Chiba, Kazuhiro 16
Chistoserdov, Andrei 48
Christodoulou, Charalampos 33
Chundawat, Shishir PS 127
Clarke, Oliver 116
Cleary, Michelle 97
Cohen, Benjamin 42
Collins, Maurice 72
Collins, Scott 217
Conti, Denise S. 91
Conway, Stephen L. 116
Cooks, Robert G. 255
Corcoran, Timothy 216
Costa, Ines Da Silva 263
Craven, Stephen 222
Crawford, Morgan 142
Crean, Abina 86, 145, 298, 306
Crick, Eric 206
Croker, Denise 61, 72, 76, 95, 115
Cronin, Michael F. 86
Cronin, Patrick 72
Crowley, Clare M. 285
Crowley, Mary Ellen 145, 298, 306
Crowther, Orlaith 97
Cryan, Philip 116
Cuitiño, Alberto 143, 277
Cuitino, Alberto M. 144
Curtis-Fisk, Jaime 289, 303, 334
Darwish, Shaza 76, 95, 115
Dave, Rajesh 315
Dave, Rajesh N. 4, 283
Davé, Rajesh N. 1, 293
Davern, Peter 285
David, Birendra K. 53
Davis, Adrian L. 81, 90
Davis, Mark 314, 316
Davis, Richey M. 245
De Beer, Thomas 68, 71, 137, 233, 239, 260
De Leersnyder, Fien 71
De Oliveira, Joshua 194
Debacq, Marie 278
Del Valle, Eva Martín 201
Deng, Vivian 126
Derrick, Andrew 161
DeSessa, Melanie 102
Deshpande, Aakash 210
Dhondt, Jens 68
Diab, Samir 74
Ding, Xiong 197
Dirion, Jean-Louis 278
Dohnal, Jiri 296
Doleski, David 27
Dominguez, Alan 43
Donnellan, Philip 113, 203
Doshi, Pankaj 239
Doss, Geny 202
Doyle, Patrick S. 284
Drazer, German 143
Dreu, Rok 338
Duarte, Íris 80, 83, 130
Dufal, Simon 40
Eickhoff, W. Mark 133, 142
Ende, David am 105
Ende, Mary T. am 239
Escotet-Espinoza, M. Sebastian 67, 125
Fan, Yi 99
Farizhandi, Amir A. K. 5
Faulhammer, Eva 65, 338
Feidl, Fabian 263
Ferguson, Rachel 219
Fisher, Robert 124
Fitzgerald, Jonathan 243
Fitzgibbon, Sean 43
Flamm, Matthew H. 117
Flanagan, Michael 209
Flynn, Nicholas 248
Fogarty, Julie 328
Forciniti, Daniel 247
Fortela, Dhan Lord 48
Foster, Matthew 128
Fradette, Louis 165
Franchville, Justin 280
Fränkel, Thomas 261
Fruchtl, McKinzie 258
Fuentes-Gari, Maria 30
Fung, Conrad 41
Futamura, Haruka 56
Galán, Miguel A. 201
Galbraith, Shaun 117
Galindo, Amparo 40
Gan, Qi 209
Gandek, Thomas 202
Ganesh, Sudarshan 96
Gao, Joe 141
Gao, Ping 333
Gao, Yijie 342
Garapaty, Anusha 189
Garcia, Elba 251
García-Muñoz, Salvador 33, 36, 79, 140, 178, 206
Gargava, Ankit 141

AUTHOR INDEX

Garikipati, SVB Janardhan 53
Garoff, Stephen 216
Gavi, Emmanuela 29
Gellman, Andrew J. 226
Gendler, Sandra J. 198
Georgakis, Christos 332
Gernaey, Krist V. 68, 71, 137, 233
Gerogiorgis, Dimitrios I. 38, 74
Gerrard, Stephen 256
Ghaderzadeh, Kanan 163
Ghadiri, Mojtaba 294
Ghijs, Michael 71, 239
Ghosh, Pranay K. 29
Gibbs, Mason 116
Gill, Harvinder 215
Gilmore, Richard 219
Giridhar, Arun 96
Glace, Andrew 42
Glasser, Benjamin 277
Glennon, Brian 111, 113, 222
Godeshala, Sudhakar 198
Godfrin, P. Douglas 284
Gogos, Costas G. 129
Goldman, Johnathan M. 205
Gong, Junbo 183, 184, 185
Gong, Zifan 213
Good, Melissa 98
Gorden, Jannick 44
Graham, Hugh 329
Gregoire, Simpson 101, 326
Gressl, Corinna 81, 90
Griffin, Daniel J. 45
Grismer, Dane A. 219
Grof, Zdenek 296
Grover, Martha A. 45
Gu, Zongyu 124
Guan, Jingjiao 214
Guo, John 50
Gupta, Balrina 224
Hackel, Benjamin J. 210
Haddad, Nizar 100
Hadjittofis, Eftychios 6, 322
Haig, Brian 268
Han, Dandan 185
Hannon, Joe 161
Hanson, Joshua 79, 138
Hao, Hongxun 186
Hapgood, Karen 8, 268
Harrison, Roger 257
Hartmanshenn, Clara 69, 82
Hartson, Steven 248
Hasan, Tayyaba 243
Hasenfus, Erik 103
Hauseder, Gregor 78
Hay, Michael 190
Hayes, Kevin 145
He, Xiaorong 141
Hebrault, Dominique 100
Hedberg, Sarah 223
Hegarty, Avril 145
Heng, Jerry 3, 6, 223, 322
Henriques, João 83
Henry, David 64
Hernandez, Rafael 48
Hervy, Martial 64
Hewlett, Kathryn 303, 334
Hickey, Anthony 91
Higgins, John 224
Hilden, Jon 280, 344
Hilger, Ryan T. 255
Hinder, Steven J. 6
Hirao, Masahiko 56, 134
Ho, Raimundo 333
Hodnett, B. Kieran 285
Hofmeister, Rudy 265
Holbrook, Debora 199
Holleran, Tim 246
Holman, James 136
Holmes, William 48
Holt, Jay 91
Hong, Donghui 116
Hong, Moo Sun 124
Hörmann, Theresa R. 60, 119
Horn, Martin 73
Horspool, Keith 141
Horton, Andrew 99
Howe, Alaina 246
Hoza, Jon 116
Hsiao, Tony 217
Hsii, Judy F. 166
Huang, Huang-Chiao 243
Huang, Wenjun 311
Huang, Zhuangrong 117
Hudson, Sarah 285
Hurley, Samantha 117
Hurst, Robert 257
Hurter, Patricia 202
Hutacharoen, Panatpong 40
Iasella, Steven 216
Ierapertritou, Marianthi 69, 262, 295
Ierapetritou, Marianthi G. 67, 82
Ikeda, Craig B. 142
Imperial, Josh 199
Ioannidis, Nicolas 129
Isbell, Mark A. 6
Islam, Tariq 30
Jackson, George 40
Jacob, Karl 99
Jahn, Niklas Henning 319
Jain, Era 209

AUTHOR INDEX

Jain, Priti ... 289
Jaladi, Tejasvi ... 102
Jaman, Zinia ... 255
Jarmer, Daniel ... 31, 112
Jayawickrama, Dimuthu A. ... 126
Jayjock, Eric ... 125
Jewell, Christopher M. ... 192
Jiang, Mo ... 181
Jo, Ami ... 245
Johnson, Sadie M. ... 210
Jolliffe, Hikaru G. ... 38
Jones, Nephi ... 217
Jones, Roderick ... 113
Kalal, Zbynek ... 218
Kalaycioglu, Gokce Dicle ... 230
Kalb, Oskar ... 29
Kalra, Ashish ... 243
Kamiya, Hidehiro ... 16
Kano, Manabu ... 54
Karim, M. Nazmul ... 102
Karlsson, Amy J. ... 213
Kastantin, Mark ... 234
Kawajiri, Yoshiaki ... 45
Kawasaki, Koji ... 56
Ke, Jiaying ... 46
Kelly, Catherine ... 61, 314, 316
Kelly, Jessica ... 249
Kendall, Richard ... 256
Kerschhaggl, Matthias ... 78
Keyvan, Golshid ... 143
Khademhosseini, Ali ... 156
Khair, Aditya S. ... 46
Khan, Saif A. ... 321
Khinast, Johannes G. ... 60, 65, 73, 78, 81, 90, 114, 119, 135, 175, 261, 338
Khoiuzam, Maria ... 224
Khosla, Chaitan ... 23
Khot, Shrikant ... 289, 303, 318, 334
Kientzler, Donald ... 136
Kiernan, Luke ... 145
Kim, Jaeyeon ... 243
Kim, Soojin ... 141
Kiziltepe, Tanyel ... 187
Klein, Lee ... 142
Kline, Billie ... 41, 208
Klingeleers, Didier ... 68
Knarr, Matthias ... 199
Kobierski, Michael ... 206
Koch, Melvin ... 19
Kodam, Madhusudhan ... 99
Kofler, Tobias ... 267
Kollamaram, Gayathri ... 61, 76
Komives, Claire F. ... 259
Koralege, Rangika Hikkaduwa ... 248
Korde, Sachin ... 316
Koscher, Gerold ... 60, 114, 119
Koswara, Andy ... 255, 320
Kovacik, Pavel ... 238
Koynov, Athanas ... 282, 317
Koynov, Sara ... 97
Kraxner, Michael ... 267
Krebs, Kathleen ... 64
Kreimer, Manuel ... 114
Krins, Brian ... 298
Krogmeier, Stephanie ... 202
Kruisz, Julia ... 73, 135
Krull, Scott M. ... 1, 293
Kucuk, Gulsad ... 2
Kuhs, Manuel ... 76, 95, 115
Kumar, Ashish ... 68
Kumar, Sunny ... 194
Kuzmission, Drew ... 202
Laemthong, Tunyaboon ... 247
Lamberto, David J. ... 52
Lampe, Kyle ... 244
Landis, Margaret ... 81, 90
Lang, Matthew J. ... 127
Langer, Robert ... 195
Langroudi, Mehrdad Kheiripour ... 133, 142
Larson, Ronald G. ... 311, 312
Laske, Stephan ... 60, 119, 175
Lau, Raymond ... 5
Lavine, Barry ... 257
Leavesley, Ian ... 138
Lee, Ann L. ... 166
Lee, Chris ... 160
Lee, Elaine ... 41
Lee, Helen ... 243
Lee, Hung Lin ... 51
Lee, Kai ... 239
Lee, Kelvin H. ... 18
Lee, May Ginn ... 8
Lee, Sau ... 26, 124
Lee, Tu ... 51
Lei, Yuguo ... 330
Leone, Anthony ... 325
Lepek, Daniel ... 114
Lesakova, Veronika ... 296
Leung, Kai ... 106
Li, Hanlin ... 173
Li, John ... 259
Li, Kangli ... 184
Li, Meng ... 1, 4, 129, 315
Li, Qiang ... 330
Li, Shuaili ... 124
Li, Yan ... 214
Li, Yongfu ... 318
Li, Yongjun ... 317
Lim, Hong Lee ... 268
Lin, Haishuang ... 330

AUTHOR INDEX

Lin, Henry 103
Lin, Hong Yu 51
Ling, Jing 281
Ling, Kevin 242
Ling, Wai Lam 98
Liu, Claire Yiqing 255
Liu, Joyce 243
Liu, Xu 130
Liu, Yingzi 16
Liu, Yumin 185
Liu, Zhen 325
Loidolt, Peter 338
Loladze, Vakhtang 220
Long, Barry 108
Looker, Adam 202
Lopez-Ribot, Jose 106
Loren, Bradley P. 255
Lower, Asher 168
Lu, Yuan 328
Luciani, Carla 169, 178, 206
Lukas, Tim 3
Ma, Ji 162
MacDonald, Hailey 173
Macias, Kevin 136
Madsen, Ben 217
Madsen, Cathy S. 198
Madurawe, Rapti 26, 250
Mahboubah, Pishnamazi 72
Mahbubani, Krishnaa 256
Manley, Leo 118
Mann, Amanda 317
Mansouri, Shahnaz 8
Mantalaris, A. 327
Mantalaris, Athanasios 20
Mantri, Rao 160
Mao, Guangzhao 197
Marks, Brian 325
Marsac, Patrick 325
Martel, Sophie 175
Martin, Doug 249
Martinetti, Jessica Lovey 175
Martinetz, Michael 135
Matic, Josip 175
Matos, Ines 235
Matsunami, Kensaku 134
Mazzei, Luca 33
McAuliffe, Michael 145
McCarthy, Joseph J. 276
McCurdy, Conor 97
McKernan, Patrick 257
Mecca, Jodi 289
Mehan, Rishi 53
Mehta, Ankit 218
Mejia, Franklin 187
Mendoza, Adlin 251
Menegatti, Stefano 194
Meng, Wei 265
Menning, Mark 297
Merkel, Alyssa 191
Mesías-Recamán, Fernando 201
Metzler, Catherine 173
Meyer, Robert F. 116, 117
Mihlbachler, Kathleen 22
Mitchell, Niall 30
Mitragotri, Samir 194
Miyano, Takuya 134
Moghadam, Soroush 312
Mohammadpour, Maryam 298, 316
Montaño, Adriana M. 209
Morales, Zuleyka 251
Morbidelli, Massimo 263
Moreira, Jorge 235
Moreno, Jacqueline 283, 315
Moreno, Mariana 96, 266
Morrison, Christopher J. 50
Mortier, Séverine T.F.C. 71, 233
Moshkelani, Delaram 171
Mott, Landon A. 109
Mukherjee, Rudra Palash 258
Mukherjee, Samrat 179
Muliadi, Ariel R. 28
Munir, Ahsan 43
Murphy, Nick P. 244
Mustafaoglu, Nur 187
Muzzio, Fernando 123, 143, 265
Muzzio, Fernando J. 67, 262, 277
Myerson, Allan S. 284
Nagy, Zoltan K. 24, 31, 96, 112, 255, 266, 281, 320
Nakagawa, Hiroshi 134
Narang, Ajit 337
Narayanan, Vibha 127
Nere, Nandkishor 320
Neves, Filipe 80
Ng, Denise Zi Ling 321
Nicolaï, Niels 137
Nie, Haichen 325
Nikfar, Faranak 205
Njeng, Alex Stéphane Bongo 278
Noicely, Kenroy 43
Nopens, Ingmar 68, 71, 137, 233, 239
Novak, Matej 238
Obando, Louis 224
O'Brien, David 333
O'Brien, Eoin 97
O'Connor, Colleen 196
O'Connor, Thomas 123, 124
O'Donnell, Kevin 318
Oka, Sarang 265
Okada, Yohei 16
Oliva, Joseph 255

AUTHOR INDEX

O'Mahony, Colm ... 169, 206
O'Mahony, Graham ... 145
Orbe, Paul ... 4
Osborne, Douglas ... 219
Owen, Andrew ... 116
Padrela, Luis ... 108
Page, Susanne ... 29
Pagire, Sudhir ... 316
Palaparthi, RaviChandra ... 53
Pan, Xinghua ... 102
Pandey, Preetanshu ... 132, 174
Panikar, Savitha ... 265
Papageorgiou, Charles D. ... 55
Papaioannou, Vasileios ... 40
Papathanasiou, Maria M. ... 20, 327
Paradkar, Anant ... 316
Paredes, Ingrid J. ... 277
Park, Keunhan ... 225
Park, Tai Hyun ... 155
Parks, Conor ... 255, 320
Partani, Yash ... 53
Patel, Shiragi ... 209
Patel, Vani ... 220
Patkar, Kshitij A. ... 171
Patrascu, Michael Shoham ... 139
Paudel, Amrit ... 65, 81, 90, 175, 338
Pawar, Pallavi ... 265
Pecheul, Marylene ... 64
Peddapatla, Raghu V. G. ... 86
Pena, Ramon ... 31, 112
Pencheva, Klimentina ... 81, 90
Peng, Ching-An ... 193
Pentz, Julie ... 116
Peppas, Nicholas ... 196
Pereira, Glinka Cathy ... 295
Petermann, Oliver ... 199
Peterson, Ryan ... 173
Petrova, Rositza ... 52
Pfefferlé, François ... 175
Phenix, Brian ... 208
Pielecha-Safira, Barbara ... 4
Pillei, Martin ... 267
Pinto, João F. ... 83
Pistikopoulos, Efstratios N. ... 20, 327
Polizzi, Mark ... 280, 344
Pollard, Jennifer ... 21
Pollock, James ... 21
Pope, Carey ... 248
Pope, Jing ... 248
Porfirio, Tiago ... 130, 235
Porter, William ... 289
Prescott, James K. ... 343
Pritchard, Justin ... 25
Przybycien, Todd M. ... 216
Purcell, Mark ... 116
Pynn, Abigail ... 331
Quach, Amy L. ... 282
Quiroga-Campano, Ana L. ... 20
Raftery, Jonathan ... 102
Raikes, Michelle ... 141
Rajniak, Pavol ... 30
Ramachandran, Rohit ... 262, 295
Ramasamy, Manoharan ... 116
Ramasubramanian, Anand ... 106
Ramesh, Rajagopal ... 257
Ramirez, Antonio ... 37
Ramkrishna, Doraiswami ... 31, 320
Ramsey, Joshua ... 248
Rane, Anil ... 94
Ranjan, Ashish ... 248
Razavi, Sonia ... 143
Razavi, Sonia M. ... 144
Reckamp, Joseph ... 110
Rege, Kaushal ... 198
Rehrl, Jakob ... 73
Reklaitis, Gintaras ... 96, 266
Remy, Brenda ... 174
Ren, Yi ... 214
Revellame, Emmanuel ... 48
Ricart, Brendon G. ... 116
Ricci, Francesco ... 141
Rijksen, Bart ... 289
Ritacco, Frank ... 126
Rizvi, Imran ... 243
Robertson, John ... 30
Robertson-Lavalle, Sophia ... 132
Robinson, John ... 221
Rodriguez-Rodriguez, Jesus ... 201
Rogers, True ... 303, 334
Román-Ospino, Andrés D. ... 295
Roush, David ... 21
Rousseau, Ronald W. ... 45
Rowe, Jasmine M. ... 205
Rowe, William ... 202
Rowland, Martin ... 3
Ruane, Anthony ... 97
Ryan, Kevin M. ... 108
Ryan, Michael ... 202, 204
Sacher, Stephan ... 73, 78, 135, 261
Sahoo, Kaustuv ... 248
Salan, Jerry S. ... 105
Saluja, Bhawana ... 91
Sammler, Robert ... 318
Sanchez, Elda E. ... 259
Sandell, Dennis ... 91
Santiago, Darlene ... 251
Santos, José L.C ... 35, 159, 207
Savard, Jeffrey ... 101, 326
Schaefer, Jan-Erik ... 103
Schaepman, Alexander ... 136

AUTHOR INDEX

Schaller, Barbara 72
Schenck, Luke 282, 317
Scherholz, Megerle 69
Scheuerle, Rebekah 256
Schmitt, Robert 303, 334
Schultz, Leon 2
Schwabe, Robert 2
Scrivens, Garry 81, 90
Seamans, T. Craig 218
Seibert, Kevin 169, 206
Seletsky, Andrew 98
Sell, Scott A. 209
Semião, Viriato 235
Sempkowski, Michelle 227
Seshadri, Divya 42
Sevimli, Sema 191
Shae, Daniel 191
Shaikh, Rahamatullah 76, 95, 115
Sharma, Ramankur 216
Sheikh, Ahmad 333
Sheth, Poonam 91
Shi, Weixian 104
Shi, Zhenqi 79, 140
Shirazian, Saeed 76
Shrestha, Uma 318
Shukla, Nisha 226
Silva, Rui C. 35, 159, 207
Singh, Ravendra 67, 262, 295
Sinko, Patrick J. 69, 82
Sivakollundu, Sivashankar 101, 326
Skliar, Dimitri 37
Slater, Nigel 256
Slaton, Joel 257
Sluggett, Gregory 81, 90
Smith, Kevin 164
Smith, Michael 190
Smrcka, David 296
Sofou, Stavroula 227, 246
Sokolnicki, Adam 43
Sokolov, Michael 263
Soledade, Maryanna De Melo 278
Soni, Nikita 295
Sorensen, Eva 33
Sotiriou, Georgios A. 228
Souquet, Jonathan 263
Spencer, Rick 107
Srcic, Stanko 338
Srinivasan, Anand 106
Staggs, Darrell 128
Stamato, Howard 160
Starkey, Derek 178
Steichen, Stephanie 196
Stepanek, Frantisek 238, 296
Sterling, Julie A. 191
Stevens, Timothy 126, 136
Stranzinger, Sandra 338
Stras, Sally 246
Strong, John C. 333
Stueber, Dirk 317
Su, Qinglin 96, 266
Subramanian, Jayashree 331
Sugiyama, Hirokazu 34, 56, 134
Suh, SeungBeum 245
Sun, Jennifer 97, 133
Sun, Li 214
Sun, Zhigang 237
Sunkara, Divya 333
Suriel, Indiana 129
Suzuki, Yasuhiro 176
Swartz, James 328
Swinney, Kelly 25
Tam, Annie 190
Tam, Hok Hei 195
Tang, Lauren 331
Tantuccio, Anthony 136
Tao, Jing 132
Teleki, Alexandra 228
Temtem, Márcio 80, 83, 130, 235
Teng, Yue 302
Theuerkauf, Joerg 99
Thirunahari, Satyanarayana 170
Thomas, John A. 164
Thompson, David H. 255
Thompson, Gina 116
Thompson, Ken 116
Thurber, Greg 212
Tian, Geng 124
Tilton, Robert D. 216
Tinke, Arjen 294
Tom, Palmerly 116
Tomassone, Silvina 229
Torres, Nelson 106
Tran, William 103
Trout, Bernhardt L. 284
Trusek-Holownia, Anna 47
Tsabet, Emir 165
Tucson, Ricardo 220
Tugcu, Nihal 21
Tuleu, Catherine 256
Tung, Hsien-Hsin 320
Turnbull, Neil 239
Uddin, Md Jasim 215
Uerpmann, Carsten 263
Undey, Cenk 43
Unsal, Hande 231
Vale, Nobel 126
Valente, Pedro 80, 130, 235
Van Hauwermeiren, Daan 71, 239
Vanarase, Aditya 136
Vanhoorne, Valérie 71

AUTHOR INDEX

VanLang, Christopher C. 329
Variankaval, Narayan 317
Varsolona, Richard 52
Veiseh, Omid 195
Velev, Orlin D. 219
Vercruysse, Jurgen 71
Verma, Anuj A. 179
Verma, Vivek 285
Verstraeten, Maxim 239, 260
Vicente, João 235
Virani, Needa 257
Visco, Donald P. 77
Vitu, Stéphane 278
Vrieze, Derek 202
Wabuyele, Busolo 282
Wahl, Patrick R. 78, 135, 261
Walerack, Corinne 64
Walker, Gavin 61, 72, 76, 95, 115, 298, 314, 316
Walls, Mackenzie T. 213
Walsh, Paul 317
Wang, Bing 264
Wang, Jingkang 182, 183, 184, 185, 186
Wang, Ting 182
Wang, Yan 183
Wang, Zhenyu 332
Wang, Zhibin 214
Wang, Zilong 67
Wassgren, Carl 177
Webber, Matthew 211
Weber, Jennifer L. 64
Welsh, John 21
Whelan, Jessica 111, 222
Willcox, David 41, 202, 204, 208
Williams, Daryl 223, 319
Wilson, John 191
Wilson, Thomas 206
Winkler, Jennifer 229
Witschnigg, Ândreas 119
Witulski, Frank D. 116
Wleklinski, Michael 255
Wolfgang, Matthias 261
Woo, Meng Wai 8
Wood, Geoffrey 81, 90
Woodman, Jessica 326
Wu, Huiquan 153, 180
Wu, Yongqi 126
Wucherpfennig, Thomas 103
Wutscher, Thomas 65
Xia, Junfei 214
Xie, Lingxiao 197
Xu, Wei 325
Ya, Naiqi 172
Yabuta, Keisho 56
Yan, Yuanwei 214
Yang, Bing-Shiou 141
Yang, Darwin 226
Yang, Xiaochuan 124
Yang, Yihui 55
Ydstie, B. Erik 46
Yeap, Eunice Wan Qi 321
Yin, Qiuxiang 182
Yohannes, Bereket 277
Yoon, Seongkyu 17, 117
Yoshizaki, Ryosuke 54
Zadrazil, Ales 238
Zagler, Michaela 119
Zakrewsky, Michael 194
Zaman, Noreen 173
Zappi, Mark 48
Zeglinski, Jacek 76
Zeiner, Tim 44
Zellnitz, Sarah 65
Zettl, Manuel 261
Zettler, Aaron 279, 344
Zhang, Anni 122
Zhang, Dina 317
Zhang, Feng 130
Zhang, Fengying 5
Zhang, Geoff G. Z. 6, 322
Zhang, Jun 193
Zhang, Lu 4
Zhang, Peipei 192
Zhang, Siying 276
Zhao, Jin 199
Zhou, George 52
Zhou, Lina 182
Zhou, Yanan 186
Zhou, Yue 64
Zhu, Xiaowen 184
Ziaee, Ahmad 72
Zimmerman, James R. 255
Zustiak, Silviya P. 209

AIChE
120 Wall Street, FL 23
New York, NY 10005-4020

ISBN 978-1-5108-3445-3

9 781510 834453

CURRAN ASSOCIATES INC.
proceedings.com